Antimicrobial Resistance in Wastewater

Antimicrobial resistance is considered one of the most pressing threats to human health worldwide and has firmly established itself as a major public health challenge. In urban environments, wastewater serves as a key reservoir of antibiotic resistance, hosting bacteria from environmental, human, and animal sources, many of which carry antibiotic resistance genes. The role of the environment in spreading antimicrobial resistance is now better recognized, along with the impact of pollutants that may foster resistance in bacteria. Contaminants released by wastewater treatment plants have become a growing concern.

This book provides a comprehensive evaluation of antimicrobial resistance in wastewater treatment, compiling insights from global experts in fields such as microbiology, environmental biotechnology, civil engineering, biochemistry, environmental science, and public health. It delves into topics including the role of the environment in the development and spread of antimicrobial resistance, chemical analysis of antibiotics in environmental samples, molecular methods to study antimicrobial resistance bacteria and genes, advanced wastewater treatment technologies, the effects of antimicrobial resistance, and the public health implications related to wastewater treatment.

Key features of this book are as follows:

1. Explores the public health impacts of antimicrobial resistance with a focus on wastewater treatment.
2. Provides detailed biochemical analyses of antibiotics in environmental samples.
3. Discusses advanced wastewater treatment processes and their role in mitigating antimicrobial resistance.
4. Highlights recent findings on the role of the environment in the emergence and spread of antimicrobial resistance.
5. Examines molecular approaches for studying antimicrobial resistance genes.

Current Trends in Antimicrobial Research
Series Editor: Arti Gupta

Antimicrobials in Pharmaceutical and Medicinal Research
Edited by Arti Gupta and Ram Prasad

Antimicrobials in Food Science and Technology
Edited by Arti Gupta and Ram Prasad

Antimicrobial Resistance in Agriculture and Its Consequences
Edited by Arti Gupta and Ram Prasad

Antimicrobials in Agriculture
Edited by Arti Gupta and Ram Prasad

Antimicrobials in Environment
Edited by Arti Gupta and Ram Prasad

Antimicrobials in Animal Husbandry
Edited by Arti Gupta and Ram Prasad

Antimicrobial Resistance in Wastewater
Edited by Arti Gupta and Ram Prasad

For more info:
https://www.routledge.com/Current-Trends-in-Antimicrobial-Research/book-series/CTAR

Antimicrobial Resistance in Wastewater

Edited by

Arti Gupta and Ram Prasad

CRC Press
Taylor & Francis Group
Boca Raton London New York

CRC Press is an imprint of the
Taylor & Francis Group, an **informa** business

Designed cover image: Getty Images [Image ID: 1413992436]

First edition published 2026
by CRC Press
2385 NW Executive Center Drive, Suite 320, Boca Raton FL 33431

and by CRC Press
4 Park Square, Milton Park, Abingdon, Oxon, OX14 4RN

CRC Press is an imprint of Taylor & Francis Group, LLC

ISBN: 978-1-032-21742-0 (hbk)
ISBN: 978-1-032-21743-7 (pbk)
ISBN: 978-1-003-26982-3 (ebk)

DOI: 10.1201/9781003269823

Typeset in Times
by codeMantra

Contents

Preface

Antibiotic resistance is an urgent global public health challenge, with numerous studies highlighting the presence of resistant bacteria and resistance genes in environmental samples. Wastewater, in particular, serves as a critical reservoir for antibiotic resistance in urban areas. The wastewater microbiome comprises bacteria from environmental, human, and animal origins, many of which carry antibiotic resistance genes. There is now greater recognition of the role the environment plays in the dissemination of antimicrobial resistance and the influence of pollutants that may drive resistance development in bacteria. Contaminants from wastewater treatment plants are a notable concern. Effective strategies, including UV treatment, ozone application, and nanofiltration, offer promising solutions to reduce antibiotic-resistant bacteria.

This book explores antimicrobial resistance in wastewater treatment processes and gathers comprehensive insights from global experts in fields such as microbial technology, environmental biotechnology, biochemistry, environmental biology, and public health. Key highlights include the role of the environment in the development and spread of antimicrobial resistance, biochemical analysis of antibiotics in environmental samples, molecular techniques for studying antimicrobial resistance bacteria and genes, cutting-edge wastewater treatment methods, the effects of antimicrobial resistance, and its public health implications with an emphasis on wastewater treatment.

This book also evaluates the impact of human water use, treatment, and conservation on the proliferation of antimicrobial resistance. Offering a multidisciplinary perspective, it addresses the interplay between antimicrobial resistance and wastewater treatment processes. This book will appeal to research scholars, scientists, policymakers, and educators.

Chapter 1, authored by Pinar et al., delves into the critical issue of "Wastewater and the Rise of Superbugs: Detection, Treatment, and Future Solutions." This chapter highlights the growing concerns surrounding superbugs—microorganisms that have developed resistance to multiple antibiotics—due to improper wastewater management. It discusses innovative detection methods and potential treatment solutions to mitigate the spread of these resistant strains.

In Chapter 2, Sushil examines "Wastewater Treatment Plants as Hubs of Antibiotic Resistance: Occurrence and Control." This chapter emphasizes the role of wastewater treatment facilities as key sites where antibiotic resistance genes proliferate, outlining both the prevalence of these genes and strategies for controlling their dissemination into the environment.

Chapter 3, authored by Ajay et al., investigates the "Impact of Wastewater on the Occurrence of Antibiotic-Resistant Bacteria," presenting data on how wastewater contributes to the proliferation of these bacteria in aquatic ecosystems and highlighting the health risks they pose to humans and wildlife.

In Chapter 4, presented by Prajapati, a comprehensive review titled "Antimicrobial Resistance in Domestic and Industrial Wastewater" is provided, detailing the differences in resistance patterns found in wastewater from various sources and the implications for public health and environmental safety.

Chapter 5, written by Nawal and Amro Abd Al Fattah Amara, addresses the emerging role of "Aquaculture Antiviral Short Peptides: Protection for Aquaculture, Humans, Animals, and Marine Water Quality." This chapter discusses the potential use of specific peptides in aquaculture systems to combat viral infections while simultaneously promoting the overall health of aquatic environments.

Chapter 6, by Rajesh et al., explores the "Mechanisms, Sources, and Treatment Methods of Antimicrobial Resistance in Wastewater." This chapter provides a detailed analysis of the biological and chemical processes that lead to the emergence of resistance as well as various treatment methods deployed to address the issue in wastewater management facilities.

In Chapter 7, Jamal et al. highlight the "Characteristics and Impact of Antimicrobial Resistance in Wastewater," focusing on how resistant strains interact with various environmental factors and their potential to disrupt aquatic ecosystems and public health.

Chapter 8, authored by Ganseh, focuses on the "Biogenic Synthesis of CuO Nanoparticles in Response to Disperse Dyes and Microbes in Wastewater Using *Strychnos potatorum* Seed." This chapter discusses the innovative application of plant-derived nanoparticles in treating wastewater contaminated by dyes and pathogenic microorganisms.

In Chapter 9, Roamana discusses "Tracking Antimicrobial Resistance in Wastewater: Emerging Threats and Mitigation Strategies," outlining the latest research methods for monitoring resistance trends and proposing practical strategies for mitigating these threats in urban settings.

Chapter 10, by Nisha et al., presents "Novel Approaches to Monitoring Antimicrobial Resistance in Wastewater: An Escalating Environmental Concern," emphasizing cutting-edge technologies like genomic sequencing and bioinformatics that enhance our ability to track resistance in real time.

In Chapter 11, also authored by Nisha, the discussion focuses on "From Waste to Resource: Mitigating Antimicrobial Resistance in Wastewater-Irrigated Agriculture." This chapter explores the dual challenge of using treated wastewater for irrigation while addressing the risk of transmitting antibiotic resistance to crops and soil microbiota.

Lastly, Chapter 12, written by Amar, delves into "Antimicrobial Resistance in Wastewater Bacteria," providing insight into the molecular mechanisms that bacteria employ to resist antibiotics and the implications of these findings for public health and wastewater treatment practices.

Arti Gupta and Ram Prasad

Acknowledgements

First and foremost, we would like to thank God for his never-ending grace, mercy, and provision during what ended up being one of the toughest times of our life.

We are indebted to the many people who helped bring this book to light. I wish to thank series editors Dr. Vijai Kumar Gupta, Dr. Marc Gutierrez, and Dr. Bonita Glanville-Morris; Senior Production Editor Dr. Jyotsna Jangra; and Commissioning Editor Dr. Renu Upadhyaya, and Taylor & Francis/CRC Press, for their generous assistance, constant support, and patience in initializing the series volume. The editors are very thankful to Vice Chancellor Faizabad University, Ayodhya and Mahatma Gandhi Central University, Motihari, for their kind support and constant encouragement. Special thanks are due to well-wishers, family members, and friends.

Dr. Arti Gupta owes a debt of gratitude to her wonderful husband, Mr. Saket Agrawal, and their son, Reyansh. Their unrelenting support and motivation have been pivotal in inspiring her to undertake this endeavor, overcome all obstacles, and assemble all the pieces into a coherent whole. Dr. Ram Prasad gives his special regards to his family for their constant invaluable support and motivation to publish. We also give special thanks to all faculty colleagues and senior faculties of our team.

Arti Gupta
Shri Avadh Raj Singh Smarak Degree College, Gonda, UP, India

Ram Prasad
Mahatma Gandhi Central University, Motihari, Bihar, India

About the Editors

Arti Gupta, Ph.D., earned her Ph.D. in Animal Science from Mahatma Jyotiba Phule Rohilkhand University in Bareilly, India, in 2010. Dr. Gupta's current research interests include animal biotechnology, molecular plant biotechnology, molecular animal biotechnology, bioprocess technology, and microbiology. Since July 2014, she has been serving at Shri Avadh Raj Singh Smarak Degree College in Gonda, Uttar Pradesh, India, where she is also engaged in editorial work.

With 13 years of teaching and research experience in Plant and Animal Biotechnology and Microbial Biotechnology, Dr. Gupta has received several awards, including the University Topper (Gold Medal) for her M.Sc. in Biotechnology, the Young Scientist Award (Gold Medal), Best Poster Presenter, and the Dr. V.P. Agarwal Award (Gold Medal).

Dr. Gupta has contributed extensively to academia, with two monographs, 17 edited books published by various international publishers, 13 book chapters, and approximately 52 national and international research papers. She holds lifetime memberships in the Indian Science Congress Association, the Biotech Research Society of India, the Zoological Society of India, and the International Consortium of Contemporary Biologists. Additionally, she serves on several editorial boards.

Previously, Dr. Gupta was a member of the Teaching Personnel at G.B. Pantnagar University in Pantnagar, Uttarakhand, India, and served as a Visiting Academic Researcher at the Biorefining and Advanced Material Research Center (BAMRC) at Scotland's Rural College (SRUC), University of Edinburgh, in the United Kingdom.

Ram Prasad, Ph.D., is associated with the Department of Botany, Mahatma Gandhi Central University, Motihari, Bihar, India. His research interests include applied and environmental microbiology, plant–microbe interactions, sustainable agriculture, and nanobiotechnology. Dr. Prasad has more than 300 publications to his credit, including research papers, review articles, and book chapters, and seven patents issued or pending, and has edited or authored several books. He has guided one postdoc, two M.Phil, nine Ph.D., and several master dissertation students. Dr. Prasad has 15 years of teaching/research experience and has been awarded the Sir CV Raman Scientist Award, the Young Scientist Award, and Prof. J.S. Datta Munshi Gold Medal by the International Society for Ecological Communications; Fellow of the World Academy of Productive Science (FWAPS); Fellowship of the Association of Fungal Biologists (FAFB); Fellow of Agricultural Technology Development Society (FATDS); Fellow of Biotechnology Research Society of India (FBRS); Fellow of the Society for Applied Biotechnology (FSAB); the American Cancer Society UICC International Fellowship for Beginning Investigators, USA; Outstanding Scientist Award in the field of Microbiology; BRICPL Science Investigator Award; Research Excellence Award, etc. He has been serving as a member of several editorial boards. Previously, Dr. Prasad served as Assistant Professor at Amity University Uttar Pradesh, India; Visiting Assistant Professor, Whiting School of Engineering, Department of Mechanical Engineering at Johns Hopkins University, Baltimore, United States, and Research Associate Professor at the School of Environmental Science and Engineering, Sun Yat-sen University, Guangzhou, China.

Contributors

Wan Nazwanie Wan Abdullah
School of Chemical Science
Universiti Sains Malaysia
Minden, Pulau Pinang, Malaysia

Amro Abd Al Fattah Amara
Protein Research Department
Genetic Engineering and Biotechnology Research Institute [GEBRI]
City of Scientific Research and Technological Applications [SRTA-City]
Universities and Research Center District
New Borg El-Arab, Alex, Egypt

Ambikapathy V
PG and Research Department of Botany
A.V.V.M. Sri Puspham College (Autonomous) (Affiliated to Bharathidasan University, Trichy)
Poondi, Thanjavur, Tamil Nadu, India

Mohd. Ayub Ansari
Department of Chemistry
Bipin Bihari College
Jhansi, Uttar Pradesh, India

Babu S
PG and Research Department of Botany
A.V.V.M. Sri Puspham College (Autonomous) (Affiliated to Bharathidasan University, Trichy)
Poondi, Thanjavur, Tamil Nadu, India

Kunal Kumar Bharti
Division of Life Sciences
School of Basic and Applied Sciences
Galgotias University
Greater Noida, Uttar Pradesh, India

Pınar Aytar Çelik
Department of Biotechnology and Biosafety
Graduate School of Natural and Applied Sciences
Eskişehir Osmangazi University
Eskişehir, Turkey
And
Environmental Protection and Control Program
Eskişehir Osmangazi University
Eskişehir, Turkey

Richa Chandra
School of Pharmaceutical Sciences
Chhatrapati Shahu Ji Maharaj University
Kanpur, Uttar Pradesh, India

Keshika Dhanesh
Department of Biotechnology
Sri Venkateswara College of Engineering
Sriperumbudur, India

Nawal Abd El-Baky
Protein Research Department
Genetic Engineering and Biotechnology Research Institute [GEBRI]
City of Scientific Research and Technological Applications [SRTA-City]
Universities and Research Center District
New Borg El-Arab, Alex, Egypt

Blaise Manga Enuh
Department of Biotechnology and Biosafety
Graduate School of Natural and Applied Sciences
Eskişehir Osmangazi University
Eskişehir, Turkey

Ganesh N
PG and Research Department of Botany
A.V.V.M. Sri Puspham College (Autonomous) (Affiliated to Bharathidasan University, Trichy)
Poondi, Thanjavur, Tamil Nadu, India

Amar P. Garg
Research & Development Cell
Swami Vivekanand Subharti University
NH-58, Subhartipuram, Meerut, India

Pratul Gaurav
Division of Life Sciences
School of Basic and Applied Sciences
Galgotias University
Greater Noida, Uttar Pradesh, India

Gomathi S
PG and Research Department of Botany
A.V.V.M. Sri Puspham College (Autonomous) (Affiliated to Bharathidasan University, Trichy)
Poondi, Thanjavur, Tamil Nadu, India

Arti Gupta
Department of Zoology
Shri Awadh Raj Singh Degree College
Ram Manohar Lohia University
Ayodhya, Uttar Pradesh, India

Prakash Chandra Gupta
School of Pharmaceutical Sciences
Chhatrapati Shahu Ji Maharaj University
Kanpur, Uttar Pradesh, India

Haider Iqbal
Department of Chemistry
Bipin Bihari College
Jhansi, Uttar Pradesh, India

Shilpa Deshpande Kaistha
Department of Biotechnology
School of Life Sciences and Biotechnology
Chhatrapati Shahu Ji Maharaj University
Kanpur, Uttar Pradesh, India

Kalpana
School of Pharmaceutical Sciences
Chhatrapati Shahu Ji Maharaj University
Kanpur, Uttar Pradesh, India

Khushboo
Department of Medical Microbiology
Subharti Medical College, Swami Vivekanand Subharti University
Meerut, India

Ajay Kumar
Division of Life Sciences
School of Basic and Applied Sciences
Galgotias University
Greater Noida, Uttar Pradesh, India

H C Lakshman
Department of Botany
Karnatak University
Dharwad, Karnataka, India

Vineet Kumar Maurya
Department of Botany and Microbiology
Hemvati Nandan Bahuguna Garhwal University (A Central University)
Srinagar (Garhwal), Uttarakhand, India

Romana Mirdhe
Department of Botany
Faculty of Sciences
Khaja Bandanawaz University
Kalaburagi, Karnataka, India

Divya Sri Muthu
Department of Biotechnology
Sri Venkateswara College of Engineering
Sriperumbudur, India

Maadhangi Muthukannan
Department of Biotechnology
Sri Venkateswara College of Engineering
Sriperumbudur, India

Monali Nanda
Bio-Resource Product Research Laboratory
Department of Botany
School of Life Science
Guru Ghasidas Vishwavidyalaya (A Central University)
Bilaspur, Chhattisgarh, India

Nur Atiqah Nasir
Faculty of Science and Technology
Universiti Sains Islam Malaysia
Nilai, Negeri Sembilan, Malaysia

Rajesh Nithyanandam
Department of Chemical Engineering
Sri Venkateswara College of Engineering
Sriperumbudur, India

Panneerselvam A
PG and Research Department of Botany
A.V.V.M. Sri Puspham College (Autonomous) (Affiliated to Bharathidasan University, Trichy)
Poondi, Thanjavur, Tamil Nadu, India

R. K. Prajapati
Department of Chemistry
D.J. College
Baraut Baghpat, Uttar Pradesh, India

Prakash P
PG and Research Department of Botany
A.V.V.M. Sri Puspham College (Autonomous) (Affiliated to Bharathidasan University, Trichy)
Poondi, Thanjavur, Tamil Nadu, India

Krishnan Ganesh Prasath
Department of Biotechnology
Sri Venkateswara College of Engineering
Sriperumbudur, India

Salmiah Jamal Mat Rosid
East Coast Environmental Research Institute
Universiti Sultan Zainal Abidin
Gong Badak Campus
Kuala Nerus, Terengganu, Malaysia

Sarina Mat Rosid
Faculty of Bioresources and Food Industry
Universiti Sultan Zainal Abidin
Besut Campus, Besut, Terengganu, Malaysia

Pragati Saini
Division of Life Sciences
School of Basic and Applied Sciences
Galgotias University
Greater Noida, Uttar Pradesh, India

Sandhya
Research & Development Cell
Swami Vivekanand Subharti University
Subhartipuram, Meerut, India
And
Unisza Science and Medicine Foundation Centre
Universiti Sultan Zainal Abidin
Gong Badak Campus
Kuala Nerus, Terengganu, Malaysia

Mamta Patra Shahi
Bio-Resource Product Research Laboratory
Department of Botany
School of Life Science
Guru Ghasidas Vishwavidyalaya (A Central University)
Bilaspur, Chhattisgarh, India

Sushil Kumar Shahi
Bio-Resource Product Research Laboratory
Department of Botany
School of Life Science
Guru Ghasidas Vishwavidyalaya (A Central University)
Bilaspur, Chhattisgarh, India

Nisha Sharma
School of Pharmaceutical Sciences
Chhatrapati Shahu Ji Maharaj University
Kanpur, Uttar Pradesh, India

Deepak Singh
ICAR-Indian Institute of Soil & Water Conservation
Dehradun, Uttarakhand, India

Nisha Singh
Department of Biochemistry
Faculty of Science
University of Lucknow
Lucknow, India

Ruchi Singh
School of Pharmaceutical Sciences
Chhatrapati Shahu Ji Maharaj University
Kanpur, Uttar Pradesh, India

Sweta
School of Pharmaceutical Sciences
Chhatrapati Shahu Ji Maharaj University
Kanpur, Uttar Pradesh, India

Belma Nural Yaman
Department of Biotechnology and Biosafety
Graduate School of Natural and Applied Sciences
Eskişehir Osmangazi University
Eskişehir, Turkey
and
Department of Biomedical Engineering
Eskisehir Osmangazi University
Eskisehir, Turkey

1 Wastewater and the Rise of Superbugs
Detection, Treatment, and Future Solutions

Blaise Manga Enuh, Belma Nural Yaman, and Pınar Aytar Çelik

1.1 INTRODUCTION

The number of bacterial pathogens with multidrug resistance phenotypes and their infections has been on the rise within the last decade. As a result, they are now considered by the World Health Organization (WHO) to be an emergent global disease and a public health problem (World Health Organization, 2021). Antimicrobial resistance (AMR) occurs when microorganisms no longer respond to previously effective treatments, making it more difficult to treat infections. Due to the inefficiency of the treatments, the disease from such infections can spread easily and cause severe illness and even death (WHO, 2020).

Antibiotics and disinfectants are generally used as treatment and for hygiene, respectively, against microorganisms. After use, some of these are released into the environment where their concentrations become less and have suboptimal effects against microorganisms, encouraging the buildup of resistance. Such resistant microorganisms can be found in environmental samples like water (surface and groundwater), soil, and wastewater (Serwecińska, 2020). The disposal of these chemical agents into the environment continues to persist and so is the increase in the proportion of resistant microorganisms.

Antibiotic resistance can be acquired vertically or horizontally, vertically from parent to offspring and horizontally in processes like conjugation, transduction, or transformation. Studies show that areas with high human activity reveal high levels of AMR, for example, agricultural areas, industrial areas, hospitals, or wastewater from inhabited areas. The runoffs of these areas are often referred to as "hotspots for AMR genes" because these contain a high density of plasmids, phages, and bacteria that encourage high levels of exchange of genetic material (Nadeem et al., 2020). Studies comparing the repertoire of AMR genes prove that effluents from humans and animals do contain AMR genes, but there is a significantly higher diversity in the AMR genes in human wastes originating from municipal wastewater treatment plants (Agga et al., 2015). This can also explain the high potential of AMR transfer across species in wastewater. The potential contribution of such dynamics to AMR in wastewater, in general, cannot be underrated.

When other emerging global issues are factored in, it is seen from a holistic point of view that the problem of AMR is becoming increasingly complex. Amidst these, there are also several knowledge gaps on the issue in question. For example, the surveillance of AMR is considered to contain the most research gaps as well as research methodologies (Bulteel et al., 2021). Effective surveillance requires the use of effective tools for the identification of resistant microbes. Traditional culture methods are limited by the scope of microorganisms that can be grown and identified. Molecular tools based on 16S and 23S ribosomal RNA enable better identification and the phylogenetic classification of bacteria and can be used in the investigation of complex microbial populations in

DOI: 10.1201/9781003269823-1

effluents. In addition, a combination of several molecular tools can be used for comparing community structures and composition across different environments (Church et al., 2020; Rytkönen et al., 2021).

In wastewater treatment plants for example, during biological treatment, the activated sludge contains very complex microbial communities (Barancheshme and Munir, 2018). This makes their investigation with traditional culture and molecular methods difficult. Advanced next-generation technologies such as pyrosequencing can be used to obtain insights into community structure and diversity, as well as to detect genes, minor populations and reveal microbial dark matter (Garner et al., 2021). Furthermore, these technologies allow for more sophisticated whole-genome sequencing (WGS), metagenomics, and computational biology analysis, which help generate insights from samples faster than had been previously done. Understanding the structure and function of bacterial communities is critical for designing effective wastewater treatment strategies, as community structure is different for different types of effluents (Wang et al., 2020). These methods could help in the long run to narrow the gap in knowledge about AMR and contribute to the discovery of effective solutions.

Wastewater is of interest as one of the most important hotspots of AMR. As a result, several methods to fight AMR in wastewater are being researched. They range from the use of carefully designed structures such as treatment reactors, activated sludge systems, and constructed wetlands. Other technologies include the use of nanomaterials, coagulation, and biochar (Alazaiza et al., 2022; Cui and Smith, 2021; Yang et al., 2021).

This chapter discusses wastewater sources, the transmission of resistance in wastewater, and microbes of importance in wastewater. It also discusses some effective methods to detect antimicrobial-resistant bacteria and antimicrobial-resistant genes. Finally, it presents some methods and technologies used to fight AMR in wastewater.

1.1.1 Antimicrobial Resistance in Different Wastewater Sources

As previously mentioned, there are trace amounts of antibiotics and other chemicals used against microorganisms in wastewater that cause a selection pressure for AMR (Karkman et al., 2018). The selection of antibiotic resistance through horizontal gene transfer is common in wastewaters and their treatment plants. Hospitals and municipal wastes are important sources of antibiotic-resistant bacteria (ARB) and antibiotic-resistant genes (ARGs) (Karkman et al., 2018). Of importance in these wastewaters are AMR genes because of the potential for transfer. Surveillance of ARGs is based on the use of indicator genes, for example, the class 1 integrase gene *intI1*, for assessing the potential for horizontal gene transfer; *blaTEM*, often common with Enterobacteriaceae and widely distributed; *vanA*, mostly in clinical isolates; and *marA*, an Enterobacteriaceae stress-response gene. Wastewater plants were observed to have limitations in removing these resistance genes, especially with the *intI1* gene, which indicated that the microbial populations have a high potential to acquire genes in general (Narciso-da-Rocha et al., 2014). Hence, treating wastewater effluents requires removing not only bacterial populations but also further genetic sterilization to ensure the destruction of potential AMR genes. A summary of the different sources and AMR routes is shown in Figure 1.1.

It is also critical to underline the importance of monitoring the presence of genes that determine antibiotic resistance in wastewater released from treatment plants since these genes can assist in detecting the public health risk that treated wastewater provides.

1.2 ANTIMICROBIALS IN WASTEWATER

The overuse of antimicrobial agents in both humans and animals has resulted in the buildup of these chemicals in the environment, and the influence of such accumulation on the emergence of antibiotic resistance should not be underestimated (World Health Organization, 2021). The antimicrobials triclosan, tetracyclines, sulfonamides, penicillin, and quinolones have been detected in wastewater,

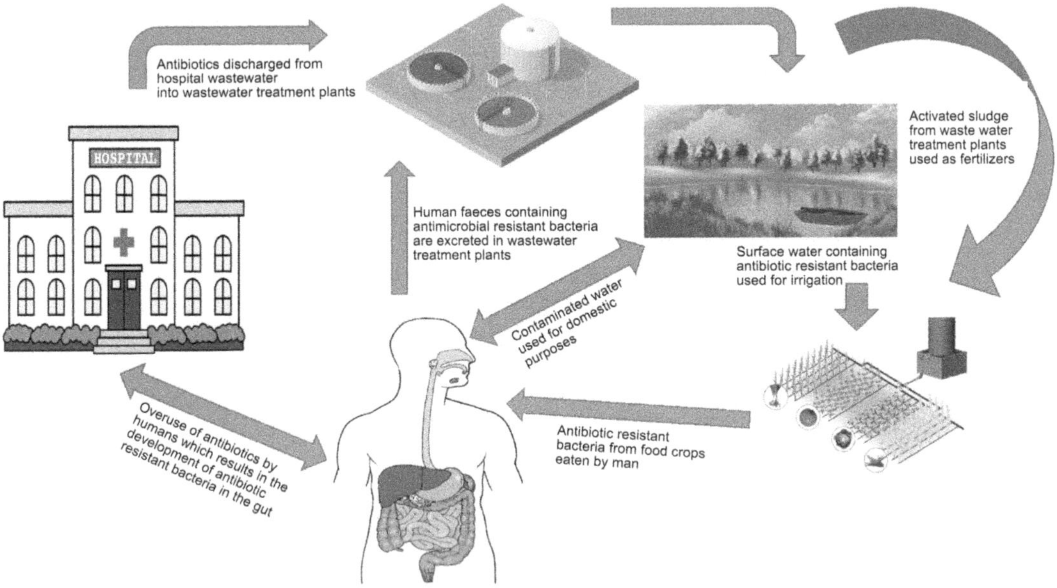

FIGURE 1.1 Different sources of antimicrobial resistance and their transmission in the environment. (Adapted from Ebomah and Okoh, 2020.)

which could in turn also influence microbial population structures and composition in wastewater treatment plants (Novo et al., 2013). This can negatively affect the outcome of biological treatment, especially as the determinants of AMR remain in treated water due to the ineffectiveness of wastewater treatment in removing them (Wang et al., 2017). For example, a study on the efficiency of biological treatment in activated sludge and membrane bioreactor systems on the removal of commonly used antimicrobials showed that membrane bioreactors exhibit better performance than activated sludge. In general, the biological wastewater treatment procedures largely eradicated the β-lactam, glycopeptide, and fluoroquinolone families, although trimethoprim and lincosamides appeared to be more persistent (Tran et al., 2016).

Most water quality legislations focus on the presence of indicator bacteria but do not address antibiotic concentrations in sewage and their treatment facilities (Roca et al., 2015; Uluseker et al., 2021). Strategies to reduce the dangers of environmental exposure can focus on improving industrial sanitation systems. Advanced treatment strategies with nanofiltration, reverse osmosis, and ionization could be improved to remove antibiotic residues in wastewater effluents (Monachan et al., 2022). Also, to provide lasting solutions to this, many novel environmental surveillance approaches have been proposed and pilot testing is being carried out by the WHO (World Health Organization, 2021), which is expected to help strengthen legislation through advocacy.

Finally, in the face of the current COVID-19 pandemic, the indiscriminate use of antibiotics in treating patients and in cleaning agents has been reported to be on the rise, which in turn accumulates in sewage effluents, affecting outcomes in treatment plants. An increase in public awareness and special attention and investments could be required during such pandemics or outbreaks (Ukuhor, 2021; Usman et al., 2020).

1.3 RESISTANT MICROBES OF IMPORTANCE IN WASTEWATER

Antibiotic residues and chemicals used in homes end up in sewage water (Table 1.1). Therefore, it is likely that bacterial species of importance will include coliforms such as *Escherichia coli*, *Klebsiella* spp., and *Aeromonas* spp., which by the way have also been shown to survive wastewater treatment (Verburg et al., 2019). In addition to these, *Acinetobacter* and *Shigella* have also been

TABLE 1.1
Resistant Microbes of Importance in Wastewater

Bacteria Species	Reference
Carbapenemase-producing *Klebsiella pneumoniae* (CKP)	Sakkas et al. (2019)
Vancomycin-resistant *Enterococci* (VRE)	
Methicillin-resistant *Staphylococcus aureus* (MRSA)	
Extended-spectrum β-lactamase (ESBL) producing *E. coli*	
Multidrug-resistant *Pseudomonas aeruginosa*	
Escherichia, Shigella, Klebsiella, and *Acinetobacter*	Ferreira da Silva et al. (2007)
Enterococcus faecium	Łuczkiewicz et al. (2010)
Enterococcus faecalis	
Enterococcus hirae	
Enterococcus casseliflavus/gallinarum	
Enterococcus durans	
Extended-spectrum β-lactamase producing *E. coli.*	

isolated (Ferreira da Silva et al., 2007). In sewage waters, bacteria such as *E. coli* acquire AMR genes more frequently or are more likely to be positively selected for resistance than in other water sources (Ogura et al., 2020).

1.4 METHODS OF DETECTION OF AMR AND ARGs IN WASTEWATER

Antibiotic resistance research has mostly concentrated on pathogens over the past 70 years. Pure culture isolation has been and continues to be the gold standard approach in clinical microbiology (Karkman et al., 2018). High-throughput-based metagenomic approaches have been used in combination with a structured database of AMR genes as a tool for a long-term comprehensive survey of different AMR genes (Lakin et al., 2017). Metagenomic approaches can be used to monitor AMR genes, drug-resistant bacteria, and mobile genetic elements such as plasmids, prophages, and phages (Yadav and Kapley, 2019). Function and sequence-based metagenomic techniques have previously enabled the identification of DNA sequences that had no resemblance to known antibiotic resistance determinants. Besides, metagenomic clones and a plasmid that showed resistance to kanamycin, ampicillin, or chloramphenicol in *E. coli* were identified (Parsley et al., 2010). Hence, effective surveillance has evolved from detecting just pathogens to detecting genetic elements that carry resistance traits.

In surveillance studies to assess the potential spread of resistance genes, fast detection of resistance mechanisms plays a significant epidemiological function (Uluseker et al., 2021; World Health Organization, 2021). As the resistance to new medicines develops quickly, the use of faster machine learning algorithms could help process the growing amount of data generated from AMR studies (Imchen et al., 2020).

Discharges from wastewater treatment plants may also contribute to AMR spread therefore should not be overlooked. Some studies using quantitative polymerase chain reaction (qPCR) have revealed the presence of antimicrobial-resistant and multidrug-resistant (MDR) *E. coli* in such samples. Findings like these are important for coming up with better environmental rules and policies to assist in regulating AMR/MDR emissions into the environment, as well as mitigation methods and effective wastewater treatment strategies (Mukherjee et al., 2021).

1.4.1 Whole-Genome Sequencing for Resistome Surveillance

Many clinical laboratories still use the traditional two-fold serial dilution and disk diffusion procedures to detect antibiotic resistance (Hendriksen et al., 2019). Traditional strategies for ARG surveillance have various drawbacks, including a lack of appropriate methodologies for several microorganisms, variances in laboratory settings, and a limit on the number of medicines analyzed (Imchen et al., 2020). Whole-genome sequencing (WGS) has aided in the detection of outbreak sources, the tracking of studies, the development of standard procedures, etc. (Hendriksen et al., 2019). In addition to predicting horizontal gene transfer and co-resistance to various antibiotics and heavy metals through analyzing chromosomes or plasmids, WGS offers several other advantages over standard resistance testing (Hendriksen et al., 2019).

1.4.2 Metagenomic Approaches for Resistome Surveillance

Although WGS is a strong tool for ARG surveillance, it requires pure cultures, which is a severe drawback because most bacteria have yet to be cultivated in the laboratory. This can be sustainably circumvented with the aid of metagenomics (Imchen et al., 2020). The metagenomic approach uses genomes from a broad population instead of specific organisms. This is suitable for wastewater samples as different microbe species are often present in communities. Continuous global surveillance and prediction of AMR via metagenomic analysis of sewage could be an ethically acceptable and economically practical method (Hendriksen et al., 2019).

1.4.3 Expression-Based Metagenomics and Computational Biology

Functional metagenomics has emerged as a promising tool for biomining of novel genes, peptides, and secondary metabolites (Johnson et al., 2016). Shotgun metagenomics has some disadvantages as compared to functional metagenomics. Because most microorganisms have yet to be isolated, this allows for the screening of specific genes over the whole microbial community. Furthermore, because of the high-throughput nature of functional metagenomics, a huge number of sequences can be screened. The library can be kept, replicated, and tested for repeatability again. It works with any kind of environmental sample. However, functional metagenomics relies on database alignment, but shotgun metagenomics is only better in that it is unaffected by prior sequence knowledge (Imchen et al., 2020).

1.5 APPROACHES TO FIGHT ANTIMICROBIAL RESISTANCE IN WASTEWATER

To prevent the spread of ARB and ARGs into the environment, wastewater must be properly treated before being discharged into rivers, lakes, or sewage systems. Low-energy anaerobic–aerobic treatment reactors, artificial wetlands, and disinfection procedures have all been reported to have high removal efficiency. Nanomaterials and biochar, in combination with other treatment methods and coagulation processes, are relatively new ARB and ARG removal solutions that require further research (Barancheshme and Munir, 2018). According to current research, based on the type of genes present and the treatment techniques utilized, a wide range of ARG removal efficiency can be accomplished; nevertheless, some gaps must be researched further (Barancheshme and Munir, 2018). As previously stated, AMR in wastewaters is a major issue, and the problem is made worse by the potential for gene transfer. However, wastewater treatment plants use some additional treatment procedures to eradicate resistant bacteria. Chlorine, UV irradiation, photocatalytic oxidation, Fenton reactions, ozonation, engineered wetlands, adsorbents, coagulants, and nanoparticles are examples of these techniques (Sharma et al., 2016).

1.5.1 Anaerobic and/or Aerobic Treatment Reactors

These are large reactors that use microorganisms in oxygen-controlled conditions to treat wastewater. It is generally considered a sustainable option for treating chemical oxygen demand because it consumes relatively less energy and is environmentally friendly. Using this process, it is possible to eliminate antibiotics, resistant bacteria, and ARGs (Christgen et al., 2015). The best removal efficiency is achieved when both anaerobic and aerobic reactors are operated in sequence. For example, pretreating the wastewater in anaerobic conditions to reduce the chemical oxygen demand initially, then proceeding with an aerobic process that can be optimized to treat the effluents to the required standards for safe discharge. On the other hand, when the reactors are used alone, they are not as effective (Barancheshme and Munir, 2018).

1.5.2 Activated Sludge Systems

The activated sludge system consists of aerobic and anaerobic digestion, composting, air drying, and bio drying where the general mechanism of removing antibiotics and ARB is by adsorption. These processes can considerably reduce the number of fecal indicators and are effective in treating wastewaters from many sources (Łuczkiewicz et al., 2010). Since the process relies on the activity of microbes, resistant microbes and genetic elements carrying resistance genes may persist and potentially spread ARGs. Meanwhile, many biosurfactant-producing bacteria have also been isolated from such wastewater treatment facilities that show broad-spectrum activity against drug-resistant *E. coli*, *Saccharomyces*, and *Staphylococcus*. The potential for this approach has been reviewed as a way that could sustainably help prevent outbreaks from such resistant bacteria populations (Çelik et al., 2020; Manga et al., 2021). However, including physicochemical pretreatment approaches such as the use of ozone, thermal hydrolysis, and microwaving could help further remove resistant microbes and ARGs (Wang et al., 2021; Xue et al., 2019).

1.5.3 Constructed Wetlands

Constructed wetlands are being touted as a long-term solution for removing antibiotics from various types of effluents. They have been described as easy-to-operate, low-cost, and energy-efficient systems for wastewater treatment (Wang et al., 2022). They can be described as small and semi-aquatic ecosystems that allow the proliferation of a large number of distinct microbial populations where various physical and chemical reactions occur. Multiple physical, chemical, and biological processes such as volatilization, plant uptake and accumulation, plant exudation, adsorption, photolysis, and microbial degradation all occur simultaneously in these green systems. These are because of interactions between soil/sediment, microorganisms, and plants. Higher plants are crucial in constructed wetlands because they ensure the substrate's hydraulic conductivity, aid in nutrient uptake, and support microbial assemblages within their roots. Even though constructed wetlands are simple to maintain and run, are inexpensive, and produce high-quality effluent with minimal energy dissipation, there is still a dearth of knowledge on removal methods, toxicity hazards, design impacts, and the impact of environmental factors on constructed wetlands' efficiency (Bôto et al., 2016; Khan et al., 2020). As with the activated sludge system, they do present a potential to enhance the spread of antibiotic resistance genes as well (Zhang et al., 2020).

1.5.4 Nanomaterials

Antimicrobial nanotechnology has been shown to be effective in protecting against antibiotic resistance, ARBs, and ARGs using various nanomaterial combinations (Prasad and Thirugnanasanbandham, 2019). Nanomaterials exhibit inhibitory and bactericidal properties against bacteria, particularly MDR pathogens (Ali et al., 2020; Ojemaye et al., 2020) but also sublethal concentrations may lead to the development and horizontal transfer of AMR (Cui and Smith, 2021). The use of nanotechnology is quite novel, and there are gaps in knowledge on the health impacts of the use of nanotechnology, natural transformation, and specific effects of different engineered nanoparticles are still to be properly evaluated, and the adsorption and desorption processes are still to be well understood (Cui and Smith, 2021).

1.5.5 Coagulation

Coagulation is an active method for treating turbidity, natural organic matter, color, and heavy metals in water (Alazaiza et al., 2022). To reuse wastewater, coagulation has long been used as a tertiary treatment process in wastewater treatment plants to remove residual suspended particles, organic debris, and phosphorus. Some studies show that coagulation can effectively remove ARGs from effluent with 0.5-log to 3.1-log reduction (Li et al., 2017). Also, results on the removal of antibiotics from hospital effluents are promising (Nonfodji et al., 2020). Chemical coagulants like alum are more often used to remove bacteria during the flocculation and sedimentation process of treatment plants, but natural and environmentally friendly alternatives like diatomite are also being researched (Sha'arani et al., 2019).

1.5.6 Biochar

Biochar is a type of active charcoal made by pyrolyzing carbon-rich biomass. It is a porous material with abundant mineral components and a large specific surface area (Yang et al., 2021), providing the greatest number of sites for pollutant sorption. Biochar can be made from various organic feedstocks, such as municipal waste, waste biomass, and agricultural and livestock waste (Rawat et al., 2023). It is created by heating biomass at 700°C in the absence of air. Aside from being cheap, they have desired physicochemical properties and can be easily modified to suit particular needs. Their mechanism of action is mainly by the adsorption of antibiotics (Krasucka et al., 2021; Yao et al., 2021) and when combined with nanocomposites can prove effective against several broad-spectrum superbugs (Gao et al., 2020; Singh et al., 2020; Chausali et al., 2021; Sonowal et al., 2023).

1.6 CONCLUSIONS AND PROSPECTS

AMR is considered a serious global public health problem by the WHO. The indiscriminate use of antibiotics in products of daily use, industrial processes, and agriculture is causing an increasingly concerning AMR problem. Besides antibiotics, other products such as sanitizers and detergents can cause AMR when disposed into sewage wastewaters. This is particularly concerning in this period of the COVID-19 pandemic when the use of these products is on the rise. Wastewater is a hotspot for AMR development and should be given special attention. Better tools for AMR detection in wastewater treatment plants are required with effective treatment options tailored toward specific types of effluents. Surveillance research is continuously needed to track spreading resistance and to detect emerging ones at early stages. Current approaches to fight AMR in sewage plants need to be evaluated and improved as resistance is easily and rapidly transferred between bacteria species while new approaches need to be researched further to understand their mechanisms of action and possible implications for human and environmental health (Figure 1.2). New tools for the detection, surveillance, and management of AMR have to be high throughput and responsive to meet up with the current rate at which AMR is increasing. Sustainable solutions will require collective and coordinated efforts from the public, research institutions, governments, and international organizations to mount effective strategies to fight AMR in wastewater.

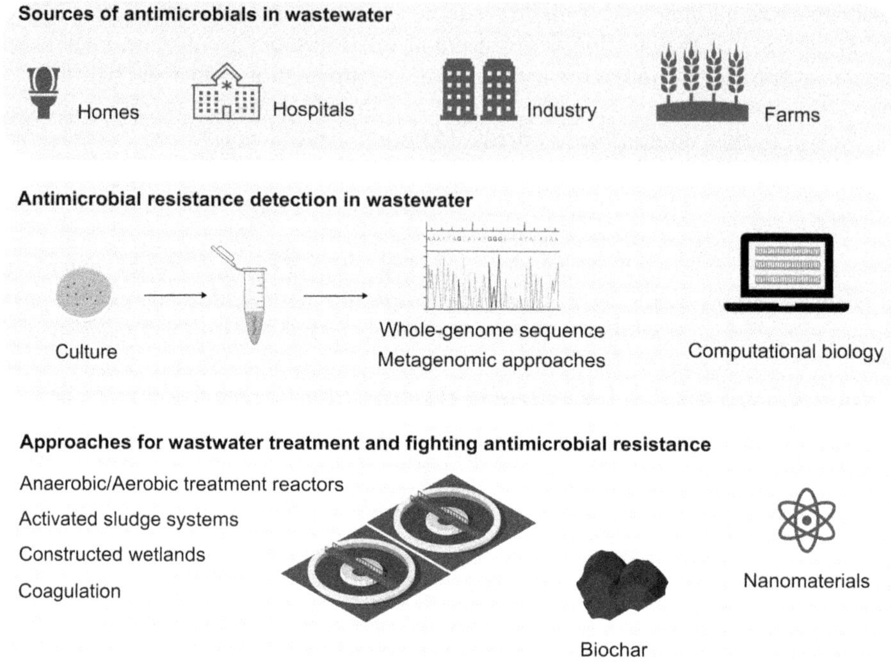

FIGURE 1.2 Antimicrobial resistance in wastewater.

REFERENCES

Agga, G.E., Arthur, T.M., Durso, L.M., Harhay, D.M., Schmidt, J.W., 2015. Antimicrobial-resistant bacterial populations and antimicrobial resistance genes obtained from environments impacted by livestock and municipal waste. *PLoS One* 10, e0132586. https://doi.org/10.1371/journal.pone.0132586

Alazaiza, M.Y.D., Albahnasawi, A., Ali, G.A.M., Bashir, M.J.K., Nassani, D.E., Al Maskari, T., Amr, S.S.A., Abujazar, M.S.S., 2022. Application of natural coagulants for pharmaceutical removal from water and wastewater: A review. *Water* 14, 140. https://doi.org/10.3390/w14020140

Ali, H.R., Emam, A.N., Koraney, N.F., Hefny, E.G., Ali, S.F., 2020. Combating the prevalence of water-borne bacterial pathogens using anisotropic structures of silver nanoparticles. *J Nanopart Res* 22, 47. https://doi.org/10.1007/s11051-020-4760-6

Barancheshme, F., Munir, M., 2018. Strategies to combat antibiotic resistance in the wastewater treatment plants. *Front Microbiol* 8, 2603. https://doi.org/10.3389/fmicb.2017.02603

Bôto, M., Almeida, C.M.R., Mucha, A.P., 2016. Potential of constructed wetlands for removal of antibiotics from saline aquaculture effluents. *Water* 8, 465. https://doi.org/10.3390/w8100465

Bulteel, A.J.B., Larson, E.L., Getahun, H., 2021. Identifying global research gaps to mitigate antimicrobial resistance: A scoping review. *Am J Infect Control* 49, 818–824. https://doi.org/10.1016/j.ajic.2020.11.024

Çelik, P.A., Manga, E.B., Çabuk, A., Banat, I.M., 2020. Biosurfactants' potential role in combating COVID-19 and similar future microbial threats. *Appl Sci* 11, 334.

Chausali, N., Saxena, J., Prasad, R., (2021). Nanobiochar and biochar based nanocomposites: Advances and future prospects. *Journal of Agriculture and Food Research 5*, 100191. https://doi.org/10.1016/j.jafr.2021.100191

Christgen, B., Yang, Y., Ahammad, S.Z., Li, B., Rodriquez, D.C., Zhang, T., Graham, D.W., 2015. Metagenomics shows that low-energy anaerobic–aerobic treatment reactors reduce antibiotic resistance gene levels from domestic wastewater. *Environ Sci Technol* 49, 2577–2584. https://doi.org/10.1021/es505521w

Church, D.L., Cerutti, L., Gürtler, A., Griener, T., Zelazny, A., Emler, S., 2020. Performance and application of 16S rRNA gene cycle sequencing for routine identification of bacteria in the clinical microbiology laboratory. *Clin Microbiol Rev* 33, e00053–19. https://doi.org/10.1128/CMR.00053-19

Cui, H., Smith, A.L., 2021. Impact of engineered nanoparticles on the fate of antibiotic resistance genes in wastewater and receiving environments: A comprehensive review. *Environ Res* 204, 112373. https://doi.org/10.1016/j.envres.2021.112373

Ebomah, K.E., Okoh, A.I., 2020. An African perspective on the prevalence, fate and effects of carbapenem resistance genes in hospital effluents and wastewater treatment plant (WWTP) final effluents: A critical review. *Heliyon* 6(5), e03899.

Ferreira da Silva, M., Vaz-Moreira, I., Gonzalez-Pajuelo, M., Nunes, O.C., Manaia, C.M., 2007. Antimicrobial resistance patterns in Enterobacteriaceae isolated from an urban wastewater treatment plant. *FEMS Microbiol Ecol* 60, 166–176. https://doi.org/10.1111/j.1574-6941.2006.00268.x

Gao, Y., Pramanik, A., Patibandla, S., Gates, K., Hill, G., Ignatius, A., Ray, P.C., 2020. Development of human host defense antimicrobial peptide-conjugated biochar nanocomposites for combating broad-spectrum superbugs. *ACS Appl Bio Mater* 3, 7696–7705. https://doi.org/10.1021/acsabm.0c00880

Garner, E., Davis, B.C., Milligan, E., Blair, M.F., Keenum, I., Maile-Moskowitz, A., Pan, J., Gnegy, M., Liguori, K., Gupta, S., Prussin, A.J., Marr, L.C., Heath, L.S., Vikesland, P.J., Zhang, L., Pruden, A., 2021. Next generation sequencing approaches to evaluate water and wastewater quality. *Water Res* 194, 116907. https://doi.org/10.1016/j.watres.2021.116907

Hendriksen, R.S., Munk, P., Njage, P., van Bunnik, B., McNally, L., Lukjancenko, O., Röder, T., Nieuwenhuijse, D., Pedersen, S.K., Kjeldgaard, J., Kaas, R.S., Clausen, P.T.L.C., Vogt, J.K., Leekitcharoenphon, P., van de Schans, M.G.M., Zuidema, T., de Roda Husman, A.M., Rasmussen, S., Petersen, B., Amid, C., Cochrane, G., Sicheritz-Ponten, T., Schmitt, H., Alvarez, J.R.M., Aidara-Kane, A., Pamp, S.J., Lund, O., Hald, T., Woolhouse, M., Koopmans, M.P., Vigre, H., Petersen, T.N., Aarestrup, F.M., 2019. Global monitoring of antimicrobial resistance based on metagenomics analyses of urban sewage. *Nat Commun* 10, 1–12. https://doi.org/10.1038/s41467-019-08853-3

Imchen, M., Moopantakath, J., Kumavath, R., Barh, D., Tiwari, S., Ghosh, P., Azevedo, V., 2020. Current trends in experimental and computational approaches to combat antimicrobial resistance. *Front Genet* 11, 563975. https://doi.org/10.3389/fgene.2020.563975

Johnson, J., Jain, K., Madamvar, D., 2016. Functional metagenomics: Exploring nature's gold mine, in: Gunasekaran, P., Noronha, S., Pandey, A., (eds) *Current Developments in Biotechnology and Bioengineering: Functional Genomics and Metabolic Engineering*. pp. 27–43, Elsevier, Amsterdam, The Netherlands. https://doi.org/10.1016/B978-0-444-63667-6.00002-X

Karkman, A., Do, T.T., Walsh, F., Virta, M.P.J., 2018. Antibiotic-resistance genes in waste water. *Trends Microbiol* 26, 220–228. https://doi.org/10.1016/j.tim.2017.09.005

Khan, S., Nawab, J., Waqas, M., 2020. Constructed wetlands: A clean-green technology for degradation and detoxification of industrial wastewaters, in: Bharagava, R., Saxena, G. (eds) *Bioremediation of Industrial Waste for Environmental Safety*. pp. 127–163, Springer, Singapore. https://doi.org/10.1007/978-981-13-3426-9_6

Krasucka, P., Pan, B., Sik Ok, Y., Mohan, D., Sarkar, B., Oleszczuk, P., 2021. Engineered biochar – A sustainable solution for the removal of antibiotics from water. *Chem Eng J* 405, 126926. https://doi.org/10.1016/j.cej.2020.126926

Lakin, S.M., Dean, C., Noyes, N.R., Dettenwanger, A., Ross, A.S., Doster, E., Rovira, P., Abdo, Z., Jones, K.L., Ruiz, J., Belk, K.E., Morley, P.S., Boucher, C., 2017. MEGARes: An antimicrobial resistance database for high throughput sequencing. *Nucleic Acids Res* 45, D574–D580. https://doi.org/10.1093/nar/gkw1009

Li, N., Sheng, G.-P., Lu, Y.-Z., Zeng, R.J., Yu, H.-Q., 2017. Removal of antibiotic resistance genes from wastewater treatment plant effluent by coagulation. *Water Res* 111, 204–212. https://doi.org/10.1016/j.watres.2017.01.010

Łuczkiewicz, A., Jankowska, K., Fudala-Książek, S., Olańczuk-Neyman, K., 2010. Antimicrobial resistance of fecal indicators in municipal wastewater treatment plant. *Water Res* 44, 5089–5097. https://doi.org/10.1016/j.watres.2010.08.007

Manga, E.B., Celik, P.A., Cabuk, A., Banat, I.M., 2021. Biosurfactants: Opportunities for the development of a sustainable future. *Curr Opin Colloid Interface Sci* 56, 101514.

Monachan, M., Dixit, N., Maliyekkal, S., Singh, S., 2022. Reverse osmosis (RO) and nanofiltration (NF) membranes for emerging contaminants (ECs) removal, in: Singh, S., Agarwal, A.K., Gupta, T., Maliyekkal, S.M. (eds) *New Trends in Emerging Environmental Contaminants*. pp. 407–425, Springer, Singapore. https://doi.org/10.1007/978-981-16-8367-1_17

Mukherjee, M., Laird, E., Gentry, T.J., Brooks, J.P., Karthikeyan, R., 2021. Increased antimicrobial and multidrug resistance downstream of wastewater treatment plants in an urban watershed. *Front Microbiol* 12, 1269. https://doi.org/10.3389/fmicb.2021.657353

Nadeem, S.F., Gohar, U.F., Tahir, S.F., Mukhtar, H., Pornpukdeewattana, S., Nukthamna, P., Moula Ali, A.M., Bavisetty, S.C.B., Massa, S., 2020. Antimicrobial resistance: More than 70 years of war between humans and bacteria. *Crit Rev Microbiol* 46, 578–599. https://doi.org/10.1080/1040841X.2020.1813687

Narciso-da-Rocha, C., Varela, A.R., Schwartz, T., Nunes, O.C., Manaia, C.M., 2014. blaTEM and vanA as indicator genes of antibiotic resistance contamination in a hospital–urban wastewater treatment plant system. *J Glob Antimicrob Resist* 2, 309–315. https://doi.org/10.1016/j.jgar.2014.10.001

Nonfodji, O.M., Fatombi, J.K., Ahoyo, T.A., Osseni, S.A., Aminou, T., 2020. Performance of *Moringa oleifera* seeds protein and *Moringa oleifera* seeds protein-polyaluminum chloride composite coagulant in removing organic matter and antibiotic resistant bacteria from hospital wastewater. *J Water Process Engin* 33, 101103. https://doi.org/10.1016/j.jwpe.2019.101103

Novo, A., André, S., Viana, P., Nunes, O.C., Manaia, C.M., 2013. Antibiotic resistance, antimicrobial residues and bacterial community composition in urban wastewater. *Water Res* 47, 1875–1887. https://doi.org/10.1016/j.watres.2013.01.010

Ogura, Y., Ueda, T., Nukazawa, K., Hiroki, H., Xie, H., Arimizu, Y., Hayashi, T., Suzuki, Y., 2020. The level of antimicrobial resistance of sewage isolates is higher than that of river isolates in different *Escherichia coli* lineages. *Sci Rep* 10, 17880. https://doi.org/10.1038/s41598-020-75065-x

Ojemaye, M.O., Adefisoye, M.A., Okoh, A.I., 2020. Nanotechnology as a viable alternative for the removal of antimicrobial resistance determinants from discharged municipal effluents and associated watersheds: A review. *J Environ Manage* 275, 111234. https://doi.org/10.1016/j.jenvman.2020.111234

Parsley, L.C., Consuegra, E.J., Kakirde, K.S., Land, A.M., Harper, W.F., Liles, M.R., 2010. Identification of diverse antimicrobial resistance determinants carried on bacterial, plasmid, or viral metagenomes from an activated sludge microbial assemblage. *Appl Environ Microbiol* 76, 3753–3757. https://doi.org/10.1128/AEM.03080-09

Prasad, R., Thirugnanasanbandham, K., 2019. Advances Research on Nanotechnology for Water Technology. Springer International Publishing. https://www.springer.com/us/book/9783030023805

Rawat, J., Sanwal, P., Saxena, J., Prasad, R. (2023). Exploring the biochar as a suitable carrier for a bioinoculant Aspergillus niger K7 and its consequence on Eleusine coracana in field studies. *Journal of Agriculture and Food Research*. https://doi.org/10.1016/j.jafr.2023.100825

Roca, I., Akova, M., Baquero, F., Carlet, J., Cavaleri, M., Coenen, S., Cohen, J., Findlay, D., Gyssens, I., Heure, O.E., Kahlmeter, G., Kruse, H., Laxminarayan, R., Liébana, E., López-Cerero, L., MacGowan, A., Martins, M., Rodríguez-Baño, J., Rolain, J.-M., Segovia, C., Sigauque, B., Tacconelli, E., Wellington, E., Vila, J., 2015. The global threat of antimicrobial resistance: Science for intervention. *New Microbes New Infect* 6, 22–29. https://doi.org/10.1016/j.nmni.2015.02.007

Rytkönen, A., Tiwari, A., Hokajärvi, A.-M., Uusheimo, S., Vepsäläinen, A., Tulonen, T., Pitkänen, T., 2021. The use of ribosomal RNA as a microbial source tracking target highlights the assay host-specificity requirement in water quality assessments. *Front Microbiol* 12, 673306.

Sakkas, H., Bozidis, P., Ilia, A., Mpekoulis, G., Papadopoulou, C., 2019. Antimicrobial resistance in bacterial pathogens and detection of carbapenemases in *Klebsiella pneumoniae* isolates from hospital wastewater. *Antibiotics* 8, 85. https://doi.org/10.3390/antibiotics8030085

Serwecińska, L., 2020. Antimicrobials and antibiotic-resistant bacteria: A risk to the environment and to public health. *Water* 12, 3313. https://doi.org/10.3390/w12123313

Singh, S., Kumar, V., Dhanjal, D.S., Datta, S., Bhatia, D., Dhiman, J., Samuela, J., Prasad, R., Singh, J., (2020). A sustainable paradigm of sewage sludge biochar: Valorization, opportunities, challenges and future prospects. *Journal of Cleaner Production*. https://doi.org/10.1016/j.jclepro.2020.122259

Sha'arani, S., Azizan, S.N.F., Md Akhir, F.N., Muhammad Yuzir, M.A., Othman, N., Zakaria, Z., Mohd Noor, M.J.M., Hara, H., 2019. Removal efficiency of Gram-positive and Gram-negative bacteria using a natural coagulant during coagulation, flocculation, and sedimentation processes. *Water Sci Technol* 80, 1787–1795. https://doi.org/10.2166/wst.2019.433

Sharma, V.K., Johnson, N., Cizmas, L., McDonald, T.J., Kim, H., 2016. A review of the influence of treatment strategies on antibiotic resistant bacteria and antibiotic resistance genes. *Chemosphere* 150, 702–714. https://doi.org/10.1016/j.chemosphere.2015.12.084

Sonowal, S., Koch, N., Sarma, H., Prasad, K., Prasad, R. (2023). A review on magnetic nanobiochar with their use in environmental remediation and high-value applications. *Journal of Nanomaterial*, article ID 4881952. https://doi.org/10.1155/2023/4881952

Tran, N.H., Chen, H., Reinhard, M., Mao, F., Gin, K.Y.-H., 2016. Occurrence and removal of multiple classes of antibiotics and antimicrobial agents in biological wastewater treatment processes. *Water Res* 104, 461–472. https://doi.org/10.1016/j.watres.2016.08.040

Ukuhor, H.O., 2021. The interrelationships between antimicrobial resistance, COVID-19, past, and future pandemics. *J Infect Public Heal* 14, 53–60. https://doi.org/10.1016/j.jiph.2020.10.018

Uluseker, C., Kaster, K.M., Thorsen, K., Basiry, D., Shobana, S., Jain, M., Kumar, G., Kommedal, R., Pala-Ozkok, I., 2021. A review on occurrence and spread of antibiotic resistance in wastewaters and in wastewater treatment plants: Mechanisms and perspectives. *Front Microbiol* 12, 717809.

Usman, M., Farooq, M., Hanna, K., 2020. Environmental side effects of the injudicious use of antimicrobials in the era of COVID-19. *Sci Tot Environ* 745, 141053. https://doi.org/10.1016/j.scitotenv.2020.141053
Verburg, I., García-Cobos, S., Hernández Leal, L., Waar, K., Friedrich, A.W., Schmitt, H., 2019. Abundance and antimicrobial resistance of three bacterial species along a complete wastewater pathway. *Microorganisms* 7, 312. https://doi.org/10.3390/microorganisms7090312
Wang, J., Long, Y., Yu, G., Wang, G., Zhou, Z., Li, P., Zhang, Y., Yang, K., Wang, S., 2022. A review on microorganisms in constructed wetlands for typical pollutant removal: Species, function, and diversity. *Front Microbiol* 13, 845725.
Wang, M., Chen, H., Liu, S., Xiao, L., 2021. Removal of pathogen and antibiotic resistance genes from waste activated sludge by different pre-treatment approaches. *Sci Tot Environ* 763, 143014. https://doi.org/10.1016/j.scitotenv.2020.143014
Wang, M., Shen, W., Yan, L., Wang, X.-H., Xu, H., 2017. Stepwise impact of urban wastewater treatment on the bacterial community structure, antibiotic contents, and prevalence of antimicrobial resistance. *Environ Pollut* 231, 1578–1585. https://doi.org/10.1016/j.envpol.2017.09.055
Wang, Q., Liang, J., Zhang, S., Yoza, B.A., Li, Q.X., Zhan, Y., Ye, H., Zhao, P., Chen, C., 2020. Characteristics of bacterial populations in an industrial scale petrochemical wastewater treatment plant: Composition, function and their association with environmental factors. *Environ Res* 189, 109939. https://doi.org/10.1016/j.envres.2020.109939
WHO, 2020. Antimicrobial Resistance. https://www.who.int/news-room/fact-sheets/detail/antimicrobial-resistance (accessed 10.24.20).
World Health Organization, 2021. Global Antimicrobial Resistance and Use Surveillance System (GLASS) Report: 2021. https://iris.who.int/handle/10665/341666.
Xue, G., Jiang, M., Chen, H., Sun, M., Liu, Y., Li, X., Gao, P., 2019. Critical review of ARGs reduction behavior in various sludge and sewage treatment processes in wastewater treatment plants. *Crit Rev Environ Sci Technol* 49, 1623–1674. https://doi.org/10.1080/10643389.2019.1579629
Yadav, S., Kapley, A., 2019. Exploration of activated sludge resistome using metagenomics. *Sci Total Environ* 692, 1155–1164. https://doi.org/10.1016/j.scitotenv.2019.07.267
Yang, Y., Ye, S., Zhang, C., Zeng, G., Tan, X., Song, B., Zhang, P., Yang, H., Li, M., Chen, Q., 2021. Application of biochar for the remediation of polluted sediments. *J Hazard Mater* 404, 124052. https://doi.org/10.1016/j.jhazmat.2020.124052
Yao, Q., Borjihan, Q., Qu, H., Guo, Y., Zhao, Z., Qiao, L., Li, T., Dong, A., Liu, Y., 2021. Cow dung-derived biochars engineered as antibacterial agents for bacterial decontamination. *J Environ Sci* 105, 33–43. https://doi.org/10.1016/j.jes.2020.12.022
Zhang, S., Lu, Y.-X., Zhang, J.-J., Liu, S., Song, H.-L., Yang, X.-L., 2020. Constructed wetland revealed efficient sulfamethoxazole removal but enhanced the spread of antibiotic resistance genes. *Molecules* 25, 834. https://doi.org/10.3390/molecules25040834

2 Wastewater Treatment Plants as Hubs of Antibiotic Resistance

Occurrence and Control

Monali Nanda, Mamta Patra Shahi, and Sushil Kumar Shahi

2.1 INTRODUCTION

Antibiotic resistance (AR) is a growing global health crisis, and wastewater treatment plants (WWTPs) have become critical hubs for the development and dissemination of antibiotic-resistant bacteria (ARB) and antibiotic resistance genes (ARGs). WWTPs receive wastewater from diverse sources, including hospitals, households, and industrial facilities, which may contain antibiotics, resistant bacteria, and other pharmaceuticals (Omuferen et al., 2022). During treatment, these plants often create conditions that support the survival and even proliferation of resistant bacteria. Conventional wastewater treatment processes—including primary, secondary, and tertiary treatments—are not specifically designed to remove antibiotics or eliminate resistant bacteria. Consequently, ARB and ARGs can persist through treatment and be discharged into natural water bodies, posing significant risks to human health, agriculture, and ecosystems (Li et al., 2022).

The presence of subinhibitory concentrations of antibiotics in WWTPs acts as a selective pressure, favoring the growth of resistant strains over susceptible ones. Additionally, WWTPs harbor diverse microbial communities, providing an environment conducive to horizontal gene transfer (HGT)—a process by which bacteria exchange genetic material, including ARGs (Wei et al., 2021a). This exchange occurs through mechanisms such as conjugation, transformation, and transduction, facilitating the rapid spread of resistance among bacterial populations. The discharged effluent, containing ARB and ARGs, contaminates surface waters, groundwater, and soil, ultimately entering the human food chain through irrigation and aquaculture. This environmental dissemination of resistance exacerbates the global antimicrobial resistance (AMR) crisis, making infections more difficult to treat and increasing the risk of disease outbreaks.

Addressing AR in WWTPs is crucial due to the central role these facilities play in the environmental spread of ARB and ARGs. Conventional treatment processes often fail to completely remove these contaminants, leading to their accumulation in natural ecosystems and potential integration into human and animal microbiomes (Amarasiri et al., 2020). This uncontrolled dissemination of resistance heightens the challenge of combating AMR, threatening public health and food security on a global scale.

To mitigate these risks, WWTP technologies must be upgraded with advanced treatment methods such as membrane bioreactors (MBRs), advanced oxidation processes (AOPs), and bioaugmentation techniques. Implementing stricter regulations on antibiotic discharge; promoting the prudent use of antibiotics across healthcare, agriculture, and industrial sectors; and enhancing surveillance of ARB and ARGs in wastewater are essential measures to curb resistance.

 DOI: 10.1201/9781003269823-2

Addressing AR in WWTPs is a vital component of the global strategy to combat AMR and safeguard public health.

2.2 ANTIBIOTIC RESISTANCE: A GROWING GLOBAL CONCERN

AR is the ability of microorganisms—including bacteria, fungi, viruses, and parasites—to survive and proliferate despite exposure to drugs that were previously effective in killing or inhibiting them. This resistance develops through genetic mutations or the acquisition of resistance genes, often facilitated by HGT. These genetic adaptations enable microorganisms to withstand antimicrobial agents, rendering treatments less effective and leading to prolonged infections, increased healthcare costs, and higher mortality rates (World Health Organization, 2020).

The widespread misuse and overuse of antibiotics in human medicine, veterinary care, and agriculture have significantly accelerated the emergence and spread of resistance. As a result, many common infections are becoming harder to treat, posing a severe threat to global public health by reducing the effectiveness of available antibiotics and limiting treatment options.

Antibiotics are classified into five major groups based on their mechanism of action:

a. **Inhibition of Cell Wall Synthesis:** Prevents bacteria from forming a protective cell wall, leading to cell death (e.g., vancomycin, cephalosporins, β-lactams, and bacitracin).
b. **Inhibition of Protein Synthesis:** Blocks bacterial ribosomes, preventing protein production necessary for growth and survival (e.g., aminoglycosides, chloramphenicol, tetracycline, and linezolid).
c. **Inhibition of Nucleic Acid Synthesis:** Disrupts bacterial DNA or RNA synthesis, preventing replication and transcription (e.g., rifampin, metronidazole, quinolones, and fluoroquinolones).
d. **Antimetabolites:** Interfere with essential metabolic pathways, inhibiting bacterial growth (e.g., trimethoprim, dapsone, and sulfonamides).
e. **Disruption of the Cell Membrane:** Damages bacterial membranes, leading to leakage of cellular contents and cell death (e.g., polymyxin and daptomycin) (Uluseker et al., 2021).

As AR continues to rise, there is an urgent need for global action, including stricter antibiotic stewardship, the development of novel antimicrobial agents, enhanced surveillance, and public awareness campaigns to promote responsible antibiotic use. Addressing this crisis requires a multidisciplinary approach involving health care, research, policymaking, and public engagement to safeguard the effectiveness of antibiotics for future generations.

2.3 MECHANISMS OF ANTIBIOTIC RESISTANCE

WWTPs play a crucial role in the spread of ARGs and ARB. They receive antimicrobial-resistant microorganisms and ARGs from various sources, including livestock farms, hospitals, pharmaceutical industries, and households. As a result, WWTPs are considered significant contributors to the dissemination of ARGs in aquatic environments. The process by which AR develops and spreads through wastewater systems is illustrated in Figure 2.1.

Bacteria have evolved multiple mechanisms to withstand the effects of antibiotics, allowing them to survive in the presence of antimicrobial agents. These resistance mechanisms can be intrinsic (naturally occurring) or acquired (through genetic mutation or HGT). The four primary mechanisms of AR include enzymatic degradation or modification of antibiotics, efflux pump activity, target modification, and reduced permeability.

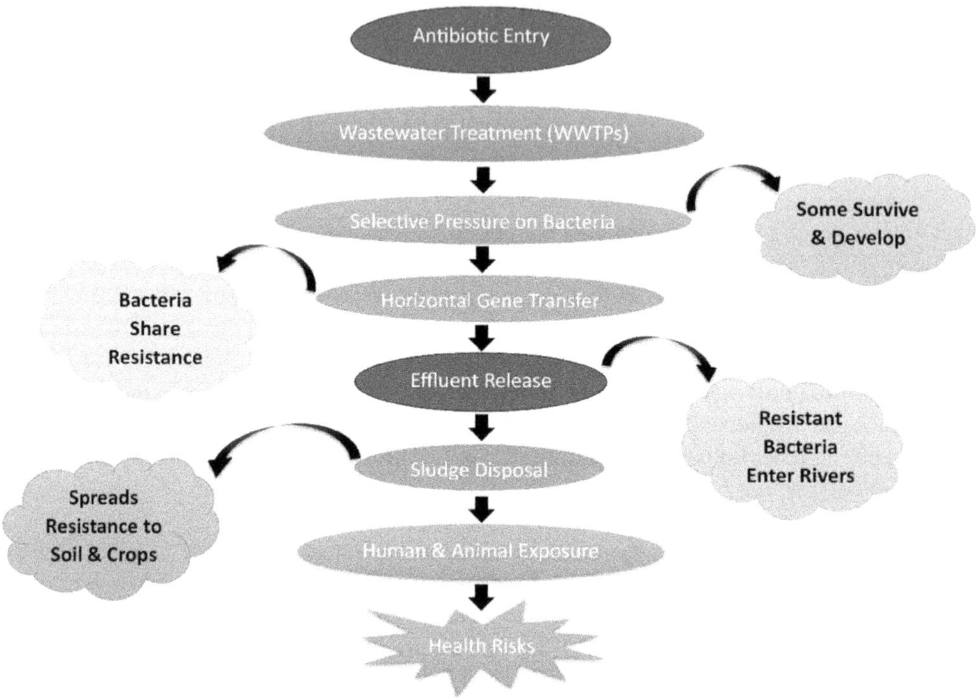

FIGURE 2.1 Pathways of antibiotic resistance development and dissemination through wastewater treatment systems.

2.3.1 Enzymatic Degradation or Modification of Antibiotics

Certain bacteria produce enzymes that chemically modify or degrade antibiotics, rendering them ineffective. These enzymes function by hydrolyzing the antibiotic structure, altering functional groups, or breaking key chemical bonds.

a. **β-Lactamases:** Enzymes that hydrolyze the β-lactam ring found in penicillins, cephalosporins, carbapenems, and monobactams. Examples include penicillinase (inactivates penicillin), extended-spectrum β-lactamases (ESBLs) (hydrolyze third-generation cephalosporins), and carbapenemases such as KPC, NDM-1, OXA-48, VIM, and IMP, which degrade carbapenems (Bush & Jacoby, 2010; Nordmann et al., 2012).
b. **Aminoglycoside-Modifying Enzymes:** Modify aminoglycoside antibiotics (e.g., gentamicin, tobramycin, and amikacin) through
   i. Acetylation (AAC enzymes)
   ii. Phosphorylation (APH enzymes)
   iii. Adenylation (ANT enzymes)

   This modification prevents aminoglycosides from binding to bacterial ribosomes (Ramirez & Tolmasky, 2010).
c. Other Antibiotic-Inactivating Enzymes:
   i. Macrolide esterases break down macrolide antibiotics.
   ii. Chloramphenicol acetyltransferase inactivates chloramphenicol by acetylation.
   iii. Rifampin ADP-ribosyltransferases modify rifampin, preventing its action against bacterial RNA polymerase (Doi & Arakawa, 2007).

2.3.2 Efflux Pumps (Active Drug Removal from Bacterial Cells)

Efflux pumps are membrane transport proteins that actively expel antibiotics from bacterial cells, reducing intracellular drug concentrations and limiting antibiotic efficacy. These pumps play a key role in multidrug resistance (MDR) by expelling a broad range of antimicrobial agents (Li & Nikaido, 2009).

2.3.2.1 Major Efflux Pump Families in Bacteria

- ATP-Binding Cassette Transporters
- Major Facilitator Superfamily
- Resistance-Nodulation-Division Family
- Small Multidrug Resistance Family
- Multidrug and Toxic Compound Extrusion Family

2.3.2.1.1 Examples of Efflux Pumps and Their Role in Resistance

- **AcrAB-TolC in *E. coli, Klebsiella pneumoniae,* and *Salmonella spp.*:** Confers resistance to fluoroquinolones, tetracyclines, β-lactams, and macrolides (Nikaido, 2009).
- **MexAB-OprM in *Pseudomonas aeruginosa*:** Provides resistance to β-lactams, aminoglycosides, and fluoroquinolones (Poole, 2001).
- **NorA in *Staphylococcus aureus*:** Involved in fluoroquinolone resistance (Hooper, 2005).

Efflux pump activity is often upregulated due to environmental stress, sublethal antibiotic exposure, or genetic mutations, significantly reducing antibiotic effectiveness (Piddock, 2006).

2.3.3 Target Modification (Altering Drug Binding Sites)

Bacteria develop resistance by modifying antibiotic binding sites, thereby decreasing the drug's ability to inhibit its target.

- **Modification of Ribosomal Targets:**
  i. Macrolide, lincosamide, and streptogramin resistance results from methylation of 23S rRNA by erm genes, reducing drug affinity for the ribosome.
  ii. Aminoglycoside resistance can result from mutations in the 16S rRNA gene, decreasing aminoglycoside binding to bacterial ribosomes (Doi et al., 2016).
- **Modification of Cell Wall Synthesis Enzymes:**
  i. Methicillin-resistant *Staphylococcus aureus* (MRSA) expresses the mecA gene, which encodes an altered penicillin-binding protein (PBP2a) with reduced β-lactam affinity (Chambers & DeLeo, 2009).
  ii. Vancomycin-resistant enterococci replace D-Ala-D-Ala in peptidoglycan with D-Ala-D-Lac, reducing vancomycin binding (Arthur & Courvalin, 1993).
- **Modification of DNA Targets:**
  i. Fluoroquinolone resistance results from mutations in DNA gyrase (gyrA, gyrB) and topoisomerase IV (parC, parE), lowering drug-binding affinity (Jacoby, 2005).
  ii. Rifampin resistance is caused by mutations in the rpoB gene, encoding the β-subunit of RNA polymerase, preventing rifampin from binding (Zhang et al., 2014).

2.3.4 Reduced Permeability (Decreased Antibiotic Uptake)

Bacteria, especially Gram-negative species, can decrease antibiotic entry by modifying outer membrane permeability.

- **Porin Channel Modifications:**
  i. Loss or mutation of porins reduces the uptake of hydrophilic antibiotics.
  ii. Example: OmpF porin downregulation in *E. coli* reduces β-lactam susceptibility (Hernandez-Allés et al., 1999).
  iii. Carbapenem resistance in *Pseudomonas aeruginosa* occurs due to the loss of the OprD porin (Quale et al., 2006).
- **Biofilm Formation:**
  i. Bacterial biofilms create a protective matrix that limits antibiotic penetration, leading to chronic infections.
  ii. Example: *Pseudomonas aeruginosa* biofilms in cystic fibrosis patients reduce antibiotic efficacy (Donlan & Costerton, 2002).

2.4 FACTORS DRIVING ANTIBIOTIC RESISTANCE

AR has emerged as a critical global health concern, primarily driven by the overuse and misuse of antibiotics, agricultural practices, inadequate infection control, and globalization. The increasing prevalence of MDR bacteria poses a severe threat to public health, leading to treatment failures, prolonged illnesses, and higher mortality rates. Understanding the key drivers behind AR is essential for developing effective control strategies and policy interventions.

One of the primary contributors to AR is the overuse and misuse of antibiotics in both human and veterinary medicine. Antibiotics are frequently prescribed for conditions that do not require them, such as viral infections, leading to unnecessary exposure and selection pressure for resistant strains (Ventola, 2015). Moreover, many patients fail to complete their prescribed antibiotic courses, allowing partially resistant bacteria to survive and proliferate (O'Neill, 2016). Self-medication and over-the-counter availability of antibiotics further exacerbate the issue, particularly in low- and middle-income countries (LMICs), where regulatory frameworks for antibiotic dispensing are weak (Llor & Bjerrum, 2014).

Another significant driver is the agricultural use of antibiotics, particularly in livestock farming. The routine administration of antibiotics for growth promotion and disease prevention in food-producing animals has been linked to the emergence of resistant bacterial strains, which can spread to humans through direct contact, food consumption, and environmental contamination (Marshall & Levy, 2011). Studies have shown that antibiotic residues persist in meat, milk, and water sources, contributing to the horizontal transfer of resistance genes among bacterial populations (Boeckel et al., 2017). The widespread use of antibiotics in aquaculture and crop production also plays a role in resistance development, as antibiotic runoff from agricultural lands can contaminate soil and waterways, creating reservoirs of resistant microbes (Cabello et al., 2013).

Poor infection control and inadequate sanitation in healthcare and community settings further accelerate the spread of ARB. Inadequate hand hygiene, improper sterilization of medical equipment, and ineffective isolation of infected patients facilitate the transmission of resistant organisms in hospitals and long-term care facilities (Laxminarayan et al., 2013). In LMICs, limited access to clean water and proper sanitation significantly increases the risk of bacterial infections, necessitating higher antibiotic use and thereby promoting resistance (Prüss-Ustün et al., 2019). Additionally, substandard and falsified antibiotics, which contain incorrect dosages or poor-quality ingredients, contribute to treatment failure and resistance evolution (Newton et al., 2011).

Globalization has also played a crucial role in the spread of resistant bacteria across international borders. The movement of people, animals, and food products enables the dissemination of resistant pathogens from one region to another, complicating efforts to contain outbreaks (Wong et al., 2016). International medical tourism, where patients seek treatments in countries with high rates of AR, further facilitates the cross-border transmission of resistant strains (Kumarasamy et al., 2010). Additionally, WWTPs and urban sewage systems serve as critical hubs for the exchange of ARGs, as untreated or partially treated effluents introduce resistant bacteria into the environment (Rizzo et al., 2013a).

The interplay of these factors highlights the complexity of AR, necessitating a multifaceted approach to mitigation. Strategies such as antimicrobial stewardship programs, stringent antibiotic regulations, improved infection control, and global surveillance systems are essential to curb resistance. International cooperation, public awareness campaigns, and research on novel antimicrobial agents also play a crucial role in addressing this growing crisis.

2.5 IMPACT OF ANTIBIOTIC RESISTANCE

The impact of AR is profound and multifaceted, affecting individuals, healthcare systems, economies, and ecosystems worldwide. It poses a significant challenge to public health, medical advancements, and environmental sustainability.

2.5.1 Increased Mortality and Morbidity

Infections caused by antibiotic-resistant pathogens are more difficult to treat, often requiring stronger or combination therapies, leading to prolonged hospital stays, intensive care, and a higher risk of complications or death. AR is already responsible for tens of thousands of deaths annually, and this number is expected to rise dramatically unless effective control measures are implemented (Barton et al., 2020).

2.5.2 Economic Burden

The economic impact of AR is immense, placing a heavy financial strain on individuals, healthcare systems, and global economies. The need for more expensive treatments, extended hospitalizations, and increased healthcare resources significantly raises medical costs. According to the World Health Organization (WHO), the global economic burden of AR could reach $100 trillion by 2050 if current trends persist (O'Neill, 2014). The loss of productivity due to prolonged illness and disability further exacerbates the economic consequences.

2.5.3 Threat to Medical Advancements

Antibiotics are crucial not only for treating infections but also for supporting various medical procedures, including surgeries, cancer treatments, and organ transplants. Without effective antibiotics, the safety and success of these procedures are compromised, increasing the risk of post-operative infections, treatment failures, and higher mortality rates (Goff et al., 2017). The rise of resistant infections may also limit the effectiveness of immunosuppressive therapies, further complicating disease management.

2.5.4 Impact on Aquatic Ecosystems

ARB in aquatic ecosystems can disrupt microbial communities and alter biodiversity. The equilibrium of microbial populations is vital for nutrient cycling, organic matter decomposition, and overall ecosystem health. The persistence of resistant bacteria can lead to the dominance of resistant strains, outcompeting susceptible ones and potentially reducing ecosystem resilience (Furtula et al., 2013). This disruption can have cascading effects on higher trophic levels, including fish and invertebrates, which rely on balanced microbial communities for their survival. Additionally, the spread of resistance genes through water systems increases the risk of human exposure and further perpetuates resistance.

2.6 OCCURRENCE OF ANTIBIOTIC RESISTANCE IN WASTEWATER TREATMENT PLANTS (WWTPs)

WWTPs are major contributors to the release of ARB and ARGs into the environment. Extensive research has been conducted worldwide to assess resistance patterns in WWTPs, highlighting their role in the dissemination of AR.

WWTPs are designed to treat municipal and industrial wastewater, ensuring it meets environmental safety standards before discharge or reuse. The primary objectives of these facilities include the removal of organic matter, nutrients, pathogens, and various contaminants. Key treatment processes involve:

- **Physical methods** (e.g., sedimentation and filtration)
- **Biological treatments** (e.g., activated sludge and biofilm reactors)
- **Chemical treatments** (e.g., chlorination and ozonation) (Xu et al., 2015; Omuferen et al., 2022).

Despite their efficiency in reducing conventional pollutants, WWTPs struggle to completely eliminate micro-pollutants, including pharmaceutical residues and antibiotics. This limitation creates selective pressures that encourage the persistence and proliferation of ARB. Consequently, WWTPs inadvertently function as reservoirs and amplification sites for AR, facilitating the spread of ARGs into natural water bodies.

2.6.1 Sources of Antibiotics and Resistant Bacteria in Wastewater

The presence of antibiotics and resistant bacteria in wastewater is a critical concern for public health and environmental sustainability. Various sources contribute to this contamination, including human and veterinary pharmaceutical use, hospital and industrial wastewater discharge, and agricultural runoff.

2.6.1.1 Human and Veterinary Pharmaceutical Usage

A major source of antibiotics in wastewater is their widespread use in human and veterinary medicine. Antibiotics consumed by humans are often excreted as active metabolites, entering municipal wastewater treatment systems. Studies suggest that up to 90% of administered antibiotics may remain unmetabolized and be excreted into wastewater streams (Kümmerer, 2009).

Similarly, veterinary antibiotics used for disease treatment and prevention in livestock are excreted into manure. This manure, when washed into nearby water systems or applied as fertilizer, contributes to the contamination of surface and groundwater (Martínez, 2008). The continuous input of antibiotic residues into aquatic ecosystems accelerates the spread of ARGs, posing long-term risks to microbial ecology and human health.

2.6.1.2 Hospital and Industrial Wastewater Contributions

Hospitals serve as significant sources of ARB and ARGs due to the high volume of antibiotics used in clinical settings. Hospital effluents contain a diverse range of antibiotics, including fluoroquinolones and β-lactams—both classified as critical for human medicine (Kümmerer, 2009). The high concentration of antibiotics in hospital wastewater creates selective pressure, promoting the emergence of MDR bacteria.

Additionally, pharmaceutical manufacturing industries contribute to antibiotic pollution through the discharge of residual antibiotics in wastewater. Effluents from these facilities often contain unmetabolized antibiotics at concentrations high enough to drive resistance evolution in environmental microbial communities (Larsson et al., 2007). In regions where pharmaceutical waste disposal is poorly regulated, these discharges can create localized hotspots of AR.

2.6.1.3 Agricultural Runoff and Its Impact

Agriculture is another major contributor to the spread of ARB and ARGs in water systems. Antibiotics are frequently used in livestock farming for disease prevention and growth promotion, leading to their accumulation in manure and slurry. When these materials are applied as fertilizers, antibiotic residues leach into soil and water, further propagating resistance genes (Sarmah et al., 2006).

Additionally, aquaculture practices, where antibiotics are directly introduced into water to prevent fish diseases, significantly contribute to antibiotic pollution in aquatic ecosystems (Cabello, 2006). Continuous exposure of environmental microorganisms to low antibiotic concentrations in agricultural runoff fosters the selection and proliferation of resistant bacteria, increasing the potential for ARG transfer across microbial populations.

2.6.2 Pathways of Antibiotic-Resistant Bacteria (ARB) and Antibiotic Resistance Genes (ARGs) in WWTPs

ARB and ARGs follow distinct pathways within WWTPs, from their sources to their eventual release into the environment. Antibiotics consumed by humans and animals are often only partially metabolized, leading to the excretion of active residues into wastewater systems. This contributes significantly to the antibiotic load entering WWTPs. Hospital wastewater, in particular, contains high concentrations of antibiotics and MDR pathogens, making it a major hotspot for the emergence and spread of AR. Similarly, discharges from pharmaceutical and veterinary industries introduce elevated antibiotic levels and potentially resistant bacteria into the wastewater stream.

Once in WWTPs, ARB and ARGs exhibit considerable resilience, often surviving conventional treatment processes despite efforts to reduce microbial loads. A major concern is their accumulation in sludge—a by-product of biological treatment—where they may become highly concentrated. Even after treatment, residual ARB and ARGs often persist in the effluent, which is subsequently discharged into aquatic ecosystems. Additionally, the use of treated sludge as agricultural fertilizer facilitates the transfer of ARGs into soils and crops, further amplifying their environmental impact.

WWTPs also act as hotspots for the proliferation and evolution of AR due to their unique microbial environments. Processes such as HGT enable the spread of ARGs among bacteria. Conjugation allows bacteria to exchange plasmids carrying ARGs, transformation enables bacteria to incorporate free DNA from lysed cells, and transduction facilitates the transfer of ARGs via bacteriophages. Furthermore, sublethal concentrations of antibiotics present in WWTPs create selective pressure, favoring the survival and proliferation of resistant strains. Co-selection with other environmental stressors, such as heavy metals and biocides, further promotes resistance. The presence of biofilms in reactor systems enhances microbial stability and fosters gene exchange, making them reservoirs for ARG retention.

A meta-analysis by Riquelme et al. (2021) revealed that ARGs such as *sul1*, *bla*, and *tet* genes are commonly detected in WWTP effluents worldwide, indicating the global scale of the issue. Karkman et al. (2018) found that hospital WWTPs exhibit significantly higher concentrations of ARGs compared to municipal facilities, emphasizing their role in resistance dissemination. Some ARGs, their associated antibiotic classes, and regulatory mechanisms are outlined in Table 2.1. Research by Rizzo et al. (2013b) demonstrated that AOPs and MBRs are more effective than conventional methods in reducing ARGs, though complete elimination remains a challenge. Zhang et al. (2020a) highlighted the persistence and proliferation of ARGs in downstream aquatic environments, underscoring the critical role of WWTPs in the broader dissemination of resistance genes.

TABLE 2.1
Most Common ARGs Found in WWTPs and Their Associated Antibiotic Class

S. No.	Antibiotic Resistance Gene (ARG)	Associated Antibiotic Class/ Associated Regulatory Systems	Reference(s)
1.	bla_{CTX-M}	Extended-spectrum β-lactamase	Ibrahim et al. (2021)
2.	*aac(6')-Ib*	Aminoglycosides	Quiroga et al. (2015)
3.	NDM-1	Carbapenems	Uppu et al. (2015)
4.	acrB	Multidrug efflux pump	Szczepanowski et al. (2009); Zhang et al. (2011)
5.	acrD	Multidrug efflux pump	Szczepanowski et al. (2009)
6.	adeA	Multidrug efflux pump	Zhang et al. (2011)
7.	amrB	Multidrug efflux pump	Yang et al. (2013)
8.	*ermB*	Macrolides	Dzyubak and Yap (2016)
9.	*gnrC*	Global nitrate and nitrite regulation	Xu et al. (2015)
10.	gyrA	Fluoroquinolones	Xu et al. (2015)
11.	*intI1*	Class 1 integrons	Chen and Zhang (2013)
12.	mdtF	Multidrug efflux pump	Zhang et al. (2011); Yang et al. (2014)
13.	mdtG	Multidrug efflux pump	Yang et al. (2014)
14.	mdtH	Multidrug efflux pump	Yang et al. (2014)
15.	mdtN	Multidrug efflux pump	Yang et al. (2014)
16.	mdtO	Multidrug efflux pump	Yang et al. (2013)
17.	*mecA*	Methicillin (β-lactams)	Hafez et al. (2009)
18.	mexB	Multidrug efflux pump	Szczepanowski et al. (2009); Yang et al. (2013)
19.	mexD	Multidrug efflux pump	Szczepanowski et al. (2009); Yang et al. (2013)
20.	mexF	Multidrug efflux pump	Szczepanowski et al. (2009); Yang et al. (2013); Zhang et al. (2011); Yang et al. (2014)
21.	mexI	Multidrug efflux pump	Szczepanowski et al. (2009)
22.	mexW	Multidrug efflux pump	Zhang et al. (2011)
23.	mexY	Multidrug efflux pump	Szczepanowski et al. (2009)
24.	norM	Multidrug efflux pump	Yang et al. (2013)
25.	orf11	Multidrug efflux pump	Szczepanowski et al. (2009)
26.	ortA	Aminoglycosides	Szczepanowski et al. (2009); Biswal et al. (2014); Harnisz et al. (2015)
27.	parC	Fluoroquinolones	Xu et al. (2015)
28.	qacB	Quaternary ammonium compounds (QACs)	Jiao et al. (2018)
29.	*qacEΔ1*	Quaternary ammonium compounds	Chen et al. (2023)
30.	qacEΔ1	Quaternary ammonium compounds (QACs)	Jiao et al. (2018)
31.	qacEΔ1–01	Quaternary ammonium compounds (QACs)	Jiao et al. (2018)
32.	qacF	Quaternary ammonium compounds (QACs)	Szczepanowski et al. (2009)
33.	qacG2	Quaternary ammonium compounds (QACs)	Szczepanowski et al. (2009)

(Continued)

TABLE 2.1 (*Continued*)
Most Common ARGs Found in WWTPs and Their Associated Antibiotic Class

S. No.	Antibiotic Resistance Gene (ARG)	Associated Antibiotic Class/ Associated Regulatory Systems	Reference(s)
34.	qacH	Quaternary ammonium compounds (QACs)	Szczepanowski et al. (2009)
35.	qnr	Fluoroquinolones	Szczepanowski et al. (2009)
36.	qnrA	Fluoroquinolones	Marti et al. (2013)
37.	qnrA3	Fluoroquinolones	Marti et al. (2013)
38.	qnrB	Fluoroquinolones	Xu et al. (2015)
39.	qnrB1	Fluoroquinolones	Szczepanowski et al. (2009)
40.	qnrB2	Fluoroquinolones	Xu et al. (2015)
41.	qnrB4	Fluoroquinolones	Xu et al. (2015)
42.	qnrB5	Fluoroquinolones	Xu et al. (2015)
43.	qnrC	Fluoroquinolones	Xu et al. (2015)
44.	qnrD	Fluoroquinolones	Biswal et al. (2014)
45.	*qnrS*	Fluoroquinolones	Xu et al. (2015)
46.	qnrS1	Fluoroquinolones	Biswal et al. (2014)
47.	qnrS2	Fluoroquinolones	Szczepanowski et al. (2009)
48.	sdeY	Multidrug efflux pump	Yang et al. (2013)
49.	smeB	Multidrug efflux pump	Zhang et al. (2011)
50.	*Sul1*	Sulfonamides	
51.	*tetA*	Tetracyclines	Yang et al. (2013)
52.	tetA	Tetracyclines	Yang et al. (2013); Jiao et al. (2018)
53.	tetA(33)	Tetracyclines	Yang et al. (2013)
54.	tetA(39)	Tetracyclines	Harnisz et al. (2015); Zhang and Zhang (2011)
55.	tetA(C)	Tetracyclines	Yang et al. (2013); Yang et al. (2014)
56.	tetA(M)	Tetracyclines	Yang et al. (2013); Zhang and Zhang (2011)
57.	tetA(P)	Tetracyclines	Yang et al. (2013); Yang et al. (2014)
58.	tetA(Q)	Tetracyclines	Yang et al. (2013); Zhang et al. (2011); Yang et al. (2014)
59.	tetB	Tetracyclines	Szczepanowski et al. (2009);
60.	tetB(P)	Tetracyclines	Laht et al. (2014)
61.	tetC	Tetracyclines	Szczepanowski et al. (2009); Zhang and Zhang (2011); Zhang et al. (2016)
62.	tetD	Tetracyclines	Xu et al. (2015)
63.	tetE	Tetracyclines	Szczepanowski et al. (2009)
64.	tetG	Tetracyclines	Szczepanowski et al. (2009); Biswal et al. (2014)
65.	tetH	Tetracyclines	Zhang and Zhang (2011); Zhang et al. (2016)
66.	tetK	Tetracyclines	Szczepanowski et al. (2009); Harnisz et al. (2015); Zhang and Zhang (2011)
67.	tetL	Tetracyclines	Szczepanowski et al. (2009)
68.	tetM	Tetracyclines	Bengtsson-Palme et al. (2016)
69.	*tetX*	Tetracyclines	
70.	*vanA*	Vancomycin	Bengtsson-Palme et al. (2016)

These findings underscore the urgent need for improved wastewater treatment technologies and stricter regulations to mitigate the spread of AR through WWTPs. Implementing advanced treatment methods, monitoring ARG levels, and developing targeted strategies for reducing antibiotic residues in wastewater are crucial steps toward controlling the global rise of AR.

2.7 ENVIRONMENTAL IMPACT OF ANTIBIOTIC RESISTANCE FROM WWTPs

AR is a growing global health and environmental crisis, with WWTPs playing a critical role in the dissemination of ARB and ARGs. These resistance determinants originate from domestic, hospital, and industrial effluents and are only partially removed during conventional wastewater treatment processes. As a result, WWTPs serve as both reservoirs and amplifiers of resistance, continuously discharging ARB and ARGs into natural water bodies. This widespread release not only contaminates aquatic ecosystems but also promotes the HGT of resistance traits among microbial communities, exacerbating the global challenge of AR (Rizzo et al., 2013a).

Once ARB and ARGs enter aquatic environments, they disrupt microbial ecology by favoring resistant strains over native microbial populations. This imbalance can negatively impact ecosystem functions, including nutrient cycling and biodiversity. Additionally, the presence of sublethal concentrations of antibiotics and other pharmaceuticals in water bodies creates selective pressure, further enhancing the survival and spread of resistant bacteria (Manaia et al., 2016). These ecological disturbances extend beyond microbial communities, affecting primary producers and altering food web dynamics.

The risks associated with ARB and ARGs in the environment are not limited to ecosystem disruptions; they also pose significant threats to human health. Contaminated water sources used for drinking, irrigation, and recreation become potential pathways for the transfer of resistance traits to human pathogens, increasing the prevalence of drug-resistant infections and complicating global disease management (Berendonk et al., 2015). The issue is particularly severe in regions with inadequate wastewater management systems, where untreated or insufficiently treated effluents directly enter water supplies.

Bioaccumulation of ARB and ARGs in aquatic organisms adds another dimension to the problem. Resistant bacteria and their genetic material can be taken up by aquatic wildlife, such as fish and shellfish, which are then consumed by higher trophic levels, including humans. This biomagnification within food chains further amplifies the risk of AR transmission, necessitating rigorous monitoring and control measures for WWTP effluents (Novo et al., 2013). Additionally, the persistent nature of resistance determinants in sediments and biofilms ensures their long-term environmental impact, as these reservoirs continue to release ARB and ARGs into ecosystems over extended periods.

Addressing the environmental impact of AR from WWTPs requires a multifaceted approach. Advanced treatment technologies, such as membrane filtration, AOPs, and biological treatments specifically targeting ARB and ARGs have shown promise in mitigating resistance spread. Furthermore, policy interventions promoting the rational use of antibiotics, enhanced wastewater management strategies, and stringent regulations on pharmaceutical discharges are essential for controlling this escalating threat. Collaborative efforts between researchers, policymakers, and industry stakeholders are crucial to developing sustainable solutions that minimize the spread of AR and protect both environmental and public health.

2.8 STRATEGIES FOR CONTROLLING ANTIBIOTIC RESISTANCE IN WASTEWATER TREATMENT PLANTS (WWTPs)

AR poses a significant public health threat, and WWTPs play a pivotal role in controlling the spread of ARB and ARGs. Effective strategies for controlling AR in WWTPs involve the implementation of advanced treatment technologies, optimization of biological processes, enhancement of monitoring and regulatory frameworks, and leveraging WWTPs as platforms for public health surveillance.

2.8.1 Advanced Treatment Technologies

Advanced treatment technologies have shown considerable promise in reducing ARB and ARGs in wastewater. MBRs, ozonation, and UV disinfection are key examples. MBRs combine biological treatment with membrane filtration, allowing for the superior retention of microbial contaminants, including ARGs. Studies have indicated that MBRs can achieve over 99% removal of certain ARBs compared to conventional activated sludge systems (Zhang et al., 2019). Ozonation, a powerful oxidation process, destroys bacterial cell walls and degrades DNA, effectively neutralizing ARGs (Rizzo et al., 2013a). UV treatment, often used alongside other disinfection methods, disrupts bacterial DNA, thereby preventing the horizontal transfer of resistance genes. These advanced technologies, particularly when used as post-treatment options, can significantly reduce the presence of ARB and ARGs after biological processing.

2.8.2 Biological Treatment Processes and Their Efficiency

Biological treatment processes, such as activated sludge systems, sequencing batch reactors, and anaerobic digestion, are fundamental components of most WWTPs. These processes rely on microbial communities to degrade organic matter and remove contaminants, including ARB. However, their ability to remove ARGs varies, as HGT can still occur within microbial communities. For instance, conventional activated sludge systems can reduce ARB populations but may not fully eliminate ARGs (Rizzo et al., 2013a).

Enhancing biological treatments, such as bioaugmentation with specific microbial consortia, holds potential for improving the degradation of antibiotics and ARGs. Recent studies suggest that combining anaerobic digestion with thermophilic conditions can achieve higher removal efficiencies by denaturing ARG-carrying plasmids (Sun et al., 2022). Optimizing factors like hydraulic retention time, temperature, and microbial diversity is crucial for improving the efficiency of biological systems in combating AR.

2.8.3 Monitoring and Regulation of Antibiotics and ARGs in Wastewater

A comprehensive framework for monitoring and regulating antibiotics and ARGs in wastewater is essential to control AR. Monitoring programs should focus on quantifying antibiotics, ARB, and ARGs at various stages of treatment to identify hotspots for resistance proliferation. Advanced techniques such as high-throughput sequencing and quantitative polymerase chain reaction are valuable tools for tracking ARGs in WWTPs (Hu et al., 2021).

Regulatory guidelines must also establish permissible limits for antibiotics and resistance markers in treated effluent to minimize environmental and human health risks. Global initiatives like the WHO and the European Union's Urban Wastewater Treatment Directive have set standards for wastewater quality. However, countries should adopt stricter regulations that address AR parameters and enforce the monitoring of ARGs and ARB in wastewater.

2.8.4 WWTPs as Public Health Surveillance Stations

WWTPs can serve as critical platforms for public health surveillance by providing insights into the prevalence of AR in communities. Wastewater-based epidemiology (WBE) can be utilized to detect AR trends and emerging threats by analyzing the genetic material of ARB and ARGs in influent and effluent samples. This approach enables real-time monitoring of AR hotspots, facilitating early detection and targeted interventions (Kitajima et al., 2020).

By integrating WBE with advanced data analytics, public health officials can correlate AR trends in wastewater with clinical settings, offering actionable insights for infection control and antibiotic stewardship programs. This proactive monitoring can guide policies for managing AR and enhance early response efforts to emerging resistance threats.

2.9 CASE STUDIES AND SUCCESS STORIES IN REDUCING ARB AND ARGs IN WASTEWATER TREATMENT PLANTS (WWTPs)

The emergence of ARB and ARGs in WWTPs presents a serious public health challenge. However, several case studies have demonstrated successful interventions in mitigating these risks. Through the adoption of advanced treatment technologies and operational strategies, several WWTPs have significantly reduced ARB and ARG concentrations. These success stories offer valuable insights and potential models for replication in other settings.

2.9.1 Upgrading to Advanced Oxidation Processes (AOPs)

In Denmark, the Egå WWTP upgraded its treatment system by implementing an AOP combined with UV radiation to treat secondary effluent. This combined approach significantly reduced the concentration of ARGs, including sul1 and tetX, which are commonly associated with sulfonamide and tetracycline resistance, respectively. Studies have shown that the use of AOPs led to a 90% reduction in detectable ARGs in the treated effluent, demonstrating the effectiveness of AOPs in addressing AMR in wastewater (Cheng et al., 2019). This case highlights the potential of advanced oxidation technologies in combating the spread of ARB and ARGs.

2.9.2 Membrane Bioreactors (MBRs) in Urban WWTPs

A WWTP in Beijing, China, successfully incorporated an MBR system to tackle the challenge of ARB and ARGs in urban wastewater. The MBR system, which combines biological treatment with membrane filtration, effectively separated bacteria from the treated water, resulting in a 99% reduction in common ARGs such as blaTEM (associated with β-lactam resistance) and qnrS (linked to quinolone resistance). This success underscores the potential of MBRs to enhance microbial safety in wastewater, particularly in densely populated urban areas where the risk of AR spread is elevated (Zhang et al., 2020a). The use of MBRs is particularly effective in environments where high-quality effluent is required for public health protection.

2.9.3 Constructed Wetlands for ARG Reduction

In Italy, constructed wetlands have emerged as an effective, low-cost, and ecofriendly solution for reducing ARB and ARGs in wastewater treatment. These systems utilize wetland vegetation alongside natural microbial communities to degrade ARGs, including sul1 and intI1 (a marker gene for integrons associated with resistance). The synergistic interaction between microbial biofilms and plant roots plays a critical role in removing genetic material linked to resistance. Research has shown that constructed wetlands can significantly reduce the prevalence of ARGs, offering a sustainable alternative for smaller or decentralized wastewater treatment systems (Barbieri et al., 2021). This case illustrates the value of integrating natural treatment processes to combat AR.

2.10 INNOVATIONS IN WASTEWATER TREATMENT AND THEIR EFFECTIVENESS

The increasing challenge of AR in wastewater has led to the development and implementation of innovative treatment technologies that effectively target ARB and ARGs. These emerging technologies offer significant promise in improving wastewater treatment efficiency, reducing public health risks, and contributing to environmental sustainability. Below are some of the latest innovations and their effectiveness:

2.10.1 Electrochemical Oxidation

Electrochemical oxidation has gained attention as a cutting-edge method for degrading ARGs in wastewater effluent. This technology uses anodic reactions to break down genetic material at the molecular level, offering a highly effective solution for mitigating resistance genes. A pilot study in the United States demonstrated that electrochemical oxidation could achieve over 95% reduction in ARG concentrations, with minimal secondary pollution, highlighting its potential as a viable method for large-scale applications (Wei et al., 2021b). The scalability and efficiency of electrochemical systems position them as a promising tool for WWTPs aiming to reduce ARGs and minimize environmental contamination.

2.10.2 Anaerobic Membrane Bioreactors (AnMBRs)

AnMBRs combine anaerobic digestion with membrane filtration, offering the dual benefits of ARG removal and energy recovery. This innovative technology has been successfully implemented at a WWTP in Switzerland, where it achieved over 90% removal of ARGs, including blaCTX-M (a β-lactam resistance gene) and mecA (associated with methicillin resistance). In addition to mitigating ARGs, the biogas produced during the anaerobic digestion process was utilized for energy generation, improving the sustainability and energy efficiency of the plant (Stoll et al., 2022). This integration of energy recovery and ARG mitigation makes AnMBRs an attractive option for wastewater treatment systems seeking both environmental and economic benefits.

2.10.3 Advanced Disinfection Technologies

Innovative disinfection methods, such as pulsed UV light and ozonation, have proven to be more effective in inactivating ARB and degrading ARGs compared to traditional chlorination. In Japan, a study demonstrated that ozonation achieved a 98% reduction in ARGs without producing harmful disinfection by-products, highlighting the advantages of ozonation as a cleaner, more efficient alternative to chlorine-based disinfection methods (Furukawa et al., 2020). Pulsed UV light, another advanced disinfection technology, uses intense bursts of UV light to disrupt microbial DNA, providing a nonchemical method for controlling ARB and ARGs. These advanced disinfection technologies underscore the importance of adopting modern strategies to address AMR in wastewater.

These innovations illustrate the potential of advanced treatment technologies to address the growing issue of ARB and ARGs in wastewater. While the initial costs of implementing systems like electrochemical oxidation, AnMBRs, and ozonation may be higher than traditional methods, their long-term benefits in reducing environmental contamination and public health risks justify the investment. Furthermore, integrating these technologies with real-time monitoring systems will be essential for adaptive management, enabling continuous improvement in the fight against AMR. Future research should focus on optimizing these technologies and exploring new combinations for even more effective wastewater treatment.

2.11 CHALLENGES AND FUTURE DIRECTIONS IN WASTEWATER TREATMENT

As the world grapples with increasing water contamination, particularly from emerging pollutants and antibiotic-resistant genes (ARGs), wastewater treatment faces several challenges. While advancements in technology have made significant strides in addressing these issues, much remains to be done. Below are some of the major challenges and future directions for improving wastewater treatment.

2.11.1 Limitations of Current Treatment Technologies

Despite the progress made with advanced wastewater treatment technologies, several limitations hinder their widespread implementation. Traditional methods like activated sludge processes and trickling filters often fail to fully remove complex contaminants such as pharmaceuticals, endocrine-disrupting compounds, and microplastics. These pollutants are persistent in treated water and can bioaccumulate in aquatic ecosystems, posing ecological and public health risks. While advanced methods, such as membrane filtration and AOPs, offer higher efficacy in removing these contaminants, they are often energy-intensive and costly, limiting their scalability and widespread use. Additionally, sludge management remains a significant challenge. Improper disposal of sludge can lead to secondary contamination, further exacerbating environmental pollution. More efficient and sustainable technologies are needed to address these challenges and ensure the protection of both public health and the environment.

2.11.2 Need for Standardized Monitoring and Reporting

Another key challenge in wastewater management is the lack of standardized protocols for monitoring and reporting contaminants, particularly emerging pollutants. Variability in sampling techniques, analytical methods, and reporting formats complicates the comparison of data across different studies and regions. This inconsistency impedes efforts to establish baseline levels of pollutants, assess the effectiveness of treatment technologies, and develop global strategies for wastewater management. Standardized monitoring protocols are essential for ensuring reliable data collection and analysis, enabling more effective decision-making. International initiatives, such as the European Water Framework Directive, have taken steps toward harmonizing water quality monitoring, but broader international collaboration and consensus are necessary to establish standardized practices worldwide.

2.11.3 Emerging Research Areas and Technological Advancements

Recent research is opening up new avenues for improving wastewater treatment, with promising innovations in both biological and technological approaches. One emerging field is the use of bioengineered microorganisms and enzymes specifically designed to target and degrade harmful contaminants like antibiotics and pesticides. This approach offers a tailored solution for treating complex pollutants that traditional methods struggle to address. Nanotechnology is also playing a significant role, with the development of nanomaterials that can adsorb pollutants or facilitate photocatalytic degradation, providing more efficient treatment options (Prasad and Thirugnanasanbandham, 2019). Additionally, the integration of artificial intelligence and machine learning is transforming wastewater management by enabling real-time monitoring, predictive maintenance, and process optimization. Furthermore, the concept of the circular economy is gaining traction in wastewater management, emphasizing the recovery of valuable resources such as water, nutrients (e.g., phosphorus and nitrogen), and energy from wastewater and sludge. This approach not only enhances sustainability but also reduces operational costs and environmental impact.

2.11.4 Policy and Regulatory Considerations

Effective wastewater treatment and management require robust policy frameworks and regulatory oversight. However, many existing policies are slow to adapt to emerging challenges, such as the spread of AR and microplastic pollution. To effectively address these issues, policies must be dynamic and responsive to the latest scientific research. They should promote the development and adoption of advanced treatment technologies while encouraging industries to implement cleaner production processes. Furthermore, there is a need for greater international collaboration, as water pollution often transcends national borders. Governments should incentivize research and infrastructure development in wastewater treatment and resource recovery technologies through targeted funding and policy support. Public awareness campaigns are also critical to educating communities on water conservation and pollution prevention, creating a culture of environmental stewardship and responsible water use.

2.12 CONCLUSION

AR is an escalating global health crisis that requires urgent, coordinated action across all sectors of society. Its impact on public health, the economy, and medical advancements is profound, and without decisive intervention, we may soon face a future where even common infections become untreatable. Tackling AR demands a comprehensive, multipronged approach that includes better stewardship of antibiotics, improved infection control practices, research into new therapeutic options, and robust public education initiatives. Only through global collaboration, sustained commitment, and the implementation of effective policies can we mitigate the threat of AR and preserve the efficacy of antibiotics for future generations.

WWTPs play a critical role in managing wastewater but also pose a significant public health risk as they act as reservoirs and conduits for ARB and ARGs. A deeper understanding of resistance pathways and mechanisms within these facilities is essential for developing effective mitigation strategies. The adoption of advanced treatment technologies, stricter regulations on antibiotic usage, and continuous monitoring efforts can help limit the spread of AR from WWTPs into the environment.

The fight against AR requires a holistic, multifaceted approach, with WWTPs playing a pivotal role. Advanced treatment technologies, optimized biological processes, and stringent regulatory frameworks, combined with innovative approaches such as using WWTPs as surveillance stations for ARB and ARGs, represent a comprehensive strategy for mitigating AR. Furthermore, continued investment in research and technology development is essential to improving the efficiency and reliability of these strategies, ensuring the long-term sustainability of public health and environmental protection.

Addressing the challenges in wastewater treatment necessitates a multifaceted approach, which includes technological innovation, the standardization of monitoring practices, and the implementation of robust policy frameworks. By bridging the gaps in research and fostering international collaboration, the future of wastewater management can transition toward more sustainable, efficient systems that protect both public health and the environment.

REFERENCES

Amarasiri, M., D. Sano, and S. Suzuki. Understanding Human Health Risks Caused by Antibiotic Resistant Bacteria (ARB) and Antibiotic Resistance Genes (ARG) in Water Environments: Current Knowledge and Questions to Be Answered. *Critical Reviews in Environmental Science and Technology* 50, no. 19 (2020): 2016–2059. https://doi.org/10.1080/10643389.2019.1692611.

Arthur, M., and P. Courvalin. Genetics and Mechanisms of Glycopeptide Resistance in Enterococci. *Antimicrobial Agents and Chemotherapy* 37, no. 8 (1993): 1563–1571.

Barbieri, M., A. Ghirardini, and G. Ruffato. Constructed Wetlands: A Natural Approach to ARG Management. *Science of the Total Environment* 772 (2021): 145423.

Barton, M. D., et al. The Economic Burden of Antibiotic Resistance. *The Lancet Infectious Diseases* 20, no. 8 (2020): 1043–1050.

Bengtsson-Palme, J., R. Hammarén, C. Pal, M. Östman, B. Björlenius, C.-F. Flach, J. Fick, E. Kristiansson, M. Tysklind, and D. G. J. Larsson. Elucidating Selection Processes for Antibiotic Resistance in Sewage Treatment Plants Using Metagenomics. *Science of the Total Environment* 572 (2016): 697–712. https://doi.org/10.1016/j.scitotenv.2016.06.228.

Berendonk, T. U., C. M. Manaia, C. Merlin, et al. Tackling Antibiotic Resistance: The Environmental Framework. *Nature Reviews Microbiology* 13 (2015): 310–317. https://doi.org/10.1038/nrmicro3439.

Biswal, B. K., A. Mazza, L. Masson, R. Gehr, and D. Frigon. Impact of Wastewater Treatment Processes on Antimicrobial Resistance Genes and Their Co-Occurrence with Virulence Genes in *Escherichia coli*. *Water Research* 50 (2014): 245–253. https://doi.org/10.1016/j.watres.2013.11.047.

Boeckel, T. P. V., C. Brower, M. Gilbert, B. T. Grenfell, S. A. Levin, T. P. Robinson, et al. Global Trends in Antimicrobial Use in Food Animals. *Proceedings of the National Academy of Sciences* 114, no. 18 (2017): E931–E941. https://doi.org/10.1073/pnas.1503141114.

Bush, K., and G. A. Jacoby. Updated Functional Classification of β-Lactamases. *Antimicrobial Agents and Chemotherapy* 54, no. 3 (2010): 969–976.

Cabello, F. C. Heavy Use of Prophylactic Antibiotics in Aquaculture: A Growing Problem for Human and Animal Health and for the Environment. *Environmental Microbiology* 8, no. 7 (2006): 1137–1144. https://doi.org/10.1111/j.1462-2920.2006.01054.x.

Cabello, F. C., H. P. Godfrey, A. Tomova, L. Ivanova, H. Dölz, A. Millanao, and A. H. Buschmann. Antimicrobial Use in Aquaculture Re-Examined: Its Relevance to Antimicrobial Resistance and to Animal and Human Health. *Environmental Microbiology* 15, no. 7 (2013): 1917–1942. https://doi.org/10.1111/1462-2920.12134.

Chambers, H. F., and F. R. DeLeo. Waves of Resistance: *Staphylococcus aureus* in the Antibiotic Era. *Nature Reviews Microbiology* 7, no. 9 (2009): 629–641.

Chen, S., J. Fu, K. Zhao, S. Yang, C. Li, P. Penttinen, X. Ao, et al. Class 1 Integron Carrying *qacEΔ1* Gene Confers Resistance to Disinfectant and Antibiotics in *Salmonella*. *International Journal of Food Microbiology* 404 (2023): 110319. https://doi.org/10.1016/j.ijfoodmicro.2023.110319.

Chen, X., Li, G.X., Zhang, H., Yuan, M., Hou, X.P., Yu, H.L., et al. Characterization of class 1 integron gene cassettes among clinical bacteria isolated from one large hospital in northern China. *Biomedical and Environmental Sciences*. 26, no. 12 (2013):1003–1007. doi: 10.3967/bes2013.038.

Cheng, W., J. Li, Y. Wu, et al. Advanced Oxidation Processes for ARG Removal in WWTPs: Efficacy and Mechanisms. *Environmental Science & Technology* 53, no. 8 (2019): 4325–4333.

Doi, Y., and Y. Arakawa. 16S Ribosomal RNA Methylation: Emerging Resistance Mechanism against Aminoglycosides. *Clinical Infectious Diseases* 45, no. 1 (2007): 88–94.

Doi, Y., J.-I. Wachino, and Y. Arakawa. Aminoglycoside Resistance: The Emergence of Acquired 16S Ribosomal RNA Methyltransferases. *Infectious Disease Clinics* 30, no. 2 (2016): 523–537.

Donlan, R. M., and J. W. Costerton. Biofilms: Survival Mechanisms of Clinically Relevant Microorganisms. *Clinical Microbiology Reviews* 15, no. 2 (2002): 167–193.

Dzyubak, E., and M.-N. F. Yap. The Expression of Antibiotic Resistance Methyltransferase Correlates with mRNA Stability Independently of Ribosome Stalling. *Antimicrobial Agents and Chemotherapy* 60, no. 12 (2016): 7178–7188. https://doi.org/10.1128/aac.01806-16.

Furtula, V., M. Choi, and A. Muckle. Environmental and Human Health Implications of Antibiotic Resistance: Risk Assessment. *Environmental Toxicology and Chemistry* 32, no. 3 (2013): 515–524. https://doi.org/10.1002/etc.2143.

Furukawa, S., T. Kondo, and H. Maeda. Effectiveness of Ozonation for ARG Removal in Wastewater. *Journal of Water Process Engineering* 34 (2020): 101201.

Goff, D. A., et al. Antimicrobial Stewardship: The Role of Clinicians in Managing Resistance. *Clinical Infectious Diseases* 65, no. 1 (2017): 1–8.

Hafez, E. E., S. A. Al-Sohaimy, and M. A. El-Saadani. The Effect of the *mecA* Gene and Its Mutant Form on the Response of *S. aureus* to the Most Common Antibiotics. *International Journal of Immunological Studies* 1, no. 1 (2009): 106–122. https://doi.org/10.1504/IJIS.2009.023564.

Harnisz, M., E. Korzeniewska, S. Ciesielski, and I. Gołaś. *tet* Genes as Indicators of Changes in the Water Environment: Relationships Between Culture-Dependent and Culture-Independent Approaches. *Science of the Total Environment* 505 (2015): 704–711. https://doi.org/10.1016/j.scitotenv.2014.10.048.

Hernández-Allés, S., V. J. Benedí, L. Martínez-Martínez, Á. Pascual, A. Aguilar, J. M. Tomás, and S. Albertí. Development of Resistance during Antimicrobial Therapy Caused by Insertion Sequence Interruption of Porin Genes. *Antimicrobial Agents and Chemotherapy* 43, no. 4 (1999): 937–939.

Hooper, B. *Integrated River Basin Governance*. IWA Publishing, London, 2005.

Hu, Y., G. Gao, C. Zhang, and F. Ju. Advanced Monitoring Techniques for Tracking Antibiotic Resistance in Wastewater. *Environmental Research* 196 (2021): 110898.

Ibrahim, M. E., T. B. Algak, M. Abbas, and B. K. Elamin. Emergence of *blaTEM, blaCTX-M, blaSHV*, and *bla-OXA* Genes in Multidrug-Resistant *Enterobacteriaceae* and *Acinetobacter baumannii* in Saudi Arabia. *Experimental and Therapeutic Medicine* 22, no. 6 (2021): 1–11. https://doi.org/10.3892/etm.2021.10885.

Jacoby, G. A. Mechanisms of Resistance to Quinolones. *Clinical Infectious Diseases* 41, no. Supplement_2 (2005): S120–S126.

Jiao, Y. N., Z. C. Zhou, T. Chen, Y. Y. Wei, J. Zheng, R. X. Gao, and H. Chen. Biomarkers of Antibiotic Resistance Genes During Seasonal Changes in Wastewater Treatment Systems. *Environmental Pollution* 234 (2018): 79–87. https://doi.org/10.1016/j.envpol.2017.11.048.

Karkman, A., T. T. Do, F. Walsh, and M. P. Virta. Antibiotic-Resistance Genes in Wastewater. *Trends in Microbiology* 26, no. 3 (2018): 220–228. https://doi.org/10.1016/j.tim.2017.09.005.

Kitajima, M., H. P. Sassi, and J. R. Torrey. Wastewater-Based Epidemiology: Current Status and Future Perspectives. *Environmental International* 131 (2020): 105017.

Kumarasamy, K. K., M. A. Toleman, T. R. Walsh, J. Bagaria, F. Butt, R. Balakrishnan, et al. Emergence of a New Antibiotic Resistance Mechanism in India, Pakistan, and the UK: A Molecular, Biological, and Epidemiological Study. *The Lancet Infectious Diseases* 10, no. 9 (2010): 597–602. https://doi.org/10.1016/S1473-3099(10)70143-2.

Kümmerer, K. Antibiotics in the Aquatic Environment: A Review – Part I. *Chemosphere* 75, no. 4 (2009): 417–434.

Laht, M., A. Karkman, V. Voolaid, C. Ritz, T. Tenson, M. Virta, and V. Kisand. Abundances of Tetracycline, Sulphonamide, and Beta-Lactam Antibiotic Resistance Genes in Conventional Wastewater Treatment Plants (WWTPs) with Different Waste Loads. *PLoS One* 9, no. 8 (2014): e103705. https://doi.org/10.1371/journal.pone.0103705.

Larsson, D. G. J., C. de Pedro, and N. Paxeus. Effluent from Drug Manufacturers Contains Extremely High Levels of Pharmaceuticals. *Journal of Hazardous Materials* 148, no. 3 (2007): 751–755.

Laxminarayan, R., A. Duse, C. Wattal, A. K. Zaidi, H. F. Wertheim, N. Sumpradit, et al. Antibiotic Resistance—The Need for Global Solutions. *The Lancet Infectious Diseases* 13, no. 12 (2013): 1057–1098. https://doi.org/10.1016/S1473-3099(13)70318-9.

Li, S., B. S. Ondon, S. H. Ho, J. Jiang, and F. Li. Antibiotic Resistant Bacteria and Genes in Wastewater Treatment Plants: From Occurrence to Treatment Strategies. *Science of the Total Environment* 838 (2022): 156544.

Li, X.-Z., and H. Nikaido. Efflux-Mediated Drug Resistance in Bacteria: An Update. *Drugs* 69 (2009): 1555–1623.

Llor, C., and L. Bjerrum. Antimicrobial Resistance: Risk Associated with Antibiotic Overuse and Initiatives to Reduce the Problem. *Therapeutic Advances in Drug Safety* 5, no. 6 (2014): 229–241. https://doi.org/10.1177/2042098614554919.

Manaia, C. M., G. Macedo, D. Fatta-Kassinos, and O. C. Nunes. Antibiotic Resistance in Urban Aquatic Environments: Can It Be Controlled? *Applied Microbiology and Biotechnology* 100, no. 4 (2016): 1543–1557.

Marshall, B. M., and S. B. Levy. Food Animals and Antimicrobials: Impacts on Human Health. *Clinical Microbiology Reviews* 24, no. 4 (2011): 718–733. https://doi.org/10.1128/CMR.00002-11.

Marti, E., J. Jofre, and J. L. Balcázar. Prevalence of Antibiotic Resistance Genes and Bacterial Community Composition in a River Influenced by a Wastewater Treatment Plant. *PLoS One* 8 (2013): e78906. https://doi.org/10.1371/journal.pone.0078906.

Martínez, J. L. Antibiotics and Antibiotic Resistance Genes in Natural Environments. *Science* 321, no. 5887 (2008): 365–367.

Newton, P. N., M. D. Green, F. M. Fernández, N. P. J. Day, and N. J. White. Counterfeit Anti-Infective Drugs. *The Lancet Infectious Diseases* 6, no. 9 (2011): 602–613. https://doi.org/10.1016/S1473-3099(06)70581-3.

Nikaido, H. Multidrug Resistance in Bacteria. *Annual Review of Biochemistry* 78, no. 1 (2009): 119–146.

Nordmann, P., L. Poirel, and L. Dortet. Rapid Detection of Carbapenemase-Producing Enterobacteriaceae. *Emerging Infectious Diseases* 18, no. 9 (2012): 1503.

Novo, A., C. M. Manaia, and R. Durán. Assessing the Role of Aquatic Environments in the Spread of Antibiotic Resistance. *Environmental Science and Pollution Research* 20, no. 3 (2013): 1572–1584.

Omuferen, L. O., B. Maseko, and J. O. Olowoyo. Occurrence of Antibiotics in Wastewater from Hospitals and Conventional Wastewater Treatment Plants and Their Impact on the Effluent Receiving Rivers: Current Knowledge between 2010 and 2019. *Environmental Monitoring and Assessment* 194, no. 4 (2022): 306.

O'Neill, J. Antimicrobial Resistance: Tackling a Crisis for the Health and Wealth of Nations. Review on Antimicrobial Resistance, 2014. https://amr-review.org.

O'Neill, J. Tackling Drug-Resistant Infections Globally: Final Report and Recommendations. The Review on Antimicrobial Resistance, Government of the United Kingdom, 2016. https://amr-review.org.

Piddock, L. J. V. Multidrug-Resistance Efflux Pumps? Not Just for Resistance. *Nature Reviews Microbiology* 4, no. 8 (2006): 629–636.

Poole, K. Multidrug Efflux Pumps and Antimicrobial Resistance in Pseudomonas Aeruginosa and Related Organisms. *Journal of Molecular Microbiology and Biotechnology* 3, no. 2 (2001): 255–264.

Prasad, R., Thirugnanasanbandham, K. 2019, Advances Research on Nanotechnology for Water Technology. Springer International Publishing. https://www.springer.com/us/book/9783030023805.

Prüss-Ustün, A., J. Bartram, T. Clasen, J. M. Colford, O. Cumming, V. Curtis, et al. Burden of Disease from Inadequate Water, Sanitation, and Hygiene for Selected Adverse Health Outcomes: An Updated Analysis with a Focus on Low- and Middle-Income Countries. *International Journal of Hygiene and Environmental Health* 222, no. 5 (2019): 765–777. https://doi.org/10.1016/j.ijheh.2019.05.004.

Quale, J., S. Bratu, J. Gupta, and D. Landman. Interplay of Efflux System, ampC, and oprD Expression in Carbapenem Resistance of *Pseudomonas aeruginosa* Clinical Isolates. *Antimicrobial Agents and Chemotherapy* 50, no. 5 (2006): 1633–1641.

Quiroga, M. P., B. Orman, L. Errecalde, S. Kaufman, and D. Centrón. Characterization of Tn 6238 with a New Allele of aac (6′)-Ib-cr. *Antimicrobial Agents and Chemotherapy* 59, no. 5 (2015): 2893–2897.

Ramirez, M. S., and M. E. Tolmasky. Aminoglycoside Modifying Enzymes. *Drug Resistance Updates* 13, no. 6 (2010): 151–171.

Riquelme, M. V., et al. The Role of WWTPs in the Removal and Spread of Antibiotic Resistance: A Global Overview. *Journal of Environmental Management* 295 (2021): 113155. https://doi.org/10.1016/j.jenvman.2021.113155.

Rizzo, L., A. Fiorentino, and A. Anselmo. Advanced Treatment of Urban Wastewater by UV Radiation: Effect on Antibiotics and Antibiotic-Resistant *E. coli* Strains. *Chemosphere* 92, no. 2 (2013a): 171–176. https://doi.org/10.1016/j.chemosphere.2013.02.045.

Rizzo, L., C. Manaia, C. Merlin, T. Schwartz, C. Dagot, M. C. Ploy, et al. Urban Wastewater Treatment Plants as Hotspots for Antibiotic Resistant Bacteria and Genes Spread into the Environment: A Review. *Science of the Total Environment* 447 (2013b): 345–360.

Sarmah, A. K., M. T. Meyer, and A. B. A. Boxall. A Global Perspective on the Use, Sales, Exposure Pathways, Occurrence, Fate, and Effects of Veterinary Antibiotics (VAs) in the Environment. *Chemosphere* 65, no. 5 (2006): 725–739.

Stoll, Z., D. Bertschi, and K. Meier. Anaerobic Membrane Bioreactors for Sustainable WWTPs. *Renewable Energy* 185 (2022): 1223–1232.

Sun, Y., X. Zhang, W. Sun, and Y. Hu. Mitigating Antibiotic Resistance in Anaerobic Digestion: Strategies and Microbial Insights. *Journal of Hazardous Materials* 423 (2022): 127015.

Szczepanowski, R., B. Linke, I. Krahn, K. H. Gartemann, T. Gutzkow, W. Eichler, A. Pühler, and A. Schlüter. Detection of 140 Clinically Relevant Antibiotic-Resistance Genes in the Plasmid Metagenome of Wastewater Treatment Plant Bacteria Showing Reduced Susceptibility to Selected Antibiotics. *Microbiology* 155 (2009): 2306–2319. https://doi.org/10.1099/mic.0.028233-0.

Uluseker, C., K. M. Kaster, K. Thorsen, D. Basiry, S. Shobana, M. Jain, and I. Pala-Ozkok. A Review on Occurrence and Spread of Antibiotic Resistance in Wastewaters and in Wastewater Treatment Plants: Mechanisms and Perspectives. *Frontiers in Microbiology* 12 (2021): 717809.

Uppu, Divakara S. S. M., G. B. Manjunath, V. Yarlagadda, J. E. Kaviyil, R. Ravikumar, K. Paramanandham, B. R. Shome, and J. Haldar. Membrane-Active Macromolecules Resensitize NDM-1 Gram-Negative Clinical Isolates to Tetracycline Antibiotics. *PLoS One* 10, no. 3 (2015): e0119422.

Ventola, C. L. The Antibiotic Resistance Crisis: Part 1: Causes and Threats. *Pharmacy and Therapeutics* 40, no. 4 (2015): 277–283.

Wei, C., D. Zhao, X. Jin, et al. Electrochemical Oxidation for ARG Elimination: A Pilot Study. *Journal of Environmental Chemical Engineering* 9, no. 2 (2021a): 104642.

Wei, Z., K. Feng, Z. Wang, Y. Zhang, M. Yang, Y. G. Zhu, and Y. Deng. High-Throughput Single-Cell Technology Reveals the Contribution of Horizontal Gene Transfer to Typical Antibiotic Resistance Gene Dissemination in Wastewater Treatment Plants. *Environmental Science & Technology* 55, no. 17 (2021b): 11824–11834.

Wong, A., D. Y. Ngu, N. Daneman, and M. Mendelson. Global Antimicrobial Stewardship: A Roadmap to Safer Antimicrobial Use. *The Lancet Infectious Diseases* 16, no. 5 (2016): 500–503. https://doi.org/10.1016/S1473-3099(16)00074-5.

World Health Organization. *Antimicrobial Resistance: Key Facts.* World Health Organization, 2020. https://www.who.int/news-room/fact-sheets/detail/antimicrobial-resistance.

Xu, J., Y. Xu, H. Wang, C. Guo, H. Qiu, Y. He, et al. Occurrence of Antibiotics and Antibiotic Resistance Genes in a Sewage Treatment Plant and Its Effluent-Receiving River. *Chemosphere* 119 (2015): 1379–1385. https://doi.org/10.1016/j.chemosphere.2014.02.040.

Yang, Y., B. Li, F. Ju, and T. Zhang. Exploring Variation of Antibiotic Resistance Genes in Activated Sludge over a Four-Year Period through a Metagenomic Approach. *Environmental Science & Technology* 47 (2013): 10197–10205. https://doi.org/10.1021/es4017365.

Yang, Y., B. Li, S. Zou, H. H. P. Fang, and T. Zhang. Fate of Antibiotic Resistance Genes in Sewage Treatment Plants Revealed by a Metagenomic Approach. *Water Research* 62 (2014): 97–106. https://doi.org/10.1016/j.watres.2014.05.019.

Zhang, Q., H. Zhao, Y. Luo, et al. Membrane Bioreactors for ARB and ARG Mitigation in Urban Wastewater. *Water Research* 168 (2020a): 115148.

Zhang, T., M. F. Shao, and L. Ye. The Role of WWTPs in the Dissemination of Antibiotic Resistance Genes. *Environmental Science & Technology* 54, no. 19 (2020b): 11412–11422. https://doi.org/10.1021/acs.est.0c04163.

Zhang, X. X., and T. Zhang. Occurrence, Abundance, and Diversity of Tetracycline Resistance Genes in 15 Sewage Treatment Plants Across China and Other Global Locations. *Environmental Science & Technology* 45 (2011): 2598–2604. https://doi.org/10.1021/es103672x.

Zhang, Y., J. Geng, Y. Zhuang, K. Xu, L. Ding, and H. Ren. Reduction of Antibiotic Resistance Genes in Municipal Wastewater Effluent by Advanced Oxidation Processes. *Science of the Total Environment* 550 (2016): 184–191. https://doi.org/10.1016/j.scitotenv.2016.01.078.

Zhang, Z., F. Dai, F. Luo, M. Zhong, Z. Huang, T. Hou, and J. Xu. Could High-Concentration Rifampicin Kill Rifampicin-Resistant M. Tuberculosis? Rifampicin MIC Test in Rifampicin-Resistant Isolates from Patients with Osteoarticular Tuberculosis. *Journal of Orthopaedic Surgery and Research* 9 (2014): 1–5.

Zhang, Y., Xiao, M., Horiyama, T., Zhang, Y., Li, X., Nishino, K., & Yan, A. The Multidrug Efflux Pump MdtEF Protects Against Nitrosative Damage during the Anaerobic Respiration in *Escherichia coli. Journal of Biological Chemistry* 286 no. 30 (2011), 26576–26584.

3 Impact of Wastewater on Occurrence of Antibiotic-Resistant Bacteria

Pratul Gaurav, Kunal Kumar Bharti, Pragati Saini, and Ajay Kumar

3.1 INTRODUCTION

We know that the majority of the population is facing serious problems regarding microorganisms, commonly bacteria. It increases morbidity and mortality in the world (Tanover, 2006) when microorganisms like fungi, bacteria, viruses, and parasites learn to adapt and grow in an environment where medication is already present (Founou et al., 2017; Dadgostar, 2019). This situation leads to antimicrobial resistance (AMR). AMR is a serious matter of concern to the public health system in the whole world (Founou et al., 2017; Prestinaci et al., 2015). In this, AMR cannot be cured with the help of antibiotics for further treatment (Chokshi et al., 2019). This will cause severe health issues. Treatment of AMR is very costly. As per the Centers for Disease Control and Prevention (CDC), in the United States, healthcare costs 20 billion dollars added for AMR (Chokshi et al., 2019). According to the CDC report, 24,000 people die every year and nearly 2.5 million people become ill with antibiotic-resistant disease (Chockshi et al., 2019). The capacity of the human immune system to fight against infectious diseases is compromised by AMR. It also causes various complications in a patient undergoing chemotherapy, joint replacement, and surgery. A high impact of AMR can be seen in a patient who is suffering from chronic diseases like rheumatoid arthritis, asthma, and diabetes. Nowadays, physicians are using polymyxins and carbapenems as medicines to reduce the symptoms of AMR. Due to its very high cost, it is rarely available in developing countries (World Health Organization, 2015).

Methicillin resistance, one of the most commonly known cases of AMR, present in *Staphylococcus aureus* (MRSA) can cause a large number of deaths every year in the world. Developing countries produce high-cost drug resistance every year. It is very costly for individual health as well as health care in developing countries. For example, in tuberculosis, only 4.2% of its cases are multidrug resistant (Chatterjee et al., 2018). The use of multidrug resistance for tuberculosis is expected to increase significantly by 2040 in countries like Russia, South Africa, and India (Friedrich, 2017).

3.1.1 Current Situation of AMR as the Major Issue in the World

The most common and effective tool used as medicine for treatment is antibiotics. In the early 1970s, the most effective drugs were fluoroquinolones, which are still used by the public. After this, no more effective AMR develops. Nowadays, the threat of AMR is increasing globally, which can cause insufficient development of antibiotics (Fouz et al., 2020). Globally, AMR is a major issue for public health and can cause death due to common infections. According to the report from the Global Antimicrobial Surveillance System under the World Health Organization (WHO), resistance to major and common infections caused by bacteria is increasing in different parts of the world. At present, approx. 7,00,000 deaths are caused due to resistant infection every year, and it is estimated that nearly 10 million deaths will be caused due to resistance in 2050 (Mott et al., 2020). As per the report from the European Center for Disease Prevention and the US CDC, around

DOI: 10.1201/9781003269823-3

TABLE 3.1
The Mortality Rate due to AMR in Different Regions by 2050

Country	Mortality
Asia	4,730,000
North America	317,000
Latin America	392,000
Oceania	22,000
Africa	4,150,000
Europe	390,000

25,000–30,000 deaths are reported every year due to AMR the countries with high income such as United States and Europe. The majority of deaths are reported in low-income countries like India and Sri Lanka, around 58,000–60,000 deaths every year (Table 3.1).

3.1.2 Reported Rate of AMR in India

The rate of AMR in India has been increasing disproportionately in the last few years (Kahn, 2017). There are some issues like the absence of proper research and data for AMR in India. This leads to the prevention of comparison nationwide. However, Indian institutions published 2,154 studies on AMR, out of which 1,042 [48.3%] were on humans, 72 [3.3%] were on animals, 11 [0.5%] were on one health, and 90 [4.2%] were on the environment. The remaining are based on editorial miscellaneous and novel agents (Gandra et al., 2017).

In 2014, India stopped the marketing of antibiotics in March, in April, the report on AMR was published by the WHO, and in July, the British Prime Minister took a look at the report. Later on, the United States implemented a plan for 5 years to tackle this problem globally. After 6 months, Obama played an important role in reducing the use of antibiotics for outpatient and inpatient by 50% and 20% using references from 2011. In 2015, a group of countries came together under a global action plan to tackle the problem of the use of antibiotics. All these were discussed during the G7 summit held in 2015. Experts from 68 countries are coming together at the World Healthcare-Association Infection Forum, in France to talk about the initiative to control the AMR in low, high, or developing countries. The current problems in India are as follows:

3.2 ANTIMICROBIAL RESISTANCE IN THE ENVIRONMENT

In India, different water sources can produce antimicrobial-resistant bacteria and their gene. Hospital waste and pharmaceutical waste are the main sources of AMR, which are released into the water bodies without proper treatment. Ganga and Yamuna receive various inlets, which contain drug-resistant bacteria. 17.4% was the extended-spectrum β-lactamase (ESBL) producer's rate among bacteria separated from Indian rivers (Azam et al., 2016).

However, we use groundwater and surface water for drinking, which is reported at the rate of *E. coli*, i.e., 18% which is resistant to cephalosporin in the third generation in the center (Kumar et al., 2013a). Study samples were collected from lakes, ponds, springs, rivers, and tube wells.

3.3 ANTIMICROBIAL RESISTANCE IN MAN

According to the "Report of Scoping on Antimicrobial Resistance," which is under the support of the government of India (Gandra et al., 2017), more than 75% of *E. coli* and *Acinetobacter baumannii* are resistant to fluoroquinolones and cephalosporins of the third generation. Colistin is frequently used by Bahmani as the last resort antimicrobial (Gandra et al., 2017). In India, the emergence of resistance to colistin takes place. 83% and 29% are the rate of resistance among *Shigella* species and *Salmonella typhi*, respectively. Overuse of the available antibiotics is the first choice of the public, which can cause AMR. Restriction policy on the use of antibiotics plays a major role in improving the situation in the healthcare system. To reduce the side effects of antibiotics, it is necessary to follow the stewardship program (Granowitz and brown, 2008). Sometimes, it is difficult to treat infections like diarrhea for which *Clostridium difficile* infection is responsible. Stewardship intervention is first targeted by prophylactic prescription of antibiotics in the hospital.

3.4 ANTIMICROBIAL RESISTANCE IN ANIMALS FOOD

India was the largest producer of fish and milk in the world in 2015 (Sekhar, 2022). Between 2000 and 2030, India is expected to rise in poultry consumption by 575% (Van Boeckel et al., 2015). India has a huge potential to produce antimicrobial agents, which increase animal food production. Uttar Pradesh with 17.6, Rajasthan with 10.6%, and Andhra Pradesh with 9.4% are the largest milk-producing states in India (Taneja and Sharma, 2019). After analyzing the sample of milk, ESBL produced in buffalo milk and cow milk Gram-negative bacilli were detected, with 50% and 48% are oxytetracycline resistant (Walia et al., 2019). India is becoming an aquaculture industry hub with 9,579 × 1,000 tons of fish production (Taneja and Sharma, 2019). For instance, in Maharashtra, tilapia fish is found, which contains 48% of Enterobacteriaceae (Marathe et al., 2016). According to the seven studies on AMR, three studies suggest that the rate of the production of ESBL varies from place to place, from 9.5% in Odisha to 87% in Madhya Pradesh.

At the international level, Denmark started the stewardship program for the veterinary sector in 1998. After the implementation of the program, the use of antibiotics for pig meat was reduced by more than 50% between 1994 and 2013 (Agersø and Aarestruo, 2013; Vickers, 2017). In the year 2011, there was a sudden increase of resistance against cephalosporin in bacteria colonizing the boiler. There are only a few countries that ban the use of antibiotics for the growth of animal meat.

In the coming days, there should be many new drugs on the market for MDR Gram-negative bacteria, but they are less effective against carbapenems. Currently, only a limited number of antimicrobial agents are available for treating carbapenem-resistant infections, with tigecycline, colistin, and imipenem among the few options considered for combination therapy.

3.4.1 The Implications of AMR by Clinical Use

The effective treatment caused by several microorganisms can be affected by the presence of AMR. Novel drugs are available in the market for the treatment of cancer through chemotherapy, transplantation of organs, and surgeries. Bacteria can resist themselves, which can affect the body to fight against the commonly known diseases such as cold and flu and can cause prolonged diseases. This long-term disease needs to be treated with very expensive drugs. These drugs are so expensive that common people cannot afford them (Krüger and Stingl, 2011).

3.4.2 Factors Advancing the AMR Rate

3.4.2.1 Overuse and Misuse of Antibiotics

In 1940, Sir Alexander Fleming warned the public about the overuse of antibiotics in the future (Zaman, 2017; Ventola, 2015). A survey across the world says that many patients are dependent on antimicrobial agents for coughs and colds or more (World Health Organization, 2015). Furthermore, there is a shortage of proper diagnostic tools in many developing countries, and the public is using antibiotics without a prescription of medicines (Chaw et al., 2018). Many antibiotics present in the market are of poor quality, and they are sold in developing countries (Chokshi et al., 2019) like India and Vietnam, where there are no laws and regulatory policies on the purchase of medicine. Because of the lack of these policies, patients are doing their self-treatment for the disease that can be treated without antibiotics (Chokshi et al., 2019; Nguyen et al., 2013; Zaman, 2017).

Moreover, the use of antibiotics unnecessarily prescribed by physicians is the main reason for the development of AMR (Spellberg, 2018). For example, many physicians in hospitals get a commission to prescribe the medicine from their pharmaceutical companies (Laxminarayan, 2014; Tang et al., 2016). Suppliers try to meet the demand for antibiotics for the patients. About 270 million prescribed antibiotics were banned in 2015, which means 30% of prescribed drugs are unnecessary. Clinicians generally suggest fewer effective drugs and do not favor prescribing first-line drugs as per the national guidelines. However, in the emergency department, nearly 50 million prescriptions are unnecessary. As per the report from CDC, around 50% of prescriptions for infections like the common cold are unnecessary. Children who are below two and adults who are above 65 are prescribed antibiotics more than others. Another important factor for resistance is self-medication. Illegal practice on the sale of drugs can cause these problems in South and Central America, Europe, and Asia.

3.4.3 Unavailability of New Novel Antibiotics

Due to the unavailability of new novel antibiotics, WHO repeatedly warned to meet the need for antibiotics to fight against antibiotic resistance (Woolhouse et al., 2016). There are only a few drugs available to treat AMR and the majority of drugs are modifications of the drugs that are already present on the market. Due to the low immunity present in the young and old, it can be life-threatening for them in the absence of drugs (Bartlett et al., 2013).

There are many obstacles in the production of antibiotics faced by pharmaceutical companies that are regulated in their production and economic obligations. Due to these obstacles, companies reduce the number of their team. Profit in these short-term antibiotics is very less, which is why the company shifted toward the production of medicine for long-term diseases, i.e., chronic diseases. The net present value (NPV) of new antibiotics is 55 million, and the NPV for diseases like neuromuscular diseases is 1.2 billion. The major interest of the pharmaceutical company is to develop drugs for methicillin-resistant *Staphylococcus aureus* (MRSA). The main reason behind this is that the pharmaceutical company wants to produce drugs for MRSA as compared to antibiotics for the Gram-negative organism (Gould and Bal, 2013). In developing countries like India and China, the high cost of antibiotics is still a major problem.

3.4.4 Antimicrobial Resistance as a Burden on Global Economics

The impact of AMR can be seen at a different level. The economic burden is one of them. Impact on the gross domestic product, global health system, and people of the country due to AMR (Golkar et al., 2014). Studies say that the AMR costs nearly 102 trillion annually because it is more dangerous than cancer. Every year around 11 million people die due to AMR. AMR can also impact morbidity and mortality including the cost of insurance, in addition to hospitals and healthcare centers also having the burden.

Various methods are being used to reduce the burdens like the rate of morbidity and mortality and treatment cost (World Health Organization, 2014). There are reasons that can complicate the burden of AMR including the lack of information on the incidence and prevalence. Data present on this issue from both public and private sectors is not proper (Thompson et al., 2016) (Table 3.2).

3.4.5 Use of Antibiotics in Agriculture

Agriculture also plays a prominent role in the cause of AMR in humans (Cassini et al., 2019; Dadgostar, 2019). For example, 75% of antibiotics are applied to food that is consumed by animals (Bartlett et al., 2013). In 2010, antibiotics used for livestock production were more than humans'

TABLE 3.2
The Mortality Rate due to Some Conditions by 2050

Name of Disease	Mortality
Cancer	8.2 million
Diabetes	1.52 million
Measles	130,000
Tetanus	60,000
Cholera	100,000–20,000
Diarrheal disease	1.4 million
Antimicrobial resistance	10 million
Road traffic accident	1.2 million

consumption of nearly 63,300 tons. Apart from the use of antibiotics to treat sick animals, antibiotics are largely used to treat healthy animals to prevent sickness. Colistin is most widely used in animal farming, which increases the production of livestock, and it is a critical last line to treat the human infection. Biocides are generally used to kill pathogens or microbes. Biocides include pesticides, fertilizers, insecticides, and disinfectants (Singer et al., 2016). Resistant organisms in the environment can be increased by the sublethal concentration of biocides (Webber et al., 2015). Between 1992 and 2007, there was a 40% growth in the biocide market in the world, but in India, the consumption of biocides cannot be defined because of the lack of proper data (Singer et al., 2016). The European Commission restricts the use of biocides that can cause AMR and also regulates the use and disposal of biocides.

Increased Level of Income: Klein et al. observe that the use of antibiotics increased by 65% in the world between 2000 and 2015 (Klein et al., 2018). As the average income of the population in the developing country increases, the overuse of antibiotics also increases (Chaw et al., 2018; Klein et al., 2018). In a developing country, animal protein consumption increases, which also increases the consumption of antibiotics in animals (Subedi et al., 2021).

The pattern of consumption of antimicrobial agents around the world changes rapidly (Klein et al., 2018; Wang et al., 2021). As per the data, in 2000, the highest consumption of antibiotics was in France, Spain, the United States, and Hong Kong; however, in low- and middle-income countries like Romania, Turkey, and Tunisia, it shows an increase in the rate of consumption of antibiotics in 2015. However, the consumption of antibiotics at this rate will surpass the consumption of antibiotics in a developed country.

Biological Factor: Bacterial infection and mutation can cause antibiotic resistance (Read and Woods, 2014). By using transposon and insertion sequences, various resistance genes can be obtained through a small fragment of DNA and plasmid (Li et al., 2019; San Millan, 2018). The transmission of these plasmids from species to bacteria also spread AMR (Sun et al., 2019).

Easy Route for Travel: A recent study suggests that the easy and modern route for traveling for goods, people, and animals leads to the circulation of AMR across the world (Castro-Sánchez et al., 2016). Humans, animals, and goods traveling can be exposed to the pathogen and infect the country (Frost et al., 2019). For example, when tourists from Europe visited India and they tested positive after returning from the trip (Ruppé et al., 2014).

3.4.6 Source of Antimicrobial Resistance

Waste of Water as the Source of Contaminated Water: Talking about pharmaceutical waste, India is the largest producer of pharmaceuticals in the world (Rehman et al., 2015). On average, Indian pharmaceuticals produce 28–31 mg/L of ciprofloxacin on two consecutive days (Larsson et al., 2007). Additionally, the pharmaceutical plant produces kilos of antibiotics every day. Classes of antimicrobial resistance (AMR), such as fluoroquinolones and sulfonium compounds, are characterized by the persistence of their residues due to high environmental stability. But some drugs like the β-lactam group of drugs can degrade faster, while the former has the potential to grow the AMR, leading to the contamination of wastewater (Ashbolt et al., 2013). All these drugs, which are produced by the Indian pharmaceutical companies, can pollute the pond of the neighboring river and sea coast. Every pharmaceutical company generates wastewater in the world, but in developed countries, these companies appropriately treat wastewater before discharge and in developing countries like India, due to the lack of proper equipment for wastewater treatment, the risk of AMR in India increases. Treatment of wastewater is a costly process, which is why developing countries cannot afford it for every pharmaceutical company (Lundborg and Tamhankar, 2017). India and China are the largest producers of wastewater in the world, which leads to a matter of concern globally.

3.4.7 Antimicrobial Waste from Hospital

Hospitals and several healthcare centers are the major sources of antimicrobial waste such as useless medicine or the secretion of the patient. The consumption of antimicrobials is high in hospitals. That is why the wastewater secreted out from hospitals favors the growth of resistance genes. Genotoxic alteration and modification of bacterial strains can be caused due to the presence of a sufficient amount of antimicrobials generated from the Indian hospital (Diwan et al., 2010). If we treat the hospital waste before its release, around 82% of antimicrobials can be removed from it (Duong et al., 2008). However, only 40% of hospitals and healthcare centers treated the wastewater released by them. In hospitals and nursing homes, the consumption of penicillin is comparatively high; due to the instability of its chemical nature, it cannot be measured. There was no use of tetracycline in the hospital, but a small concentration of it can be present in the waste from the hospital. This small concentration of tetracycline can be produced by the patient who visits the hospital (Verburg et al., 2019).

3.4.8 Wastewater of Municipal

Most of the municipal waste is placed in the ground without treatment, which favors the environment for the growth of resistant organisms and their genes. Only 20%–25% of municipal waste is treated but this 20% is not so functional for treatment (Lundborg et al., 2017). Municipal water waste, which is AMR, is released into the water bodies. Mutha River, which flows in India, carries an AMR whose concentration is 30 times more than municipal and domestic waste (Lata et al., 2016).

3.4.9 Extended-Spectrum β-Lactamases (ESBL) in Wastewater

As per the report, Enterobacteriaceae produced by ESBL can co-resist trimethoprim, fluoroquinolones, and a few aminoglycosides. The concentration of ESBL-EC colonies is lower in the other wastewater as compared to the wastewater from the nursing homes. According to the study, the requirement for cephalosporins is very low in nursing homes.

Isolated *E. coli* can be obtained from unknown media in which limited strains are present to confirm the presence of ESBL-EC phenotype. There is a major difference in finding out the phenotype of ESBL-EC with known and unknown media which can cause difficulty in determining the phenotype with the help of known media.

3.4.10 Other Sources of Antimicrobial Resistance

Discarded products from livestock: excreta of animals, with the help of antimicrobial and resistant organisms, can contaminate the surrounding (Wichmann et al., 2014). The sample collected from the fecal matter of cows and pigs contains a few antibiotics; out of these, nine antibiotics are from the fecal matter of cattle and three from a pig.

3.4.11 Manure and Sludge from Agriculture

Sludge is nothing but the slurry that can be produced during wastewater treatment and is rich in microorganisms, whereas manure is a nature-based product that helps in the growth of crops (McClellan and Halden, 2010). Antimicrobial content is different for the sources that produce differently. Various drugs like doxycycline, sulfadiazine, and oxytetracycline are present in the manure (Berendsen et al., 2015), and drugs like ciprofloxacin, norfloxacin, and trimethoprim are contained in sludge that is less water-soluble (Clarke and Smith, 2011). A group of gene called "resistome" has the potential to resist for a long period even after the decomposition of sludge or manure (Dantas and Sommer, 2012; Yang et al., 2014). India does not have proper data regarding this, so a study from China helps us to tackle this problem. According to the report from China, about 156 new resistant genes have been found in the manure and sludge.

3.4.12 Heavy Metals

Metals follow the same process to resist themselves as that of antimicrobials followed by decreasing the permeability of the membrane, sequestration of intracell, and the availability of the same plasmid (Ardestani et al., 2015). The study suggests that when the concentration of metal changes in the soil toxicities, there is no change in the process of selection of resistant genes.

3.4.13 Aquaculture

Aquaculture is the leading industry in India and other countries, which is why the use of antimicrobial agents increases to increase the productivity of seafood that harms the environment. Henriksen et al. (2015) suggest that the AMR through aquaculture can spread globally through international water. Due to the lack of data, it is not confirmed but antimicrobial residues are found in aquatic shrimps in India (Swapna et al., 2012).

3.4.14 Importance of Education for Antimicrobial Resistance

Medical students need to know about the prescription for antibiotics. It needs to amplify the strength of training for nursing, pharmacists, and dentists for graduate and undergraduate students. All are in a position to prescribe the antibiotics.

There are various techniques in the field of education and training, which are very valuable for students (Little et al., 2013). Various online courses called massive open online courses (MOOCs) are available globally. It can be available on the smartphones, and it provides the stage for antimicrobial stewardship of application. MOOCs provide good quality education without any hindrance.

3.4.15 Mechanism of Action of Antimicrobial Agents

Now we will learn about the working of antimicrobial agents. The function of the microbe is first targeted by the antimicrobial agent. A different antimicrobial agent works in a different way to kill or inhibit the bacteria. There are some different mechanisms for the action of antimicrobial agents.

3.4.16 Synthesis of Cell Wall

The cell wall of bacteria helps maintain the cell and also prevents the bacteria from lysis due to osmotic pressure (Johnson et al., 2014; Cho et al., 2014). The cell wall of bacteria contains peptidoglycan. The structure of peptidoglycan is arranged in a cross-linked manner with the help of two chains of N-acetylmuramic acid and N-acetylglucosamine using two peptidases, transpeptidase and carboxypeptidase. A range of antibiotics, including β-lactams and glycopeptides, are commonly employed for their potent antimicrobial activity (Nikolaidis et al., 2014).

3.4.17 Glycopeptides and B-Lactams Are the Drugs that Prevent the Synthesis of Bacterial Cell Walls

Various types of bacteria and penicillin-binding proteins (PBPs) can be attracted by various antibiotics like β-lactam. Structurally, it is locked like a D-Ala-D-Ala forming dipeptide bond and this binds covalently with the active site present in PBPs, which prevents the synthesis of the bacterial cell wall (Zapun et al., 2008).

Vancomycin and teicoplanin are two major glycopeptide antibiotics that are used by healthcare centers. Transpeptidase acts with the help of glycopeptides because it does not allow the formation of the link between the structural layers of peptidoglycan. The outcome of this causes the lysis of the cell as the content is released from the cell (Cole and Riordan, 2013).

3.4.18 Synthesis of Protein

The synthesis of a specific process from the cell is a complex biological process. Protein synthesis can be achieved by transcription and translation, which consist of four major steps: initiation, elongation, termination, and recycling.

Protein synthesis can be inhibited by antibiotics because of the structural change in the ribosome and eukaryotic ribosomes. In this way, bacterial growth is inhibited by using antibiotics. Antibiotics can inhibit the synthesis of protein by altering the process of both the units of ribosomes like the 30S and 50S of 70S bacterial ribosome.

Blockage of 30S Subunit: Blockage of protein synthesis at the 30S subunit of ribosomes can be initiated by macrolides, tetracycline, and aminoglycosides. The entry of positively charged carbohydrates into the bacterial cell is through diffusion. In this process, positively charged carbohydrate binds with the plasma membrane, which is negatively charged, and this further leads to diffusion. Inside the bacterial cell, the diffused particle is attacked at the A-site of 30S ribosomes. This attraction leads to a change in translation and can form mRNA–tRNA incorrectly.

Blockage of 50S Subunit: At the peptidyl transferase center, ribosomes of bacteria and amino acids of the 50S subunit attach to form a polypeptide chain. Polypeptide chains are allowed to leave the ribosomes through the nascent peptide exit tunnel of the 50S subunit. Chloramphenicol does not allow a new polypeptide to move through the tunnel. That is why disturbance in the elongation process can inhibit the synthesis of protein.

3.4.19 Resistance to Antibiotics by Dissemination of Resistance Gene

3.4.19.1 Transformation

In in vitro conditions or the environment, the transformation of naked DNA takes place. A sophisticated process consists of type 2 and 4 secretion system and a type 4 pili. In this process, the DNA present at the surface of the chromosome is taken up by the bacteria with the help of a highly preserved channel present in the membrane of the cytoplasm (Krüger and Stingl, 2011).

3.4.19.2 Transduction

It is a process in which bacteria like *Staphylococcus aureus* are required through which DNA transfers between them with the help of bacteriophages. Cells of bacteria are used by bacteriophages for replication. In this process, small pieces of DNA may enter the bacteriophage. For further replication, the DNA of the bacteriophage gets into the new bacteria. That is why the new DNA of bacteria forms and has resistance properties.

3.4.19.3 Conjugation

In this process, the horizontal movement of the gene to the recipient cell is with the help of T4SSs (transferase). It is necessary to form a DNA-binding complex from the start of the conjugation at the DNA transfer site (relaxosome) and a connection (coupling protein) for the attachment of these two.

3.4.20 Resistance Mechanism of Antibiotics

3.4.20.1 Resistance to β-Lactam Antibiotics

The resistance of the bacteria is developed with the help of two processes against the β-lactams. One of the processes is to reduce the attraction between the PBPs and antibiotics by changing the structure of PBPs. Another process is the destruction of the β-lactam ring with the help of an enzyme that can lead to the inactivation of antibiotics.

3.4.21 Resistance to Glycopeptide Antibiotics

The major matter of concern in clinical health is the presence of *Enterococci* resistance to glycopeptides and a decrease in the susceptibility of antibiotics in *S. aureus*. This happened due to the change in the sequence of transpeptidase enzymes (Marsot et al., 2012). The horizontal movement of the resistant gene (Tn1546 transposon) just after the in vivo treatment can separate the vancomycin-resistant *Staphylococcus aureus* strain in 2002. Inactivating the mechanism of vancomycin can be achieved by using several proteins, which are encoded with transposons that are mobile in the nature to recreate peptidoglycan, restrict the binding of vancomycin, and gain the stability of the cell wall.

Resistant to trimethoprim and sulfonamide (TMP) antibiotics

Substitution of an amino acid (Phe98 to Tyr98) for the gene called dhfr can cause resistance to TMP in *S. aureus* and *S. pneumonia*. With the help of these two amino acids that can change the structure of the enzyme, it can resist the sulfonamide in *S. pneumonia*.

3.4.22 Multidrug Resistance (MDR)

Bacteria gain resistance against multidrug resistance due to several reasons; one is an alteration in the membrane, inactivation of drugs, and modification of drugs. Spreading of multidrug resistance through bacterial infection is a major problem for treatment. One of the MDR is known as Gram-negative bacteria, which reduces the effectiveness of clinically used antibiotics. There are a total of five protein-encoded pumps present in the bacteria: resistance nodulation division family, major facilitator superfamily, multidrug and toxic compound extrusion families, and ATP-binding cassette MRD transporter.

In *E. coli* and Gram-negative bacteria, the RNA family is found. This contains protein that is used to transport in the inner membrane, like a periplasmic accessory protein and a channel for a protein present at the outer membrane.

3.4.23 Antimicrobial Resistance in Wastewater

3.4.23.1 Environmental Reservoir of AMR from Wastewater

AMR can spread among people and animals through various routes. The environment plays a major role as a reservoir for AMR and also helps mix mobile genetic elements, and it involves animals and humans as a host.

These reservoirs of different natures can impact the AMR gene, which leads to the development of AMR organisms (Wellington et al., 2013). AMR has been present in the environment for a long time, and there is some evidence in the environment suggesting the transfer of the AMR gene. However, the movement of ARGs between the two similar species is well organized. According to the study, the transfer of ARGs can occur phenotypically with distant species or organisms of different species. This can affect the development of new AMR organisms through evolution. Effluent collected from pharmaceutical companies and hospitals contains a major number of antimicrobial residues, ARGs, and AMR. According to the study, wastewater is the most important reservoir for AMR in the environment. This provides a suitable condition for the proper growth of AMR and ARGs. ARG is nondegradable in nature and can be spread in the environment by using the horizontal gene transfer technique. Through studies, researchers can analyze how public gathering events like the Kumbh Mela, in holy places, transmit the AMR in the environment.

3.4.24 Population of ARGs and ARB due to Its Dissemination and Wastewater from the Hospital

A recent study suggests that the hospital waste and clinically ARGs and ARB population can critically increase the level of AMR bacteria released into the environment through the hospital, which are untreated. Various AMR bacteria are used clinically, such as MDR and vancomycin and ampicillin-resistant *Enterococcus* spp., which can be detected in the sample of hospital waste. That is why the total influent from the hospital is very small. The presence of the AMR gene is the greatest amount even after the treatment. That is why the discharge of treated water from hospitals does not have a major impact on the concentration of AMR in the environment (Figure 3.1).

3.4.25 Impact on AMR Dissemination by the Water Treatment Process

In today's scenario, there is a major development in the wastewater treatment process, which can be done chemically, physically, and biologically to reduce the waste from water and make it suitable to reuse and back to the environment. The amount of the removal of AMR bacteria from untreated wastewater can depend on the process used for its treatment (Hiller et al., 2019).

From the final release of the hospital after wastewater treatment, the presence of MDR in *E. coli* strain is related to neonatal meningitis and extraintestinal serotypes. Resistance in BFG bacteria removed from treated wastewater is more as compared to the bacteria present in human feces (Niestępski et al., 2019). However, the amount of ARGs in the wastewater from the hospital after treatment is the same as before treatment.

According to these studies, the ARGs' development can occur in the WWTP with the help of technology. ARB that have high biomass and biodiversity can regulate the development of

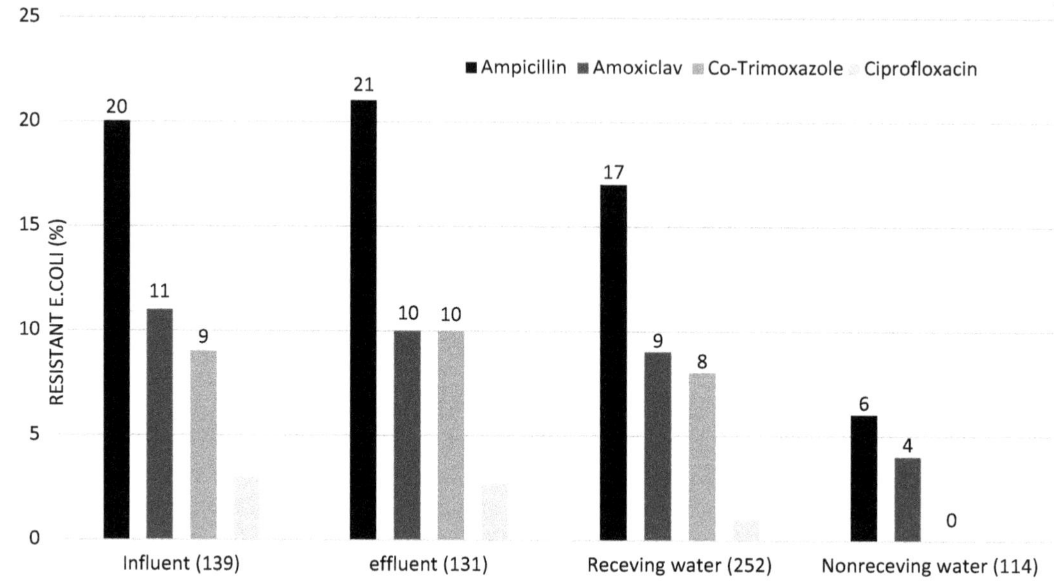

FIGURE 3.1 Resistance of antimicrobes in surface water and WWTP. *E. coli* percentage in the treatment plant against the four antimicrobials within the surface water and the control surface water. In brackets, it shows the number of isolates that are tested per location.

the ARB's population. There is an effective process for treating wastewater that reduces the presence of AMRs in it. These processes can be the membrane filtration and ultraviolet (UV) radiation.

India is a holy place for many religions. One of its examples is the Kumbh Mela. Millions of Hindu pilgrims gathered here to celebrate at the banks of the holy rivers Ganga and Yamuna (Sridhar et al., 2015). Pilgrims used to bathe in these rivers because they believed that this will clean their sins. But they are unaware of its harmful effects. This can seriously affect their health. A recent study suggests that many ARB, ARGs, and AMR are found in the Ganga river (Reddy and Dubey, 2019). Airborne and foodborne diseases are at high risk in the hajj. Studies suggest that the gathering of pilgrims in such large numbers can cause the transmission of AMR bacteria into the body. Fresh water and surface water play a major role in the dissemination of AMR pathogens. In this complicated evolutionary network and ecological system, the development of AMR takes place because of the excessive use of antimicrobial drugs. The concentration of ARGs is higher even after wastewater treatment, which can develop AMR bacteria and can be a matter of concern for public health.

3.4.26 Steps Were Taken to Curb AMR in the Environment

3.4.26.1 Political Commitments

International Efforts: AMR was sighted as a serious threat to public health by the WHO (World Health Organization, 2011). The southeast Asian regional office (SEARO) organized various events and meetings to plan and adopt measures to implement AMR in the region. In India (Jaipur), a landmark meeting was held in the year 2011 in which the Health Ministers of all countries participated and committed to adopting the Jaipur declaration (World Health Organization, 2011) by taking intensive measures on AMR. It is a complex intersectional problem, so a cure is not easy.

An increasing rate of AMR has to be controlled by taking strong political liability that initiates the locally based solution, which is context-specific (Bhatia et al., 2019). The approach for the whole of society, while containing AMR, plays a major role in political, financial, technical, regulatory, etc. The government has to cooperate and collaborate with the private sector, academia, pharmaceutical industries, and international partners to fight AMR and raise the issue on national development agendas. The national authorities influence when the pharmaceutical industry and academia collaborate and develop a new class of antimicrobials to keep in step with the exposure/emergence of resistance. In so many countries, the private sector plays an important role in the supply of health care at an individual or population level.

More than two-thirds of the health care is provided to poor people in different countries including India by private as well as nonstate actors. According to the report of the WHO in 2014, the extraordinary rates of AMR prevalent come into focus, and then India and together make joint efforts (Taneja and Sharma, 2019). Afterwhile, AMR is labeled as a flagship priority area for SEARO given by WHO (2016). A vigilant program was started by the Indian Medical Association and the aim is to train physicians and aware the general public about AMR (Taneja and Sharma, 2019). In 2015, the WHO started an action plan globally, which decided to include the concept of one health in the fight against AMR (Bhattacharjee, 2022).

National Efforts: In India, AMR is one of the highest rates of infectious disease burden in the world, and this burden is made worse by poor sanitation and malnutrition (Ganguly et al., 2011). The pathogen range of resistance varies in different regions. To use antibiotics properly, one must be aware of local patterns of resistance (Raghunath, 2008). According to some hospital-based data, antibiotic resistance is on the rise and will become a bigger issue if it is not properly addressed under current needs, as it is in other developed nations (Behera and Mathur, 2011).

At a regional level, the requirement for and a challenge in taking responsible action, including for health, is primarily a national issue (Lakshmi, 2008). The Indian Health Ministry jointly with the Family Welfare Ministry issued the national policy for the spread of antimicrobial resistance in 2011, which is a positive development. It addresses the necessary intervention strategies and the procedure for formulating and putting into practice a standard antibiotic policy (Kumar et al., 2013a).

This comprehensive plan, which incorporates the WHO's GAP's essential components, was developed in 2017 by India's Ministry of Health and Family Welfare (Gupte, 2012). It focuses on the following six strategic priority areas:

1. People have to be aware and understand that education, communication, and training are required.
2. Enhancing information and proof through surveillance;
3. Prevention and management of infections;
4. Improved application of antimicrobials in food, medicine, and animals;
5. Research and innovation in AMR;
6. Enhanced responsibility as well as governance at the international, domestic, and sectoral levels.

Notably, India`s plan outlines the importance of managing AMR across different areas like public health, animal husbandry, agriculture, or the ecosystem, concerning the true spirits of the "One Health" approach. Even though encouraged as a noble plan, some experts have challenged how it will be carried out on the ground levels (Godman et al., 2021).

3.4.27 Cost of Antimicrobial Resistance

The outcomes of a study of fiction show that the costs associated with AMR are quite expensive and vary greatly between nations (O'Neill, 2014; Utt and Wells, 2016). AMR is believed to cost the US economy \$55 billion annually, with \$20 billion going toward health care and about \$35 billion going toward lost productivity, following the CDC (Dixit et al., 2019; Dadgostar, 2019). A World Bank study recently supports that this AMR would lead to increasing poverty and have a wider influence on low-income countries than on the rest of the globe (Toops et al., 2021).

According to research findings, the average world gross domestic product could drop by around 1%, with developing nations having to lose 5%–7% by 2050 (Anderson et al., 2019; Utt and Wells, 2016). This percentage equates to \$100–210 trillion in total. By 2050, multidrug resistance tuberculosis could cost the globe \$16.7 trillion (Dadgostar, 2019; O'Neill, 2014). Additionally, AMR will widen the disparity between developing and developed nations, which will lead to a significant rise in inequality. In particular, people from low-income nations will represent the vast majority of the people who are forced into high poverty as the outcome of AMR (Dadgostar, 2019). This focuses on how the poorest people in the world will ultimately be the most influenced because their economies depend more on workers, which will suffer from such a high frequency of communicable diseases (Dadgostar, 2019).

AMR has a larger effect on employment by the decrease in income, which is caused by illness and higher mortality, as well as the direct impact on the economy (Naylor et al., 2018). Loss of life due to AMR reduces the capacity for work, which causes a negative impression on the mass of the population and also the labor productivity in the country (Naylor et al., 2018; Taylor et al., 2014). To calculate the possible financial effects of AMR on the labor force, Taylor et al. developed a hypothetical framework. In their experiments, they have related a benchmark (the lack of AMR) with the latest fad in AMR along with the worst outcome that may occur if proper care is not taken. With the result of those experiments, the current pattern of AMR has remained unchanged for 10 years, and the world's working population has been reduced by 2 years. In comparison to the rest of the world, the above transformation will be much more widespread in Europe and Asia (Taylor et al., 2014). The nation will lose approx. \$28 billion in 10 years in terms of yearly gross domestic product (GDP) reduction, if the trend of AMR is unchanged. This hypothesis indicates that the European Union and the Association for Economic Co-operation nations will suffer larger losses than the rest of the globe with such a \$20 billion decrease in GDP (Taylor et al., 2014).

AMR will have a serious impact on global trade if current AMR trends continue (Lekagul et al., 2019). By 2050, global exports may have significantly decreased, according to a World Bank assessment on the consequences of resistant bacteria on slave labor industries (Dadgostar, 2019). Due to AMR's long-term impacts on the economy, it is possible to say that its negative effects on the global financial system are initially expected to be worse than the international economic slump (Dadgostar, 2019).

The effects of AMR on livestock production will also be substantial (Hao et al., 2014). Animals would be similarly affected by AMR as humans in terms of mortality and morbidity. Animal treatment options will no longer be effective due to an increase in AMR, and infections might well become worse (Dadgostar, 2019). In the end, this will result in less livestock being produced and traded, which will raise the price of protein as a result of a reduction in sources of protein like milk, eggs, and meat (Dadgostar, 2019). Given that animal proteins are becoming more popular on a global scale, protein shortages will be a major problem (Subedi et al., 2021). The production of livestock in low- and intermediate-income countries will be severely impacted by AMR, according to the World Bank. If the current AMR tendencies do not slow down, predictions state that the productivity of cattle would fall by 11% by the year 2050 (Dadgostar, 2019). The economic situation will worsen as a result of this significant loss in animal production, which would diminish revenue-generating (Dadgostar, 2019).

3.4.28 Challenges of Antimicrobial Resistance

Successful monitoring and systematic research regarding the impact and worldwide epidemiology are unavoidable because AMR remains a serious challenge that puts a successful treatment approach in a precarious position. The main problem that is mentioned in recent studies is estimating the pressure of AMR as well as the occurrence of AMR. These studies show a large communication gap regarding mortality rate, treatment failures, disease burden, and different kinds of resistance in multiple cities, particularly in low- and intermediate-income states, which represent the biggest barriers to analyzing the outcome of antibiotic resistance. A successful plan has been started at the national level for the monitoring of the national situation, adding a new diagnostic technique first, the growth of antimicrobials and vaccines, as well as controlling the further arrival of antibiotic resistance (Väänänen et al., 2006; Wright, 2014; Safdari et al., 2018).

3.4.29 Procedures for Acknowledging AMR that Projects

3.4.29.1 Infectious Diseases Avoidance

The finest method to control infectious diseases is by the use of antibiotics that stop the growth as well as the reproduction of resistant bacteria. By checking medication patterns and resistance, AMR can be avoided, and also growing and applying effective infection control procedures, by controlling healthcare services like the primary care centers on the best antibiotics use. For prevention, the main source of antibiotic resistance features must be known and followed to examine changes in the resistance pattern. According to how surveillance and measuring food contamination, searching into separate materials, teaching people about safe food behavior tools, recognizing a high-risk set of people, and trying to encourage the proper way of handwashing, by using this, we can slow down the transmission of infections.

Contact tracing is a good method to find cases and contact people who are in danger. While keeping in mind that people who are at high risk has to be found and provided with effective treatment. This is a source-exhaustive method, and the achievement is restricted to achieving diseases like gonorrhoea, tuberculosis, and meningococcus. Some bacteria, for example, *S. pneumonia* can be prevented through vaccination. After the discovery of the first vaccine in the year 2000 the rate of antimicrobial infection falls; but these vaccines cannot show their effect on serotype 19A of pneumonia. Later in the year 2010, a new type of vaccine was discovered with a significant effect on serotype 19A (Abushaheen et al., 2020; Fridkin et al., 2014).

3.4.30 Prevention of Misuse

The antimicrobial remedy standard should involve the necessary application of an evidence-relied approach. There should be an avoidance of unwanted antibiotic recommendations by clinicians. They should ensure responsibility for advancement in patient-centric safety techniques. Healthcare providers have a major role in this domain, which is why they monitor the healthcare workers frequently and provide assessments based on that essential task. The affected people and clinicians must be assured of educational aid to achieve the maximum utilization of prescriptions. A fair example of this approach is the keen program instigated by CDC—it assists state-based programs and brings forward ways to make patients, lawmakers, and providers conscious regarding the wrong use of antibiotics (Fleming-Dutra et al., 2016; Abushaheen et al., 2020; Fridkin et al., 2014). One such method of antibiotics stewardship is the appropriate use of antibiotics in sample amounts. Additionally, it ensures better health and financial outcomes and acts as a hindrance to the spread of disease (Fleming et al., 2016; Abushaheen et al., 2020; Fridkin et al., 2014). Relatively, it saves people from the result of resistance and makes an important contribution to the fight against resistance. New antibiotics are necessary to track the progression of resistance; though, it is an evolving process that takes place over a period.

3.4.31 Promotion of Education and Innovation

To make the people aware of the different types of stakeholder groups, a modern educational program is conducted to maximize and spread return to the public health plans. Besides that, the education system has to keep the focus on the reasons, influences, as well as challenges of bacterial resistance on human and animal health. Even when promotional materials—such as specification sheets, banners, and video content—are exchanged between relevant parties, it remains critical to examine the role of early intervention in curbing the development of tolerance. (Fleming et al., 2016; Abushaheen et al., 2020; Fridkin et al., 2014). As a result, the unit of data about that topic must be provided to hospitals, doctors, state agencies, and health insurers. Likewise, early treatment needs to help with the creation of splitting advanced technology, such as research institutions, studies, and techniques to find new types of bacteria or pathogens. Citizen collaborations may be planned to support the creation of new medicines. One example is the cooperation between the American government and the Biomedical Advanced Research Development Authority. To try to solve the judicial mistakes made by the drug company industry in the search for new opiates, the improvement in public safety can be constructed. To facilitate drug makers, plans such as giving extra money for research or simplifying permission may also be fulfilled (Gould and Bal, 2013; Piddock, 2012).

3.4.32 Introduction of the Checkpoint

To stop the sales of antibiotics on the black market and for consciousness, a certain phase will be built. Due to the availability of opiates in a medical shop with any suggestion of doctor meds, this takes place. To stop the unlicensed market in the field of opioids, government action must be taken (Michael et al., 2014; Alam et al., 2019). Several parameters, along with the behavior and body language of the affected person, may change the mind of the doctor to treat the patient. Because of this, expertise must lead to the doctor's note instead of the desires of the affected persons. During the first meeting, a doctor advised the affected people to take antibiotics; similarly, the affected people are not allowed to take antibiotics if the symptoms are not seen and it is a good strategy. While the treatment of acute respiratory tract infections in Spain, New Zealand, and America, it is seen that the intake of antibiotics is decreased in those affected people (Sargent et al., 2016). The crucial fact in identifying the responsibility, as a key factor in promoting the budget, is gaps in the records of incidence and mortality rate of public opposition. For internet-based data entry on pathogens, bacterial resistance, and data planning at the government level, the CDC has launched the National Healthcare Safety Network.

By using network, we enter data and start searching for the microorganisms that affect the healthcare pathogen. It is necessary to make sure that the network works properly in the healthcare group, for example, hospitals and doctor's clinics. Physicians should be aware of the crucial tests to identify antibiotic susceptibility and resistant microbes, which may have a massive effect on a patient's health condition (Abushaheen et al., 2020; Fridkin et al., 2014).

3.4.33 Recommendations for the Clinical Management of Antimicrobial Resistance

AMR can be controlled on a large scale by the efforts of the doctor, affected person, and their parents, healthcare provider, etc. at the governmental, country, and municipal levels. As a result, communication among the different stakeholders for AMR diagnosis and management has the potential to seriously influence the pattern of antimicrobial needs.

3.4.34 Recommendations to Clinicians

1. Follow the instructions for prescribing drugs by the doctors such as the dosages, timetable, and period of recommendations, in the memo.

2. The prescribing of antibiotic treatment for the affected person's unique health problem and the bad aspects of overuse like antibiotics agents, adverse reactions, and other types of infections will be described to their mother, father, or by their family members.
3. Encourage the affected person that their health problems can be treated without any requirement of antiviral drugs and promote a straightforward orders strategy in case their diagnosis does not improve.
4. Accept the most contemporary development in exposure, locate new self-conferring resistance microbes, and utilize the information at hand to make decisions.
5. Develop the guidelines and advice for infection prevention and control to stop infections.
6. Clear and concise responsibilities designed to highlight the need for proper broad-spectrum antibiotics can be posted in the discussion halls to help affected persons and doctors.

3.4.35 Recommendations to Patients and Their Families

The commitment to stopping the spread of disease and AMR falls equally on medical practitioners, sick people, and their communities.

1. Be familiar with the information about antibiotic-resistant bacteria.
2. Admit the information and guidance of the medical personnel related to your medication management, opioid symptoms, contraindications, antibacterial drugs period, and toxicities.
3. Follow the rules of basic grooming and appropriate hand sanitary conditions.
4. Once you are sick, keep your distance from the public.
5. Need proper immunization, for example, the flu virus, and also nag your family member to get their doses at the right time.

3.4.36 Recommendations to Health Organizations and Health Agencies

1. Recognize meds patterns in healthcare facilities such as community health centers, rehabilitation centers, and hospices to indicate and discuss the cause of abusive behavior.
2. Health professionals and the general community should be made aware of the importance of utilizing antibiotics as efficiently as possible and the dangers of doing so.
3. Construct techniques, instruments, as well as software that will focus on making responsibility measurement effective and help prioritize observations.
4. Create standard operating procedures that are efficient for the detection and treatment of communicable diseases.
5. Offer possibilities for professional growth and management of youth development.
6. Create commonsense laws and expert, government, and geographical policies that guarantee their development.
7. To enable the most effective use of antibiotics, encourage technological innovation, medications, and scientific studies.

3.4.36.1 Wastewater Treatment

Wastewater treatment is known as sewage treatment, the process of cleaning up sewage or wastewater before it enters reservoirs or other natural water sources such as river systems, ponds, marshes, and deep oceans. The difference between clean water and contaminated water relies on the type and concentration of contaminants found in the water as well as on its destined use because fresh water cannot be found in nature (i.e., outside of chemical laboratories). In general, water is considered contaminated when it has too many dirt particles and it has to be proper for activities like drinking, scuba diving, and catching fish. Even though external factors have an impact on water quality, the term "pollution" typically indicates that human interaction can be the cause of groundwater pollution. Thus, sewage treatment is a key component of water pollution control. Water pollution is usually accompanied by the dewatering of industrial wastewater in and out of either groundwater or surface water.

A combination of underground aquifers, seawater, and stormwater, along with liquid- or water-borne waste removed from houses and apartments, organizations, and residential and retail establishments, can be referred to as sewage. In general, it has a high concentration of wastes that require oxygen to break them down, harmful or disease-causing pathogens, organic substances, essential minerals that encourage plant growth, inorganic chemicals, nutrients, and sediments. It might also include poisonous substances (Metcalf et al., 1991).

3.4.36.2 Wastewater Can Be Divided into the Following Four Groups

1. **Domestic:** Wastewater is discharged from homes, businesses, and other similar facilities.
2. **Industrial:** Wastewater with a heavy industrial waste element.
3. **Infiltration/Inflow:** Unnecessary extra water that enters the sewage system either directly or indirectly, such as via porous medium, leaking water joints, or other means. Stormwater inflow refers to water entering the sewage system through panels, groundwork, crawl space drainages, ceiling crossbars, and subsurface drainage interconnection.
4. **Stormwater:** Landslide rising sea levels brought on by rain.

Due to the availability of a wide range of biological, physiological, and biochemical mechanisms that can be used to improve consistently, WWTPs hire various techniques to comply with or even surpass environmental requirements relating to wastewater discharges. The three different types of water treatment processes are (1) primary, (2) secondary, and (3) advanced tertiary, irrespective of the strategies utilized at any specific WWTP. Both the initial investment required to build or improve wastewater treatment facilities and operation and administration costs are affected by the level of technology used within each treatment section. In the United States, secondary treatment processes were inspected by Muga and Mihelcic. For WWTPs with up to 3,800 m^3/day of ability, building costs varied from \$260 to \$2,770/m^3/day [2008]. At WWTPs in Louisiana with a potential of 3,800 m^3/day, the expected price for supplemental tertiary treatment excluding the cost for primary or secondary treatment was \$502/m^3/day for development and \$120,116 for yearly operation and maintenance (Ko et al., 2004).

3.4.36.3 Primary Treatment

Massive object is eliminated from arriving sewage during primary treatment using floating, securing, and screening methods. The primary treatment procedure eliminated 25%–50% of the incoming biochemical oxygen demand, 50%–70% of the total suspended soil particles, and 65% of the grease and other contaminants. Grit chambers are used to remove smaller inorganic solids like sand, and sedimentation basins, also known as primary clarifiers, are used to eliminate both inorganic and organic suspended solids. Sludge is the term for the solids that are managed to gather and separated in sedimentation basins and relocated to anaerobic digesters for additional operations. Primary sedimentation also removes some organic nitrogen, organic phosphorus, and heavy metals that are linked to solids, but colloidal and dissolved constituents are unaffected. Primary effluent is the term used to describe the effluent from the primary sedimentation unit.

3.4.36.4 Secondary Treatment

After the primary treatment stage, secondary treatment focuses on extra organic matter in sewage water and may start by removing nutrient solutions. Industrial effluents, which are made of pathogens' stone blocks that break down the organic contaminants in the liquid waste, are one secondary treatment approach. At the secondary treatment stage, organic compounds are also eliminated using synthetic wastewater, which contains a lot of pathogens. The number of organic contaminants that are consumed by microbial species in the sludge increases when primary treatment wastewater is released into the environment. Secondary clarifiers enable the withdrawal of solid particles and/or municipal wastewater from the liquid waste by letting them settle.

After standard secondary treatment, delayed and soluble substances could remain available in the liquid waste, and enhanced tertiary treatment may be required to comply with regulatory standards providing protection to accepting waters. So many chemical and biological nutrient removal (BNR) technologies are available to decrease the untreated sewage concentration levels before release. Metals and nutrients may be aimed at during additional treatment (Yeoman et al., 1988; Morse et al., 1998; Pagilla et al., 2006). BNR first came into existence about 30 years ago, and because of its natural benefits over conventional chemical methods, it has typically been a crucial component of conventional wastewater treatment. Tchobanoglous and others compare BNR to chemical modification and point out that BNR uses microscopic organisms in aerobic or anaerobic environments to eliminate nutrition by dissolved and/or connected cell growth, consuming less energy, waste, and pesticide application (Carey and Migliaccio, 2009).

Wastewater treatment plants (WWTPs) use chlorine or UV radiation as a final option before removal to decrease harmful microorganisms that may be visible in the liquid waste as a result of health problems that are associated with wastewater (Ongerth and Ongerth, 1982). Major efforts have been made to amplify and broaden sewage treatment in the United States due to widespread concern about the preservation of water resources for a purpose that transcends the healthcare system, along with aesthetics and aquatic wellness. The number of Americans served by WWTPs that offer additional sewage treatment rose from 7.8 to 108.5 million between 1972 and 2004. More than 70% of the U.S. population, or roughly 220 million persons, are now continued to serve by wastewater treatment systems that include at least secondary options for treatment.

3.4.36.5 Nutrient Removal from WWTP Effluent Discharge

Given the substantial sewage treatment until discharges, WWTPs can still greatly affect obtaining waterways. Given that rising algal blooms brought on by nutrient cycling are one of the main causes of aquatic injury, it is clear that WWTPs need to increase the efficiency of their nutrient removal processes (Eggers et al., 1991; Zhou and Smith, 2002; Pagilla et al., 2006; Correll, 1998; Magnien et al., 1992; Rabalais, 2002; Walsh et al., 2005). For WWTPs, dietary withdrawal is a special challenge because of the extra charges related to sophisticated treatment methods required to generate pollutants with poor nutrient concentration levels (Ko et al., 2004; Olivieri et al., 2005; Muga et al., 2009).

From Table 3.3, variations in nutritional extraction efficiency at WWTPs are due to various factors including (1) the use of remediation technology; (2) input sewage; (3) business and structural

TABLE 3.3
Nutrient Concentration Ranges in Untreated Sewage Wastewater (Tchobanoglous et al., 2003; Asano et al., 2007)

Constituent (mg/L)	Conventional Activated Sludge[a]	i. Untreated Wastewater	ii. Activated Sludge with BNR[b]	iii. Activated Sludge with BNR, Microfiltration, and Reverse Osmosis[c]
Total nitrogen	15–35	20–70	3–8	Less and equal to 1
NO_3–N	10–30	0–trace	2–8	Less and equal to 1
NH_3–N	1–10	12–45	1–3	Less and equal to 0.1
Total Phosphorus	4–10	4–12	1–2	Less and equal to 0.5

[a] Activated sludge with a nitrification stage is used as a secondary treatment.
[b] Tertiary treatment: Biological nutrient removal of nitrogen and phosphorus using activated sludge.
[c] Tertiary treatment: Advanced treatment combined with biological nutrient removal and activated sludge.

errors; and (4) technical facility design restrictions (Asano et al., 2007). Even though seasonal trends in sewage outflow may be present (Olivieri et al., 2005), each head of population stakeholder items in sewage can stay constant despite variations in water flow rate (Tchobanoglous et al., 2003).

3.4.37 Nitrogen Removal

The biological reactions of nitrification (the oxidation of NH_4-N and organic-N in an aerobic phase) and denitrification can remove nitrogen from inflowing wastewater (reduction of NO_3-N to gaseous nitrogen in an anoxic atmosphere). To adequately release nitrogen, such reactions are utilized to connect or dissolve cellular activities. The separation of nitrogen could be merged with that of the separation of other target components, including the separation of compounds that start raising the wastewater's biochemical oxygen demand (BOD). Although the initial stage of separating nitrogen, nitrification, requires autotrophic bacteria, tactical changes can be made to compensate them. Heterotrophic bacteria are utilized to reduce biochemical oxygen demand (BOD) through the breakdown of organic matter. Although several bacteria have been discovered for this purpose (Painter, 1970; Teske et al., 1994), *Nitrosomonas* and *Nitrobacter* are commonly used for the removal of nitrogen from sewage treatment. Payne (1981) and Gayle et al. (1989) recognize a diverse range of denitrifying autotrophic and heterotrophic bacteria that are effective at WWTPs. Biological nitrogen is completely removed through the denitrification processes (WWTPs).

3.4.38 Phosphorus Removal

Exhaustive analyses of the amount of phosphorus in sewage have been decreased using a diverse range of biochemical processes (Bowker and Stensel, 1990; Yeoman et al., 1988; Morse et al., 1998). A popular method to successfully lower untreated sewage concentration levels produced in groundwater is chemical phosphorus precipitation. Chemical precipitation is based on the idea that certain salts, when added to sewage, yield insoluble phosphates, which can then be isolated from discharged effluent through activated sludge procedures. Calcium, iron (Fe_2^+ and Fe_3^+), and aluminum are frequently found as salt precipitants (lime). During primary and secondary water treatment plants, the molecules aluminum sulfate and iron (III) chloride are frequently used as the phosphorus is covalently linked in the liquid waste that is created.

The process of biological phosphorus removal (BPR) from sewage water makes use of activated sludge, which usually contains specific bacteria that accumulate phosphorus and can comprehend the overabundance of phosphorus as polyphosphates (Wentzel et al., 1991). BPR techniques at WWTPs gained attention after Srinath (1959) found that activated sludge absorbed phosphorus at levels that were far higher than those required for biomass accumulation. To further develop the ideas of BPR, Barnard (1974) proposed that to successfully remove phosphorus, activated sludge has to be used in the anaerobic and then in the aerobic zones. To increase BPR, WWTP technicians could ensure that the pH, dissolved oxygen (DO), and temperature requirements are met for the microorganisms that are essential to blood pressure (BP).

REFERENCES

Abushaheen, M.A., Fatani, A.J., Alosaimi, M., Mansy, W., George, M., Acharya, S., Rathod, S., Divakar, D.D., Jhugroo, C., Vellappally, S., and Khan, A.A. 2020. Antimicrobial resistance, mechanisms and its clinical significance. *Disease-a-Month* 66(6):100971.

Agersø, Y., and Aarestrup, F.M. 2013. Voluntary ban on cephalosporin use in Danish pig production has effectively reduced extended-spectrum cephalosporinase-producing *Escherichia coli* in slaughter pigs. *Journal of Antimicrobial Chemotherapy* 68(3):569–572.

Alam, M.M., Islam, M., Wahab, A., and Billah, M. 2019. Antimicrobial resistance crisis and combating approaches. *Journal of Medicine* 20(1):38–45.

Anderson, M., Clift, C., Schulze, K., Sagan, A., Nahrgang, S., and Mossialos, E. 2019. *Averting the AMR Crisis: What are the Avenues for Policy Action for Countries in Europe?* Copenhagen, Denmark: European Observatory on Health Systems and Policies.

Ardestani, M.M., Van, Straalen, N.M., and Gestel, C.A. 2015. Biotic ligand modeling approach: Synthesis of the effect of major cations on the toxicity of metals to soil and aquatic organisms. *Environmental Toxicology and Chemistry* 34(10):2194–2204.

Asano, T., Burton, F., and Leverenz, H. 2007. *Water Reuse: Issues, Technologies, and Applications*. New York: McGraw-Hill Education.

Ashbolt, N.J., Amézquita, A., Backhaus, T., Borriello, P., Brandt, K.K., Collignon, P., Coors, A., Finley, R., Gaze, W.H., Heberer, T., and Lawrence, J.R. 2013. Human health risk assessment (HHRA) for environmental development and transfer of antibiotic resistance. *Environmental Health Perspectives* 121(9):993–1001.

Azam, M., Jan, A.T., and Haq, Q.M. 2016. Bla CTX-M-152, a novel variant of CTX-M-group-25, identified in a study performed on the prevalence of multidrug resistance among natural inhabitants of river Yamuna, India. *Frontiers in Microbiology* 7(176):1–13.

Barnard, J.L. 1974. Cut P and N without chemicals. *Water & Wastes Engineering*, Part 1, 11(7):33–36; Part 2, 11(8):41–43.

Bartlett, J.G., Gilbert, D.N., and Spellberg, B. 2013. Seven ways to preserve the miracle of antibiotics. *Clinical Infectious Diseases* 56(10):1445–1450.

Behera, B., and Mathur, P. 2011. High levels of antimicrobial resistance at a tertiary trauma care centre of India. *The Indian Journal of Medical Research* 133(3):343.

Berendsen, B.J., Wegh, R.S., Memelink, J., Zuidema, T., and Stolker, L.A. 2015. The analysis of animal faeces as a tool to monitor antibiotic usage. *Talanta* 132:258–268.

Bhatia, R., Katoch, V.M., and Inoue, H. 2019. Creating political commitment for antimicrobial resistance in developing countries. *The Indian Journal of Medical Research* 149(2):83.

Bhattacharjee, M.K. 2022. Global action plan and antibiotic stewardship. In *Chemistry of Antibiotics and Related Drugs*, Cham: Springer, pp. 247–262.

Bowker, R.G., and Stensel, H.D. 1990. *Phosphorus Removal from Wastewater*. Park Ridge, NJ: Noyes Data Corporation.

Carey, R.O., and Migliaccio, K.W. 2009. Contribution of wastewater treatment plant effluents to nutrient dynamics in aquatic systems: A review. *Environmental Management* 44(2):205–217.

Cassini, A., Högberg, L.D., Plachouras, D., Quattrocchi, A., Hoxha, A., Simonsen, G.S., Colomb-Cotinat, M., Kretzschmar, M.E., Devleesschauwer, B., Cecchini, M., and Ouakrim, D.A. 2019. Attributable deaths and disability-adjusted life-years caused by infections with antibiotic-resistant bacteria in the EU and the European Economic Area in 2015: A population-level modelling analysis. *The Lancet Infectious Diseases* 19(1):56–66.

Castro-Sánchez, E., Moore, L.S., Husson, F., and Holmes, A.H. 2016. What are the factors driving antimicrobial resistance? Perspectives from a public event in London, England. *BMC Infectious Diseases* 16(1):1–5.

Chatterjee, S., Poonawala, H., and Jain, Y. 2018. Drug-resistant tuberculosis: Is India ready for the challenge? *BMJ Global Health* 13(4): e000971.

Chaw, P.S., Höpner, J., and Mikolajczyk, R. 2018. The knowledge, attitude and practice of health practitioners towards antibiotic prescribing and resistance in developing countries—A systematic review. *Journal of Clinical Pharmacy and Therapeutics* 43(5):606–613.

Cho, H., Uehara, T., and Bernhardt, T.G. 2014. Beta-lactam antibiotics induce a lethal malfunctioning of the bacterial cell wall synthesis machinery. *Cell* 159(6):1300–1311.

Chokshi, A., Sifri, Z., Cennimo, D., and Horng, H. 2019. Global contributors to antibiotic resistance. *Journal of Global Infectious Diseases* 11(1):36.

Clarke, B.O., and Smith, S.R. 2011. Review of 'emerging'organic contaminants in biosolids and assessment of international research priorities for the agricultural use of biosolids. *Environment International* 37(1):226–247.

Cole, T.S., and Riordan, A. 2013. Vancomycin dosing in children: What is the question? *Archives of Disease in Childhood* 98(12):994–997.

Correll, D.L. 1998. The role of phosphorus in the eutrophication of receiving waters: A review. *Journal of Environmental Quality* 27(2):261–266.

Dadgostar, P. 2019. Antimicrobial resistance: Implications and costs. *Infection and Drug Resistance* 12:3903.

Dantas, G., and Sommer, M.O. 2012. Context matters—the complex interplay between resistome genotypes and resistance phenotypes. *Current Opinion in Microbiology* 15(5):577–582.

Diwan, V., Tamhankar, A.J., Khandal, R.K., Sen, S., Aggarwal, M., Marothi, Y., Iyer, R.V., Sundblad-Tonderski, K., and Stålsby-Lundborg, C. 2010. Antibiotics and antibiotic-resistant bacteria in waters associated with a hospital in Ujjain, India. *BMC Public Health* 10(1):1–8.

Dixit, A., Kumar, N., Kumar, S., and Trigun, V. 2019. Antimicrobial resistance: Progress in the decade since emergence of New Delhi metallo-β-lactamase in India. *Indian Journal of Community Medicine* 44(1):4–8.

Duong, H.A., Pham, N.H., Nguyen, H.T., Hoang, T.T., Pham, H.V., Pham, V.C., Berg, M., Giger, W., and Alder, A.C. 2008. Occurrence, fate and antibiotic resistance of fluoroquinolone antibacterials in hospital wastewaters in Hanoi, Vietnam. *Chemosphere* 72(6):968–973.

Eggers, E., Dirkzwager, A.H., and Van der Honing, H. 1991. Full-scale experiences with phosphate crystallization in a Crystalactor. *Water Science and Technology* 23(4–6):819–824.

Fleming-Dutra, K.E., Hersh, A.L., Shapiro, D.J., Bartoces, M., Enns, E.A., File, T.M., Finkelstein, J.A., Gerber, J.S., Hyun, D.Y., Linder, J.A., and Lynfield, R. 2016. Prevalence of inappropriate antibiotic prescriptions among US ambulatory care visits. *The Journal of the American Medical Association* 315(17):1864–1873.

Founou, R.C., Founou, L.L., and Essack, S.Y. 2017. Clinical and economic impact of antibiotic resistance in developing countries: A systematic review and meta-analysis. *Public Library of Science* 12(12): e0189621.

Fouz, N., Pangesti, K.N., Yasir, M., Al-Malki, A.L., Azhar, E.I., Hill-Cawthorne, G.A., Abd, E.L., and Ghany, M. 2020. The contribution of wastewater to the transmission of antimicrobial resistance in the environment: Implications of mass gathering settings. *Tropical Medicine and Infectious Disease* 5(1):33.

Fridkin, S., Baggs, J., Fagan, R., Magill, S., Pollack, L.A., Malpiedi, P., Slayton, R., Khader, K., Rubin, M.A., Jones, M., and Samore, M.H. 2014. Vital signs: Improving antibiotic use among hospitalized patients. *Morbidity and Mortality Weekly Report* 63(9):194.

Friedrich, M.J. 2017. Drug-resistant tuberculosis is predicted to increase in high-burden countries. *The Journal of the American Medical Association* 318(3):231

Frost, I., Van Boeckel, T.P., Pires, J., Craig, J., and Laxminarayan, R. 2019. Global geographic trends in antimicrobial resistance: The role of international travel. *Journal of Travel Medicine* 26(8):taz036.

Gandra, S., Joshi, J., Trett, A., Lamkang, A.S., and Laxminarayan, R. 2017. *Scoping Report on Antimicrobial Resistance in India*. Washington, DC: Center for Disease Dynamics, Economics & Policy, pp. 1–30.

Ganguly, N.K., Wattal, C., Chandy, S.J., Arora, S.K., Gupta, U., and Kotwani, A. 2011. Situation analysis antibiotic use and resistance in India. Public Health Foundation of India, and Center for Disease Dynamics, *Economics and Policy* 10(9):209–212.

Gayle, B.P., Boardman, G.D., Sherrard, J.H., and Benoit, R.E. 1989. Biological denitrification of water. *Journal of Environmental Engineering* 115:930–943.

Godman, B., Egwuenu, A., Haque, M., Malande, O.O., Schellack, N., Kumar, S., Saleem, Z., Sneddon, J., Hoxha, I., Islam, S., and Mwita, J. 2021. Strategies to improve antimicrobial utilization with a special focus on developing countries. *Life* 7:11(6):528.

Golkar, Z., Bagasra, O., and Pace, D.G. 2014. Bacteriophage therapy: A potential solution for the antibiotic resistance crisis. *The Journal of Infection in Developing Countries* 8(02):129–136.

Gould, I.M., and Bal, A.M. 2013. New antibiotic agents in the pipeline and how they can help overcome microbial resistance. *Virulence* 4(2):185–191.

Granowitz, E.V., and Brown, R.B. 2008. Antibiotic adverse reactions and drug interactions. *Critical Care Clinics* 24(2):421–442.

Gupte, N. 2012. Antimicrobial resistance: Time to fight it out. *Pediatric Infectious Diseases* 4(1): 15–17.

Hao, H., Cheng, G., Iqbal, Z., Ai, X., Hussain, H.I., Huang, L., Dai, M., Wang, Y., Liu, Z., and Yuan, Z. 2014. Benefits and risks of antimicrobial use in food-producing animals. *Frontiers in Microbiology* 12(5):288.

Henriksson, P.J., Troell, M., and Rico, A. 2015. Antimicrobial use in aquaculture: Some complementing facts. *Proceedings of the National Academy of Sciences* 112(26):E3317.

Hiller, C.X., Hübner, U., Fajnorova, S., Schwartz, T., and Drewes, J.E. 2019. Antibiotic microbial resistance (AMR) removal efficiencies by conventional and advanced wastewater treatment processes: A review. *Science of the Total Environment* 685:596–608.

House, O.W. 2015. *President's 2016 Budget Proposes Historic Investment to Combat Antibiotic-Resistant Bacteria to Protect Public Health*. Washington, DC: The Watergate.

Johnson, J.W., Fisher, J.F., and Mobashery, S. 2013. Bacterial cell-wall recycling. *Annals of the New York Academy of Sciences* 1277(1):54–75.

Kahn, L.H. 2017. Antimicrobial resistance: A one health perspective. *Transactions of the Royal Society of Tropical Medicine and Hygiene* 111(6):255–260.

Klein, E.Y., Van Boeckel, T.P., Martínez, E.M., Pant, S., Gandra, S., Levin, S.A., Goossens, H., and Laxminarayan, R. 2018. Aumento global increase and geographic convergence in antibiotic consumption between 2000 and 2015. *Proceeding of National Academy of Science, USA* 115(15):E3463–E3470.

Ko, J.Y., Day, J.W., Lane, R.R., and Day, J.N. 2004. A comparative evaluation of money-based and energy-based cost–benefit analyses of tertiary municipal wastewater treatment using forested wetlands vs. sand filtration in Louisiana. *Ecological Economics* 49(3):331–347.

Krüger, N.J., and Stingl, K. 2011. Two steps away from novelty–principles of bacterial DNA uptake. *Molecular Microbiology* 80(4):860–867.

Kumar, S., Tripathi, V.R., and Garg, S.K. 2013a. Antibiotic resistance and genetic diversity in water-borne Enterobacteriaceae isolates from recreational and drinking water sources. *International Journal of Environmental Science and Technology* 10(4):789–798.

Kumar, S.G., Adithan, C., Harish, B.N., Sujatha, S., Roy, G., and Malini, A. 2013a. Antimicrobial resistance in India: A review. *Journal of Natural Science, Biology and Medicine* 4(2):286.

Lakshmi, V. 2008. Need for national/regional guidelines and policies in India to combat antibiotic resistance. *Indian Journal of Medical Microbiology* 26(2):105–107.

Larsson, D.J., de Pedro, C., and Paxeus, N. 2007. Effluent from drug manufactures contains extremely high levels of pharmaceuticals. *Journal of Hazardous Materials* 148(3):751–755.

Lata, P., Ram, S., and Shanker, R. 2016. Multiplex PCR based genotypic characterization of pathogenic vancomycin resistant *Enterococcus faecalis* recovered from an Indian river along a city landscape. *Springer Plus* 5(1):1199.

Laxminarayan, R. 2014. Antibiotic effectiveness: Balancing conservation against innovation. *Science* 345(6202):1299–1301.

Lekagul, A., Tangcharoensathien, V., and Yeung, S. 2019. Patterns of antibiotic use in global pig production: A systematic review. *Vet Animal Science* 7:100058.

Li, Q., Chang, W., Zhang, H., Hu, D., and Wang, X. 2019. The role of plasmids in the multiple antibiotic resistance transfer in ESBLs-producing *Escherichia coli* isolated from wastewater treatment plants. *Frontiers in Microbiology* 10:633.

Little, P., Stuart, B., Francis, N., Douglas, E., Tonkin-Crine, S., Anthierens, S., Cals, J.W., Melbye, H., Santer, M., Moore, M., and Coenen, S. 2013. Effects of internet-based training on antibiotic prescribing rates for acute respiratory-tract infections: A multinational, cluster, randomised, factorial, controlled trial. *The Lancet* 382(9899):1175–1182.

Lundborg, C.S., Tamhankar, A.J. 2017. Antibiotic residues in the environment of South East Asia. *BMJ Clinical Research* 5:358.

Magnien, R.E., Summers, R.M., and Sellner, K.G. 1992. External nutrient sources, internal nutrient pools, and phytoplankton production in Chesapeake Bay. *Estuaries* 15(4):497–516.

Marathe, N.P., Gaikwad, S.S., Vaishampayan, A.A., Rasane, M.H., Shouche, Y.S., and Gade, W.N. 2016. Mossambicus, tilapia (Oreochromis mossambicus) collected from water bodies impacted by urban waste carries extended-spectrum beta-lactamases and integron-bearing gut bacteria. *Journal of Biosciences* 41(3):341–346.

Marsot, A., Boulamery, A., Bruguerolle, B., and Simon, N. 2012. Vancomycin: A review of population pharmacokinetic analyses. *Clinical Pharmacokinetics* 51(1):1–3.

McClellan, K., and Halden, R.U. 2010. Pharmaceuticals and personal care products in archived US biosolids from the 2001 EPA national sewage sludge survey. *Water Research* 44(2):658–668.

Metcalf, L., Eddy, H.P., and Tchobanoglous, G. 1991. *Wastewater Engineering: Treatment, Disposal, and Reuse*. New York: McGraw-Hill Education.

Michael, C.A., Dominey-Howes, D., and Labbate, M. 2014. The antimicrobial resistance crisis: Causes, consequences, and management. *Frontiers in Public Health* 16(2):145.

Morse, G.K., Brett, S.W., Guy, J.A., and Lester, J.N. 1998. Phosphorus removal and recovery technologies. *Science of the Total Environment* 212(1):69–81.

Mott, D.J., Hampson, G., Llewelyn, M.J., Mestre-Ferrandiz, J., and Hopkins, M.M. 2020. A multinational European study of patient preferences for novel diagnostics to manage antimicrobial resistance. *Applied Health Economics and Health Policy* 18(1):69–79.

Muga, H.E., Mihelcic, J.R., Reents, N.W., Gemio, G., and Morales, S. 2009. Performance of Lagoon wastewater treatment system in Sapecho, Bolivia. *Proceedings of the Water Environment Federation* 2009(8):6936–6949.

Naylor, N.R., Atun, R., Zhu, N., Kulasabanathan, K., Silva, S., Chatterjee, A., Knight, G.M., and Robotham, J.V. 2018. Estimating the burden of antimicrobial resistance: A systematic literature review. *Antimicrobial Resistance & Infection Control* 7(1):1–7.

Nguyen, K.V., Thi Do, N.T., Chandna, A., Nguyen, T.V., Pham, C.V., Doan, P.M., Nguyen, A.Q., Thi Nguyen, C.K., Larsson, M., Escalante, S., and Olowokure, B. 2013. Antibiotic use and resistance in emerging economies: A situation analysis for Viet Nam. *BMC Public Health* 13(1):1.

Niestępski, S., Harnisz, M., Korzeniewska, E., Aguilera-Arreola, M.G., Contreras-Rodríguez, A., Filipkowska, Z., and Osińska, A. 2019. The emergence of antimicrobial resistance in environmental strains of the Bacteroides fragilis group. *Environment International* 124:408–419.

Nikolaidis, I., Favini-Stabile, S., and Dessen, A. 2014. Resistance to antibiotics targeted to the bacterial cell wall. *Protein Science* 23(3):243–259.

O'Neill, J. 2014. *Antimicrobial Resistance: Tackling a Crisis for the Health and Wealth of Nations*. London: Review on Antimicrobial Resistance.

Olivieri, A.W., Soller, J.A., Olivieri, K.J., Goebel, R.P., and Tchobanoglous, G. 2005. Seasonal tertiary wastewater treatment in California: An analysis of public health benefits and costs. *Water Research* 39(13):3035–3043.

Ongerth, H.J., and Ongerth, J.E. 1982 Health consequences of wastewater reuse. *Annual Review of Public Health* 3(1):419–444.

Pagilla, K.R., Urgun-Demirtas, M., and Ramani, R. 2006. Low effluent nutrient technologies for wastewater treatment. *Water Science and Technology* 53(3):165–172.

Painter, H.A. 1970. A review of literature on inorganic nitrogen metabolism in microorganisms. *Water Research* 4(6):393–450.

Payne, W.J. 1981. *Denitrification*. New York: John Wiley & Sons Inc.

Piddock, L.J. 2012. The crisis of no new antibiotics—what is the way forward? *The Lancet Infectious Diseases* 12(3):249–253.

Prestinaci, F., Pezzotti, P., and Pantosti, A. 2015. Antimicrobial resistance: A global multifaceted phenomenon. *Pathogens and Global Health* 109(7):309–318.

Rabalais, N.N. 2002. Nitrogen in aquatic ecosystems. *AMBIO: A Journal of the Human Environment* 31(2):102–112.

Raghunath, D. 2008. Emerging antibiotic resistance in bacteria with special reference to India. *Journal of Biosciences* 33(4):593–603.

Read, A.F., and Woods, R.J. 2014. Antibiotic resistance management. *Evolution, Medicine, and Public Health* 2014(1):147.

Reddy, B., and Dubey, S.K. 2019. River Ganges water as reservoir of microbes with antibiotic and metal ion resistance genes: High throughput metagenomic approach. *Environmental Pollution* 246:443–451.

Rehman, M.S., Rashid, N., Ashfaq, M., Saif, A., Ahmad, N., and Han, J.I. 2015. Global risk of pharmaceutical contamination from highly populated developing countries. *Chemosphere* 138:1045–1055.

Ruppé, E., Armand-Lefèvre, L., Estellat, C., El-Mniai, A., Boussadia, Y., Consigny, P.H., Girard, P.M., Vittecoq, D., Bouchaud, O., Pialoux, G., and Esposito-Farèse, M. 2014. Acquisition of carbapenemase-producing Enterobacteriaceae by healthy travellers to India, France, February 2012 to March 2013. *Euro Surveillance* 19(14):20768.

Safdari, R., Saeedi, M.G., Masoumi-Asl, H., Rezaei-Hachesu, P., Mirnia, K., Mohammadzadeh, N., and Samad-Soltani, T. 2018. National minimum data set for antimicrobial resistance management: Toward global surveillance system. *Iranian Journal of Medical Sciences* 43(5):494.

San Millan, A. 2018. Evolution of plasmid-mediated antibiotic resistance in the clinical context. *Trends in Microbiology* 26(12):978–985.

Sargent, L., McCullough, A., Del Mar, C., and Lowe, J. 2016. Is Australia ready to implement delayed prescribing in primary care? A review of the evidence. *Australian Family Physician* 45(9):688–690.

Sekhar, M. 2022. *Technologies for Enhancing Water Productivity in Irrigated Agriculture. InFood, Energy, and Water Nexus*. Cham: Springer, pp. 67–95.

Singer, A.C., Shaw, H., Rhodes, V., and Hart, A. 2016. Review of antimicrobial resistance in the environment and its relevance to environmental regulators. *Frontiers in Microbiology* 7:1728.

Spellberg, B. 2018. The maturing antibiotic mantra: "Shorter is still better". *Journal of Hospital Medicine* 13(5):361–362.

Sridhar, S., Gautret, P., and Brouqui, P. 2015. A comprehensive review of the Kumbh Mela: Identifying risks for spread of infectious diseases. *Clinical Microbiology and Infection* 21(2):128–133.

Srinath, E.G., Sastry, C.A., and Pillai, S.C. 1959. A rapid removal of phosphorus from sewage by activated sludge. *Experientia* 15(9):339–340.

Subedi, D., Giri, A.K., and McDonald, T. 2021. *Commercial Farms Received the Most Government Payments in 2019*. Washington, DC: Amber Waves: The Economics of Food, Farming, Natural Resources, and Rural America.

Sun, D., Jeannot, K., Xiao, Y., and Knapp, C.W. 2019. Horizontal gene transfer mediated bacterial antibiotic resistance. *Frontiers in Microbiology* 10:1933.

Swapna, K.M., Rajesh, R., and Lakshmanan, P.T. 2012. Incidence of antibiotic residues in farmed shrimps from the southern states of India. *Indian Journal of Geo-Marine Sciences* 41(4):344–347.

Taneja, N., and Sharma, M. 2019. Antimicrobial resistance in the environment: The Indian scenario. *Indian Journal of Medical Research* 149(2):119–128.

Tang, Q., Song, P., Li, J., Kong, F., Sun, L., and Xu, L. 2016. Control of antibiotic resistance in China must not be delayed: The current state of resistance and policy suggestions for the government, medical facilities, and patients. *Bioscience Trends* 10(1):1–6.

Taylor, J., Hafner, M., Yerushalmi, E., Smith, R., Bellasio, J., Vardavas, R., Bienkowska-Gibbs, T., and Rubin, J. 2014. *Estimating the Economic Costs of Antimicrobial Resistance: Model and Results*. Santa Monica, CA: RAND Corporation.

Tchobanoglous, G., Burton, F.L., and Stensel, H.D. 2003. *Wastewater Engineering: Treatment and Reuse*, 4th edn. New York: McGraw-Hill.

Tenover, F.C. 2006. Mechanisms of antimicrobial resistance in bacteria. *The American Journal of Medicine* 119(6): S3–10.

Teske, A., Alm, E., Regan, J.M., Toze, S., Rittmann, B.E., and Stahl, D.A., 1994. Evolutionary relationships among ammonia-and nitrite-oxidizing bacteria. *Journal of Bacteriology* 176(21):6623–6630.

Thompson, N.D., LaPlace, L., Epstein, L., Thompson, D., Dumyati, G., Concannon, C., Quinlan, G., Witten, T., Warnke, L., Lynfield, R., and Maloney, M. 2016. Prevalence of antimicrobial use and opportunities to improve prescribing practices in US nursing homes. *Journal of the American Medical Directors Association* 17(12):1151–1153.

Toops, S., Peterson, M.A., Vanderbush, W., Sackeyfio, N., and Anderson, S. 2021. *International Studies: An Interdisciplinary Approach to Global Issues*. New York: Routledge, Taylor & Francis, p. 27.

USEPA, 2004. *Clean Watersheds Need Survey 2004—Report to Congress*. Washington, DC: USEPA.

Utt, E., and Wells, C. 2016. The global response to the threat of antimicrobial resistance and the important role of vaccines. *Pharmaceuticals Policy and Law* 18(1–4):179–197.

Väänänen, M.H., Pietilä, K., and Airaksinen, M. 2006. Self-medication with antibiotics—Does it really happen in Europe? *Health Policy* 77(2):166–171.

Van Boeckel, T.P., Brower, C., Gilbert, M., Grenfell, B.T., Levin, S.A., Robinson, T.P., Teillant, A., and Laxminarayan, R. 2015. Global trends in antimicrobial use in food animals. *Proceedings of the National Academy of Sciences* 112(18):5649–5654.

Ventola, C.L. 2015. The antibiotic resistance crisis: Part 1: Causes and threats. *Pharmacy and Therapeutics* 40(4):277.

Verburg, I., García-Cobos, S., Hernández, Leal. L., Waar, K., Friedrich, A.W., and Schmitt, H. 2019. Abundance and antimicrobial resistance of three bacterial species along a complete wastewater pathway. *Microorganisms* 7(9):312.

Vickers, N.J. 2017. Animal communication: When I'm calling you, will you answer too? *Current Biology* 27(14): R713–R715.

Walia, K., Sharma, M., Vijay, S., and Shome, B.R. 2019. Understanding policy dilemmas around antibiotic use in food animals & offering potential solutions. *The Indian Journal of Medical Research* 149(2):107.

Walsh, C.J., Roy, A. H., Feminella, J.W., Cottingham, P.D., Groffman, P.M., and Morgan, R.P. 2005. The urban stream syndrome: Current knowledge and the search for a cure. *Journal of the North American Benthological Society* 24(3):706–723.

Wang, Y., Zhu, K., Dai, R., Li, R., Li, M., Lv, X., and Yu, Q. 2021. Specific interleukin-1 inhibitors, specific interleukin-6 inhibitors, and GM-CSF blockades for COVID-19 (at the edge of sepsis): A systematic review. *Frontiers in Pharmacology* 12:804250.

Webber, M.A., Whitehead, R.N., Mount, M., Loman, N.J., Pallen, M.J., and Piddock, L.J. 2015. Parallel evolutionary pathways to antibiotic resistance selected by biocide exposure. *Journal of Antimicrobial Chemotherapy* 70(8):2241–2248.

Wellington, E.M., Boxall, A.B., Cross, P., Feil, E.J., Gaze, W.H., Hawkey, P.M., Johnson-Rollings, A.S., Jones, D.L., Lee, N.M., Otten, W., and Thomas, C.M. 2013. The role of the natural environment in the emergence of antibiotic resistance in Gram-negative bacteria. *The Lancet Infectious Diseases* 13(2):155–165.

Wentzel, M.C., Lötter, L.H., Ekama, G.A., Loewenthal, R.E., and Marais, G.V. 1991. Evaluation of biochemical models for biological excess phosphorus removal. *Water Science and Technology* 23(4–6):567–576.

Wichmann, F., Udikovic-Kolic, N., Andrew, S., and Handelsman, J. 2014. Diverse antibiotic resistance genes in dairy cow manure. *Mustang Bio, Inc* 5(2): e01017–e01013.

Woolhouse, M., Waugh, C., Perry, M.R., and Nair, H. 2016. Global disease burden due to antibiotic resistance–state of the evidence. *Journal of Global Health* 6(1):010306.

World Health Organization. 2011. *Regional office for South-East Asia. Jaipur Declaration on Antimicrobial Resistance*. Geneva, Switzerland: World Health Organization.
World Health Organization. 2014. *Antimicrobial Resistance: Global Report on Surveillance*. Geneva, Switzerland: World Health Organization.
World Health Organization. 2015. *Global Action Plan on Antimicrobial Resistance. WHO Report*. Geneva, Switzerland: World Health Organization.
World Health Organization. 2016. *Situation Analysis on Antimicrobial Resistance in the South-East Asia Region*. Geneva, Switzerland: World Health Organization.
Wright, G.D. 2014. Something old, something new: Revisiting natural products in antibiotic drug discovery. *Canadian Journal of Microbiology* 60(3):147–154.
Yang, Y., Li, B., Zou, S., Fang, H.H., and Zhang, T. 2014. Fate of antibiotic resistance genes in sewage treatment plant revealed by metagenomic approach. *Water Research* 62:97–106.
Yeoman, S., Stephenson, T., Lester, J.N., and Perry, R. 1998. The removal of phosphorus during wastewater treatment: A review. *Environmental Pollution* 49(3):183–233.
Zaman, S.B. 2017. A review on antibiotic resistance: Alarm bells are ringing. *Cureus* 9:e1403.
Zapun, A., Contreras-Martel, C., and Vernet, T., 2008. Penicillin-binding proteins and β-lactam resistance. *FEMS, Microbiology Reviews* 32(2):361–385.
Zhou, H., and Smith, D.W. 2002. Advanced technologies in water and wastewater treatment. *Journal of Environmental Engineering and Science* 1:247–264.

4 Antimicrobial Resistance in Domestic and Industrial Wastewater

A Review

Ramesh Kumar Prajapati, Mohd. Ayub Ansari, and Haider Iqbal

4.1 INTRODUCTION

Antimicrobial resistance (AMR) has received widespread attention due to the negative effects on the environment caused by the use and misuse of antimicrobial agents, as well as the consequences for public health and agriculture. AMR, which is mainly due to the resistance developed in bacteria against antibiotics, is currently a major emerging health issue in the public domain, causing environmental pollution to soil and water that needs immediate care and attention in developed and developing countries around the globe. Already, the World Health Organization (WHO) has expressed its concern and has mentioned that antibiotic resistance expressed by microbes is a severe threat to human health worldwide. Data associated with anti-microbial resistance indicate that nearly five million deaths are directly occurring because of drug-resistant infections and that number is increasing (Figure 4.1). It has been estimated that by 2050, this number of casualties might reach up to ten million, which, economically, will cost more than USD 100 trillion (Murray et al. 2022).

At the subcontinental level, it has been estimated that the death rate in all age groups contributed by resistance to drugs is the maximum in western sub-Saharan Africa, 27.3 deaths per 10^4 (20.9–35.3), and the minimum in Australia, 6.5 deaths (4.3–9.4) per 10^4. The six leading bacteria causing loss of life due to AMR are found to be Gram-positive bacteria: Streptococcus pneumonia and Staphylococcus aureus, and Gram-negative bacteria: *E. coli, Acinetobacter baumannii, Klebsiella pneumonia,* and *Pseudomonas aeruginosa.* In 2019, due to AMR, methicillin-resistant *S.* aureus caused more deaths per 10^4 while six other virulent drug-resistant pathogens each caused 50,000–1,00,000 deaths annually by developing multidrug-resistant tuberculosis. In fact, patients suffering from cancer are highly affected by AMR. Cancer patients are more susceptible to infections due to the weakening of their immune defence system. During surgery, radiotherapy, chemotherapy, and treatments, bone marrow transplants further force the immune system under immense pressure (O'Neill 2016; Rafraf et al. 2016; de Kraker, Stewardson, and Harbarth 2016; Naylor et al. 2018; Polianciuc et al. 2020). The objectives of this chapter are to summarize and to show the concern on environment which are faced by mankind because of resistance developed in microbes against a series of antibiotics with the techniques employed for its detection and separation. So various types of human activities cause pollution, which is due to the appearance of AMR against antibiotics in different components of the environment. Due to AMR developed in the air, water, and soil, it has led to an increase in disease occurring in animal/plant diseases and also affects the soil biodiversity. This loss further increases the use of antimicrobials, leading to a negative feedback from initial use and ultimately increasing the pressure selectively further. The impact of AMR in the environment is

DOI: 10.1201/9781003269823-4

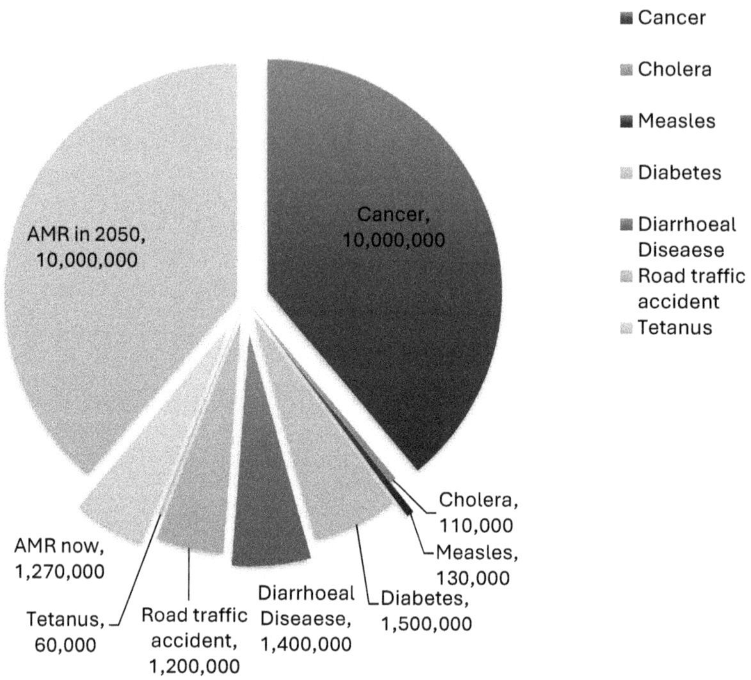

FIGURE 4.1 Death caused by antimicrobial resistance (AMR) diseases.

visualized by its complexities, cyclic dynamics, and interrelationships with microbes and antibiotics, resulting in multiple causalities.

In 1929, A Fleming made the first antibiotic discovery with the discovery of penicillin, and later on, other classes of antimicrobial drugs which are discovered naturally and synthesized have effectively reduced the infectious diseases caused by bacteria, fungi, and other pathogens in humans, animals, and plants, by saving lives and enhancing productivity (Gatimu and Kimani 2021). But since mid-nineteen, their effectiveness is in jeopardy because genetic evolution in microbes makes them resistant to antimicrobial drugs and antimicrobial treatments are found to be less effective (Figure 4.2).

When these antimicrobial compounds are thrown into the environment by anthropogenic activities, along with other factors, they selectively exert pressure on natural microbial communities in which inherited or acquired resistance has evolved and proliferated.

Atmospheric pollution containing minimal selective concentration of such antimicrobial drugs significantly contributes to the development of AMR (Sabri et al. 2020; Larsson and Flach 2021; Andersson et al. 2020).

An antibiotic helps our bodies by acting as either a killer, i.e., bactericidal antibiotics, or by inhibiting the growth of bacteria, known as bacteriostatic against the bacteria that cause so many types of diseases in humans and animals, hence are prescribed for their treatment. They are continuously reaching different components of environment through urine and excreta; hence, it is important to optimize the use of antibiotics and their control by which their concentrations could be minimized in the environment. Many of the bacteria that cause many types of disease become resistant by defending themselves against antibiotics, which means resistant bacteria are harder to kill. This is described as antibiotic resistance. The resistant bacteria start multiplying in the host, which makes antibiotics less effective. The main cause of resistance to antibiotics developed by the bacteria depends on so many factors of which some are as-antibiotic resistance develops when bacteria changes its shape and in its gene in response to overuse/over-prescription of antibiotics, lack of hygiene and poor sanitation, poor control of infections in hospital, and most importantly lack of new

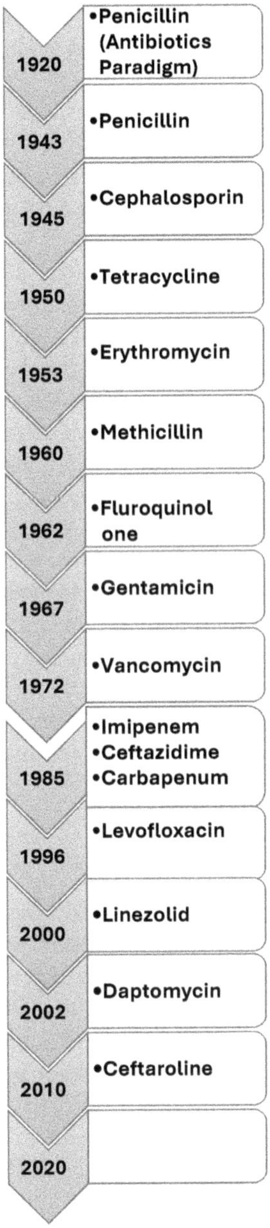

FIGURE 4.2 Chronological discovery of different antibiotics.

antibiotics being developed. Depending on the class of effectiveness and pharmacological effects of antibiotic drugs, nearly 40%–90% of antibiotic doses are excreted in the feces and via the excretory system in the active form or as their metabolites, which eventually go into the environment, where they contaminate the soils, waters, plants, livestock, and aquatic life. The excessive use of antibiotics as contaminated manure in animal farming also leads to contamination in agro-ecosystems. The improper and casual dumping of unused drugs and hospital accessories into the sewage systems is also a major polluter of water. Generally, conventional wastewater treatment plants (WWTPs) are used in the purification of wastewater contaminated by antibiotic residues, but they are not completely removed by this process. The sludge, as well as the final effluent coming out of WWTPs, still contains residues of antibiotics. The sludge mostly contains manure and fertilizers obtained from

WWTP. These are generally used by farmers on soil to increase the production of crops, whereas the effluent is directly drained into sewage and water bodies (Devarajan et al. 2016). Activated sludge and anaerobic digestion processes in conventional WWTPs are the most favorable sites for stimulating and promoting the growth of antibiotic resistance genes (ARGs) by providing a suitable trophic environment for horizontal gene transfer (HGT), a method of genetic material transmission, evolution, and adaptation in bacteria. HGTs are facilitated by the intermediacy of movable genetic vectors, e.g., with the help of plasmids, bacteriophages, and transposons (Wohde et al. 2016; Durso et al. 2015).

As soon as antibiotic residues reach in the environment, they produce adverse impacts on organisms at different trophic levels and finally affect the health of human beings when they consume contaminated food and polluted water, which increases the population of resistant bacteria. This also selectively increases the pressure by encouraging the development of antibiotic resistance and bacterial proliferation in various environmental components (Cycoń, Mrozik, and Piotrowska-Seget 2019; Kuppusamy et al. 2018). Apart from the danger of enhancement of antibiotic microbial resistance, the residues of antibiotics and their metabolites are also absorbed by plants, which interfere with physiological processes, leading to potential ecotoxicological effects. The disadvantages of antibiotic residues in plants are studied by numerous chronic and acute toxicity tests, and it has been found that antibiotics affect the physiology of plants during processes such as photosynthesis, cell division, and respiration (oxidative stress response in plants). Delay in germination of seeds and reduced biomass are also reported due to the higher quantity of antibiotics present in agricultural soils, which later on may yield fewer crops by absorbing nutrients from soils adulterated by residues of antibiotics (Wang et al., n.d.). It also affects negatively the microbiomes in human guts, causing health distress, most commonly as allergic reactions. Its prolonged exposure develops chronic toxic effects and disturbances in the normal functioning of digestive systems (Calero-Cáceres et al. 2017; Berendonk et al. 2015; Minden et al. 2017).

4.2 ENVIRONMENT-SPECIFIC RESISTANCE GENES

It has been found that so much work has done in the last few decades encompassing the knowledge of the dangers arising due to environment-specific resistance genes by the excessive use of antimicrobial agents and it is studied by the availability of detection techniques such as qPCR, conventional PCR, metagenomic methods, or culture-dependent techniques (Ju et al. 2018; Graham et al. 2016).

The most favorable environments for ARGs and ARB are found to be in hospitals, in AFOs, during aquaculture operations, and in treatments of sewage at WWTPs, where repeated high doses of antibiotics are readily available for bacteria and their growth rates are very high due to an abundance of nutrients (Atashgahi et al. 2018; Baquero, Martínez, and Cantón 2008). Hospitals, sanitary systems, and air flow systems, where various human members of microbiomes associated with ARGs are the prominent sites for spreading hospital-acquired infections with resistant bacteria, and so many studies for the advancement and multiplication of antibiotic resistance have been conducted in the ambience of hospitals (Berendonk et al. 2015; Pruden et al. 2012; Breathnach et al. 2012). In hospitals, the use of antibiotics over extended time spans by patients leads to de novo evolution of resistance against the antibiotics by pathogens, which starts the evolution of ARGs, leading to the spread of HGT in the patients (Awad et al. 2014; Kizny Gordon et al. 2017; Awad et al. 2016).

Sewage from hospitals constitutes nutrient-rich environments that support the growth of different species of bacteria and are continuously enriched with new ARB, ARGs. Different antibiotics present in sub-minimum inhibitory concentrations certainly encourage HGT (Bengtsson-Palme et al. 2016; López-Causapé et al. 2013). Sewage from municipal is treated at WWTPs by the use of different treatment processes; it minimizes the pathogens and removes inorganic and organic nitrogen residues and phosphorus before it is poured into surface water. Hence, sewage and WWTPs are the core places for the development, recombination, and spread of AMR. During WWTP, total pathogen concentration has been found to be reduced but enhanced selectivity. In most of the treatment processes, by and large, the population of resistant bacteria and hence the effective quantity of ARGs and ARB are substantially dropped in the treated effluent of hospitals, sewage, and WWTPs (Carter et al. 2014; Hocquet, Muller, and Bertrand 2016; Chessa, Jechalke, et al. 2016; Bürgmann et al. 2018). A major shortcoming of WWTP treatment is its poor performance in minimizing higher concentration of resistance genes in biosolids, which are used in landfills and as fertilizers for crops (Chessa, Pusino, et al. 2016). These bio-solids, in due course of time, pollute the soils, crops, livestock, and human-resistant pathogens. Various sewage treatment methods used in various types of WWTPs may also increase ARGs and biosolids in effluent to different concentrations as per the treatment method (Cycon and Piotrowska-Seget 2016; Cycoń et al. 2018).

The various hotspots, such as hospitals and WWTPS, are the centers of major concern for health because the resistant pathogens directly cross species barriers, spreading infections in humans. It is also very common here for bacteria to evolve resistance against antibiotics, which will directly lessen their ability to fight infections occurring in the general populace. The continuous use of a broader spectrum of antibiotics for prolonged periods may contribute to the development of a cyclic feedback of resistance between the general population and the environment. Antibiotics are also widely used in animal feed in the form of AFOs for the production of livestock, livestock farming, and aquaculture at large-scale concentrated animal feeding operations (CAFOs). It is difficult to estimate the exact magnitude of antibiotics used in AFOs and CAFOs, but it is found to be much higher than the amount used in human health (DeVries et al. 2015; Diehl and Lapara 2010; Gudeta et al. 2016a; Ventola 2015; Woolhouse et al. 2015). The data obtained from WHO reflects that in the United States, almost 80% of antibiotics are used in veterinary practices (Kuppusamy et al. 2018), whereas in Canada, however, the concentration of antibiotics in biomass was found to be at least two times higher than in humans (Diehl and Lapara 2010).

The methods of treatment of waste in AFOs are different as compared to those mostly used in municipal solid waste sewage and at WWTPs. Here, formed manure is concentrated as a waste lagoon and finally subjected to its bacterial degradation. In manure, the total amount of cultured bacteria is reduced as time passes, but lagoons and manure are found to be highly enriched in ARB (Ju et al. 2018; Graham et al. 2016; Woolhouse et al. 2015; Grenni, Ancona, and Barra Caracciolo 2018). Manure is consumed as fertilizer for crops where ARGs and surviving ARB in crop land intermix with microbiomes present in soil and, ultimately here, by dissolving in water, ARGs and ARB reach water bodies (DeVries et al. 2015).

Less attention has been paid to the resistance shown by bacteria, which is caused by the release of antibiotics into the environment from industrial sources as compared to the controlled release of antibiotics from AFOs and hospitals. The effluent contains the residues of antibiotics such as ciprofloxacin and its analogs, whose concentrations are surprisingly found above the average concentrations of the drugs in wastewater (D. Li et al. 2009), which increases antibiotic resistance in biomes (Johnning et al. 2013; Koba et al. 2017).

4.3 CARRIERS OF ARB TRANSMISSION

The ARB and ARGs are transported by means of various carriers (vectors) to the different components of environments from their origin, which takes place with the help of different vectors shown in Figure 4.3. This figure explains how antibiotics are connected to the food chain of humans and also explains the interconnectivity among the circulation of antibiotics, the development of AMR, and its side effects. It starts when antibiotics are taken during an infection period and unutilized antibiotics/metabolites are released into the environment as waste products (feces and urine), which contaminates the nutrients and potable water, and in the last causes various infections to human beings by bacteria that are not easy to treat. Vectors are generally involved in transport and also sometimes provide a means to offer a reservoir where ARB/ARGs are evolved and multiplied. ARGs are transported among the different vectors of environments within ARB, which has been investigated indirectly for the presence of resistance genes.

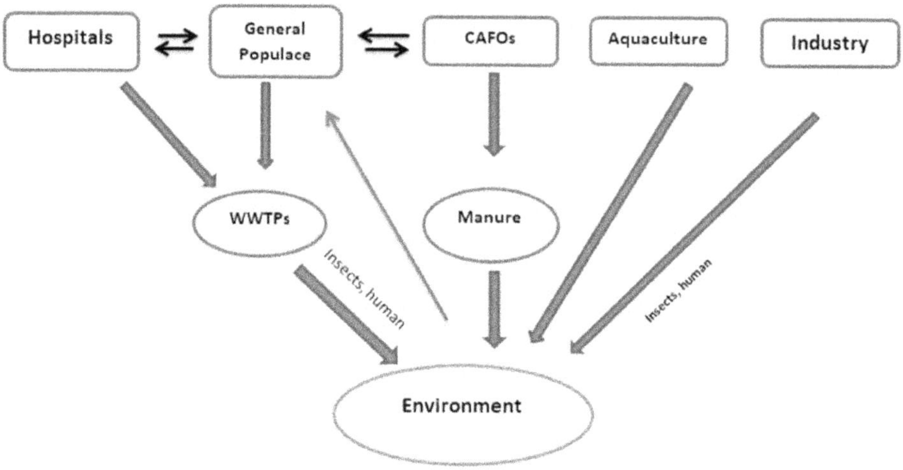

FIGURE 4.3 The evolution and transmission of antibiotic resistance-carrying bacteria (ARB) and ARGs by vectors in different components.

4.3.1 Surface Waters

Effluent from WWTPs, pharmaceutical industries, and hospitals and runoff from manure-fertilized fields are directly discharged into surface waters. Hence, AFOs are key places where transportation and development of ARB occur (Bilal et al. 2020; Pei et al. 2006). The rivers contain higher ARGs and are directly influenced by the sources of contaminants, e.g., WWTPs or AFOs. ARGs found in rivers facilitate a better medium for the transfer of genetic elements to bacteria to change their genetic materials (Kristiansson et al. 2011).

4.3.2 Air

Researchers have shown a keen interest in studying air as a potential site for dispersal of ARB, and the presence of ARGs against the seven most commonly prescribed antibiotics is found in the shoot particles of air in various concentrations (J. Li et al. 2018). Floating bacteria species like *P. aeruginosa* and *P. syringae* present in air particulates contain ARGs and get precipitated (Hu et al. 2018; Dueker et al. 2017; Kachhawaha et al. 2017).

The presence of ARGs has also been detected in a nearby area of WWTPs downwind (Tolmacheva et al. 2016), which is a main concern about occupational hazards for workers. ARGs have also been developed by the AFOs, specifically CAFOs, and in a minimum quantity, airborne transmission of ARB and ARGs is also reported in hospital areas (Subedi et al. 2015; J. Li et al. 2016; Gao et al. 2017).

4.3.3 Animal Vectors

Animal vectors, specifically microbiomes, are better carriers for ARB colonies by providing nutrient-rich environments. In animal vectors, which are a better site for antibiotic resistance, ARGs can easily multiply and modify their genetic materials. Humans and farmers, who are directly in contact with livestock, are the most studied animal vectors. They have been found to increase resistance against antibiotics (Papageorgiou, Kosma, and Lambropoulou 2016). Other animals, rodents, groups of vertebrates, e.g., insects, specifically houseflies, and cockroaches, are also better carriers of ARB vectors in association with ARB from AFOs to CAFOs (Serra-Compte et al. 2021), to WWTPs (Brown and Wong 2018). Whereas in hospitals, both active proliferation of ARB (Price et al. 2007; Lesser et al. 2018) and HGT of ARGs in flies have been observed.

TABLE 4.1
Timeline of Techniques Used in the Detection of Antibiotic Residues in Environment

S. No.	Time Period	Technique Applied
1	Prior to 1970	Adsorption chromatography – thin layer chromatography (TLC)
2	1970–1980	GC coupled with Electron Capture Detector (GC-ECD) (HPLC)
3	1980–1990	GC-MS and GC-MS coupled with (SIM) (selected ion monitoring) HPLC in association of (DAD-UV) (diode array detection-ultraviolet
4	1990–2000	G C- M S in the presence of ECD Partition chromatography—mass spectrometry
5	2000–2010	UPLC - Ultraperformance liquid chromatography alone and also in association of time of flight
6	Onwards 2010	UPLC coupled with mass spectrometry (Electrospray Ionization)

4.4 PROCESS FOR TREATMENT OF WASTEWATER CONTAINING ARGS AND ARB

The wastewater treatment process is a suitable place for HGT and other modern membrane separation processes are mostly used within their limits to treat the waste material present in effluents of hospitals, public places, and pharmaceutical industries' wastewater. The different techniques used in the analysis of the contaminants are given in Table 4.1 as per their discoveries.

Various methods used in the elimination of pollutants and antibiotics in water are shown in Table 4.2.

4.5 CONCLUSIONS

Large-scale antibiotic pollution and AMR are increasing and pose a significant environmental concern affecting the health of humans and animals, which requires a wide study of the sources and processes involved in their development and upsurge. The presence of ARB and ARGs in large amounts in municipal WWTPs, which are generated in activated sludge treatment sewage, directly and in fewer instances, indirectly, depends on the applied modification of the sewage treatment system, which is a suitable place for the growth and dissemination of AMR in bacteria. Researchers have taken a large interest in recent years in addressing the concern about the reservoir of resistance genes in various components of the environment. Currently, researchers are focusing on environmental dimensions of resistance against antibiotics, which proves the presence of resistance genes in microbes within the environment. In this study, little attention has been paid to the threats, which are inherently linked with the transmission of ARGs, and thus, side effects on human health are a major concern (Han et al. 2020; de Ilurdoz, Sadhwani, and Reboso 2022; Kraemer, Ramachandran, and Perron 2019).

Antibiotic pollution contributed by the presence of antibiotic resistance has the capacity to spread infection among humans and livestock and directly affects ecosystem health. Antibiotic pollution adversely affects the environments of living aquatic organisms by inhibiting the functions of the aquatic ecosystem and, ultimately, the entire life cycle of organisms. Antibiotics present in the environment cause several diseases in humans, such as obesity, diabetes, and asthma. Because of it, many separation and analysis processes have been studied to solve the problem of antibiotic pollution in ecosystems, especially in surface and groundwater. It has been found that mostly conventional physical processes, viz., coagulation and flocculation, are less suitable for the removal of antibiotic pollutants from industrial wastes. Biological methods for the separation of antibiotics from surface water usually have fewer negative effects on the ecosystems and generate fewer by-products.

TABLE 4.2
Method Used in the Elimination of Antibiotic Pollutants in Water

Chemical method (Gautam, Kumar, and Sabumon 2007; Suarez, Lema, and Omil 2009) It involves coagulation and flocculation.	Adsorption		Physiochemical advanced oxidation process (AOP). It is classified into four subclasses	Membrane Technology	Biological		Constructed wetlands An artificial wastewater treatment process achieved by a combined action of soil, plants, and microorganisms is used to purify wastewater entering the wetland (Lindmark et al. 2014; Huang et al. 2021)
	Activated carbon methods are of two types 1. Powder activated carbon (PAC) (Kovalova et al. 2013) 2. Granular activated carbon (Cole 1972)	Other cellulose membranes	1. Electrochemical oxidation (Huang et al. 2021). 2. Oxidation process (Kasprzyk-Hordern, Ziółek, and Nawrocki 2003) 3. Fenton process 4. UV/H_2O_2 method (Subedi et al. 2015)	1. Microfiltration (MF) (de Jesus Gaffney et al. 2015) 2. Ultrafiltration (UF) (Rivera-Jaimes et al. 2018) 3. Nanofiltration (NF) 4. Reverse osmosis (RO) (Yu et al. 2022)	Aerobic It involves a biologically aerated filter (Chen et al. 2021)	Anaerobic It involves the following subclasses: anaerobic digestion (Cheng et al. 2018), anaerobic baffled reactor (ABR) (Mendoza-Espinosa and Stephenson 2009)	

REFERENCES

Andersson, D. I., N. Q. Balaban, F. Baquero, P. Courvalin, P. Glaser, U. Gophna, R. Kishony, S. Molin, and T. Tønjum. 2020. Antibiotic Resistance: Turning Evolutionary Principles into Clinical Reality. *FEMS Microbiology Reviews* 44 (2): 171–188. https://doi.org/10.1093/FEMSRE/FUAA001.

Atashgahi, S., I. Sánchez-Andrea, H. J. Heipieper, J. R. Van Der Meer, A. J. M. Stams, and H. Smidt. 2018. Prospects for Harnessing Biocide Resistance for Bioremediation and Detoxification. *Science* 360 (6390): 743–746. https://doi.org/10.1126/SCIENCE.AAR3778.

Awad, Y. M., S. C. Kim, S. A. M. Abd El-Azeem, K. H. Kim, K. R. Kim, K. Kim, C. Jeon, S. S. Lee, and Y. S. Ok. 2014. Veterinary Antibiotics Contamination in Water, Sediment, and Soil near a Swine Manure Composting Facility. *Environmental Earth Sciences* 71 (3): 1433–1440. https://doi.org/10.1007/S12665-013-2548-Z/FIGURES/4.

Awad, Y. M., Y. S. Ok, A. D. Igalavithana, Y. H. Lee, Y. K. Sonn, A. R. A. Usman, M. I. Al-Wabel, and S. S. Lee. 2016. Sulphamethazine in Poultry Manure Changes Carbon and Nitrogen Mineralisation in Soils. *Chemistry and Ecology* 32 (10): 899–918. https://doi.org/10.1080/02757540.2016.1216104.

Baquero, F., J. L. Martínez, and R. Cantón. 2008. Antibiotics and Antibiotic Resistance in Water Environments. *Current Opinion in Biotechnology* 19 (3): 260–265. https://doi.org/10.1016/J.COPBIO.2008.05.006.

Bengtsson-Palme, J., M. Angelin, M. Huss, S. Kjellqvist, E. Kristiansson, H. Palmgren, D. G. Joakim Larsson, and A. Johansson. 2015. The Human Gut Microbiome as a Transporter of Antibiotic Resistance Genes between Continents. *Antimicrobial Agents and Chemotherapy* 59 (10): 6551–6560. https://doi.org/10.1128/AAC.00933-15/SUPPL_FILE/ZAC010154477SO1.PDF.

Berendonk, T. U., C. M. Manaia, C. Merlin, D. Fatta-Kassinos, E. Cytryn, F. Walsh, H. Bürgmann, H. Sørum, M. Norström, M. N. Pons, N. Kreuzinger, P. Huovinen, S. Stefani, T. Schwartz, V. Kisand, F. Baquero, and J. L. Martinez. 2015. Tackling Antibiotic Resistance: The Environmental Framework. *Nature Reviews Microbiology* 13 (5): 310–317. https://doi.org/10.1038/nrmicro3439.

Bilal, M., S. Mehmood, T. Rasheed, and H. M. N. Iqbal. 2020. Antibiotics Traces in the Aquatic Environment: Persistence and Adverse Environmental Impact. *Current Opinion in Environmental Science & Health* 13 (Feb): 68–74. https://doi.org/10.1016/J.COESH.2019.11.005.

Breathnach, A. S., M. D. Cubbon, R. N. Karunaharan, C. F. Pope, and T. D. Planche. 2012. Multidrug-Resistant Pseudomonas Aeruginosa Outbreaks in Two Hospitals: Association with Contaminated Hospital Waste-Water Systems. *Journal of Hospital Infection* 82 (1): 19–24. https://doi.org/10.1016/J.JHIN.2012.06.007.

Brown, A. K., and C. S. Wong. 2018. Distribution and Fate of Pharmaceuticals and Their Metabolite Conjugates in a Municipal Wastewater Treatment Plant. *Water Research* 144 (Nov): 774–783. https://doi.org/10.1016/J.WATRES.2018.08.034.

Bürgmann, H., D. Frigon, W. H. Gaze, C. M. Manaia, A. Pruden, A. C. Singer, B. F. Smets, and T. Zhang. 2018. Water and Sanitation: An Essential Battlefront in the War on Antimicrobial Resistance. *FEMS Microbiology Ecology* 94 (9): 101. https://doi.org/10.1093/FEMSEC/FIY101.

Calero-Cáceres, W., J. Méndez, J. Martín-Díaz, and M. Muniesa. 2017. The Occurrence of Antibiotic Resistance Genes in a Mediterranean River and Their Persistence in the Riverbed Sediment. *Environmental Pollution* 223 (Apr): 384–394. https://doi.org/10.1016/J.ENVPOL.2017.01.035.

Carter, L. J., E. Harris, M. Williams, J. J. Ryan, R. S. Kookana, and A. B. A. Boxall. 2014. Fate and Uptake of Pharmaceuticals in Soil-Plant Systems. *Journal of Agricultural and Food Chemistry* 62 (4): 816–825. https://doi.org/10.1021/JF404282Y/SUPPL_FILE/JF404282Y_SI_001.PDF.

Chen, L., Z. Chen, Y. Wang, Y. Mao, and Z. Cai. 2021. Effective Treatment of Leachate Concentrate Using Membrane Distillation Coupled with Electrochemical Oxidation. *Separation and Purification Technology* 267 (July): 118679. https://doi.org/10.1016/J.SEPPUR.2021.118679.

Cheng, D. L., H. H. Ngo, W. S. Guo, Y. W. Liu, J. L. Zhou, S. W. Chang, D. D. Nguyen, X. T. Bui, and X. B. Zhang. 2018. Bioprocessing for Elimination Antibiotics and Hormones from Swine Wastewater. *Science of the Total Environment* 621 (Apr): 1664–1682. https://doi.org/10.1016/J.SCITOTENV.2017.10.059.

Chessa, L., A. Pusino, G. Garau, N. P. Mangia, and M. V. Pinna. 2016. Soil Microbial Response to Tetracycline in Two Different Soils Amended with Cow Manure. *Environmental Science and Pollution Research* 23 (6): 5807–5817. https://doi.org/10.1007/S11356-015-5789-4/TABLES/4.

Chessa, L., S. Jechalke, G. C. Ding, A. Pusino, N. P. Mangia, and K. Smalla. 2016. The Presence of Tetracycline in Cow Manure Changes the Impact of Repeated Manure Application on Soil Bacterial Communities. *Biology and Fertility of Soils* 52 (8): 1121–1134. https://doi.org/10.1007/S00374-016-1150-4/FIGURES/8.

Cole, K. S. 1972. *Membranes, Ions and Impulses: A Chapter of Classical Biophysics*. https://books.google.com/books?hl=en&lr=&id=ZCAIOKhZMWUC&oi=fnd&pg=PA5&dq=related:ontgUmc-Rx8J:scholar.google.com/&ots=EuhMrpfAZL&sig=9Whfu9ryk9NJReeDGxDkqero2sQ.

Cycoń, M., A. Mrozik, and Z. Piotrowska-Seget. 2019. Antibiotics in the Soil Environment—Degradation and Their Impact on Microbial Activity and Diversity. *Frontiers in Microbiology* 10 (Mar): 338. https://doi.org/10.3389/FMICB.2019.00338/FULL.

Cycon, M., and Z. Piotrowska-Seget. 2016. Pyrethroid-Degrading Microorganisms and Their Potential for the Bioremediation of Contaminated Soils: A Review. *Frontiers in Microbiology* 7 (Sep): 1463. https://doi.org/10.3389/FMICB.2016.01463.

Cycoń, M., K. Orlewska, A. Markowicz, A. Zmijowska, J. Smoleń-Dzirba, J. Bratosiewicz-Wasik, T. J. Wasik, and Z. Piotrowska-Seget. 2018. Vancomycin and/or Multidrug-Resistant Citrobacter Freundii Altered the Metabolic Pattern of Soil Microbial Community. *Frontiers in Microbiology* 9 (May): 1047. https://doi.org/10.3389/FMICB.2018.01047/BIBTEX.

de Ilurdoz, M. S., J. J. Sadhwani, and J. V. Reboso. 2022. Antibiotic Removal Processes from Water & Wastewater for the Protection of the Aquatic Environment - A Review. *Journal of Water Process Engineering* 45 (Feb): 102474. https://doi.org/10.1016/J.JWPE.2021.102474.

de Jesus Gaffney, V., C. M. M. Almeida, A. Rodrigues, E. Ferreira, M. J. Benoliel, and V. V. Cardoso. 2015. Occurrence of Pharmaceuticals in a Water Supply System and Related Human Health Risk Assessment. *Water Research* 72 (Apr): 199–208. https://doi.org/10.1016/J.WATRES.2014.10.027.

de Kraker, M. E. A., A. J. Stewardson, and S. Harbarth. 2016. Will 10 Million People Die a Year Due to Antimicrobial Resistance by 2050? *PLoS Medicine* 13 (11): e1002184. https://doi.org/10.1371/JOURNAL.PMED.1002184.

Devarajan, N., A. Laffite, C. K. Mulaji, J. P. Otamonga, P. T. Mpiana, J. I. Mubedi, K. Prabakar, B. W. Ibelings, and J. Poté. 2016. Occurrence of Antibiotic Resistance Genes and Bacterial Markers in a Tropical River Receiving Hospital and Urban Wastewaters. *PLoS One* 11 (2): e0149211. https://doi.org/10.1371/JOURNAL.PONE.0149211.

DeVries, S. L., M. Loving, X. Li, and P. Zhang. 2015. The Effect of Ultralow-Dose Antibiotics Exposure on Soil Nitrate and N_2O Flux. *Scientific Reports* 5 (1): 1–8. https://doi.org/10.1038/srep16818.

Diehl, D. L., and T. M. Lapara. 2010. Effect of Temperature on the Fate of Genes Encoding Tetracycline Resistance and the Integrase of Class 1 Integrons within Anaerobic and Aerobic Digesters Treating Municipal Wastewater Solids. *Environmental Science and Technology* 44 (23): 9128–9133. https://doi.org/10.1021/ES102765A/SUPPL_FILE/ES102765A_SI_001.PDF.

Dueker, M. E., G. D. O'Mullan, J. M. Martínez, A. R. Juhl, and K. C. Weathers. 2017. Onshore Wind Speed Modulates Microbial Aerosols along an Urban Waterfront. *Atmosphere* 8 (11): 215. https://doi.org/10.3390/ATMOS8110215.

Durso, L. M., D. A. Wedin, J. E. Gilley, D. N. Miller, and D. B. Marx. 2015. Assessment of Selected Antibiotic Resistances in Ungrazed Native Nebraska Prairie Soils. *Wiley Online Library* 45 (2): 454–462. https://doi.org/10.2134/jeq2015.06.0280.

Gao, M., R. Jia, T. Qiu, M. Han, and X. Wang. 2017. Size-Related Bacterial Diversity and Tetracycline Resistance Gene Abundance in the Air of Concentrated Poultry Feeding Operations. *Environmental Pollution* 220 (Jan): 1342–1348. https://doi.org/10.1016/J.ENVPOL.2016.10.101.

Gatimu, S. M., and R. W. Kimani. 2021. Does Mass Drug Administration of Azithromycin Reduce Child Mortality? *The Lancet Global Health* 9 (11): e1485–1486. https://doi.org/10.1016/S2214-109X(21)00417-4.

Gautam, A. K., S. Kumar, and P. C. Sabumon. 2007. Preliminary Study of Physico-Chemical Treatment Options for Hospital Wastewater. *Journal of Environmental Management* 83 (3): 298–306. https://doi.org/10.1016/J.JENVMAN.2006.03.009.

Graham, D. W., C. W. Knapp, B. T. Christensen, S. McCluskey, and J. Dolfing. 2016. Appearance of β-Lactam Resistance Genes in Agricultural Soils and Clinical Isolates over the 20th Century. *Scientific Reports* 6 (1): 1–8. https://doi.org/10.1038/srep21550.

Grenni, P., V. Ancona, and A. B. Caracciolo. 2018. Ecological Effects of Antibiotics on Natural Ecosystems: A Review. *Microchemical Journal* 136 (Jan): 25–39. https://doi.org/10.1016/J.MICROC.2017.02.006.

Gudeta, D. D., V. Bortolaia, S. Pollini, J. D. Docquier, G. M. Rossolini, G. C. Gregory, E. M. Elizabeth, and L. Guardabassi. 2016a. Expanding the Repertoire of Carbapenem-Hydrolyzing Metallo-β-Lactamases by Functional Metagenomic Analysis of Soil Microbiota. *Frontiers in Microbiology* 7 (Dec): 1985. https://doi.org/10.3389/FMICB.2016.01985/FULL.

Han, Y., L. Yang, X. Chen, Y. Cai, X. Zhang, M. Qian, X. Chen, et al. 2020. Removal of Veterinary Antibiotics from Swine Wastewater Using Anaerobic and Aerobic Biodegradation. *Science of the Total Environment* 709 (Mar): 136094. https://doi.org/10.1016/J.SCITOTENV.2019.136094.

Hocquet, D., A. Muller, and X. Bertrand. 2016. What Happens in Hospitals Does Not Stay in Hospitals: Antibiotic-Resistant Bacteria in Hospital Wastewater Systems. *Journal of Hospital Infection* 93 (4): 395–402. https://doi.org/10.1016/J.JHIN.2016.01.010.

Hu, J., F. Zhao, X. X. Zhang, K. Li, C. Li, L. Ye, and M. Li. 2018. Metagenomic Profiling of ARGs in Airborne Particulate Matters during a Severe Smog Event. *Science of the Total Environment* 615 (Feb): 1332–1340. https://doi.org/10.1016/J.SCITOTENV.2017.09.222.

Huang, A., M. Yan, J. Lin, L. Xu, H. Gong, and H. Gong. 2021. A Review of Processes for Removing Antibiotics from Breeding Wastewater. *International Journal of Environmental Research and Public Health* 18 (9): 4909. https://doi.org/10.3390/IJERPH18094909.

Johnning, A., E. R. B. Moore, L. Svensson-Stadler, Y. S. Shouche, D. G. J. Larsson, and E. Kristiansson. 2013. Acquired Genetic Mechanisms of a Multiresistant Bacterium Isolated from a Treatment Plant Receiving Wastewater from Antibiotic Production. *Applied and Environmental Microbiology* 79 (23): 7256–7263. https://doi.org/10.1128/AEM.02141-13/SUPPL_FILE/ZAM999104895SO1.PDF.

Ju, F., K. Beck, X. Yin, A. Maccagnan, C. S. McArdell, H. P. Singer, D. R. Johnson, T. Zhang, and H. Bürgmann. 2018. Wastewater Treatment Plant Resistomes Are Shaped by Bacterial Composition, Genetic Exchange, and Upregulated Expression in the Effluent Microbiomes. *The ISME Journal* 13 (2): 346–360. https://doi.org/10.1038/s41396-018-0277-8.

Kachhawaha, A. S., P. M. Nagarnaik, M. Jadhav, A. Pudale, P. K. Labhasetwar, and K. Banerjee. 2017. Optimization of a Modified QuEChERS Method for Multiresidue Analysis of Pharmaceuticals and Personal Care Products in Sewage and Surface Water by LC-MS/MS. *Journal of AOAC International* 100 (3): 592–597. https://doi.org/10.5740/JAOACINT.17-0060.

Kasprzyk-Hordern, B., M. Ziółek, and J. Nawrocki. 2003. Catalytic Ozonation and Methods of Enhancing Molecular Ozone Reactions in Water Treatment. *Applied Catalysis B: Environmental* 46 (4): 639–669. https://doi.org/10.1016/S0926-3373(03)00326-6.

Kizny Gordon, A. E., A. J. Mathers, E. Y. L. Cheong, T. Gottlieb, S. Kotay, A. Sarah Walker, T. E. A. Peto, D. W. Crook, and N. Stoesser. 2017. The Hospital Water Environment as a Reservoir for Carbapenem-Resistant Organisms Causing Hospital-Acquired Infections—A Systematic Review of the Literature. *Clinical Infectious Diseases* 64 (10): 1435–1444. https://doi.org/10.1093/CID/CIX132.

Koba, O., O. Golovko, R. Kodešová, M. Fér, and R. Grabic. 2017. Antibiotics Degradation in Soil: A Case of Clindamycin, Trimethoprim, Sulfamethoxazole and Their Transformation Products. *Environmental Pollution* 220 (Jan): 1251–1263. https://doi.org/10.1016/J.ENVPOL.2016.11.007.

Kovalova, L., H. Siegrist, U. Von Gunten, J. Eugster, M. Hagenbuch, A. Wittmer, R. Moser, and C. S. McArdell. 2013. Elimination of Micropollutants during Post-Treatment of Hospital Wastewater with Powdered Activated Carbon, Ozone, and UV. *Environmental Science and Technology* 47 (14): 7899–7908. https://doi.org/10.1021/ES400708W/SUPPL_FILE/ES400708W_SI_001.PDF.

Kraemer, S. A., A. Ramachandran, and G. G. Perron. 2019. Antibiotic Pollution in the Environment: From Microbial Ecology to Public Policy. *Microorganisms* 7 (6): 180. https://doi.org/10.3390/MICROORGANISMS7060180.

Kristiansson, E., J. Fick, A. Janzon, R. Grabic, C. Rutgersson, B. Weijdegård, H. Söderström, and D. G. J. Larsson. 2011. Pyrosequencing of Antibiotic-Contaminated River Sediments Reveals High Levels of Resistance and Gene Transfer Elements. *PLoS One* 6 (2): e17038. https://doi.org/10.1371/JOURNAL.PONE.0017038.

Kuppusamy, S., D. Kakarla, K. Venkateswarlu, M. Megharaj, Y. E. Yoon, and Y. B. Lee. 2018. Veterinary Antibiotics (VAs) Contamination as a Global Agro-Ecological Issue: A Critical View. *Agriculture, Ecosystems & Environment* 257 (Apr): 47–59. https://doi.org/10.1016/J.AGEE.2018.01.026.

Larsson, D. G. J., and C. F. Flach. 2021. Antibiotic Resistance in the Environment. *Nature Reviews Microbiology* 20 (5): 257–269. https://doi.org/10.1038/s41579-021-00649-x.

Lesser, L. E., A. Mora, C. Moreau, J. Mahlknecht, A. Hernández-Antonio, A. I. Ramírez, and H. Barrios-Piña. 2018. Survey of 218 Organic Contaminants in Groundwater Derived from the World's Largest Untreated Wastewater Irrigation System: Mezquital Valley, Mexico. *Chemosphere* 198 (May): 510–521. https://doi.org/10.1016/J.CHEMOSPHERE.2018.01.154.

Li, D., M. Yang, J. Hu, J. Zhang, R. Liu, X. Gu, Y. Zhang, and Z. Wang. 2009. Antibiotic-Resistance Profile in Environmental Bacteria Isolated from Penicillin Production Wastewater Treatment Plant and the Receiving River. *Environmental Microbiology* 11 (6): 1506–1517. https://doi.org/10.1111/J.1462-2920.2009.01878.X.

Li, J., J. Cao, Y. G. Zhu, Q. L. Chen, F. Shen, Y. Wu, S. Xu, et al. 2018. Global Survey of Antibiotic Resistance Genes in Air. *Environmental Science and Technology* 52 (19): 10975–10984. https://doi.org/10.1021/ACS.EST.8B02204/ASSET/IMAGES/LARGE/ES-2018-02204J_0006.JPEG.

Li, J., L. Zhou, X. Zhang, C. Xu, L. Dong, and M. Yao. 2016. Bioaerosol Emissions and Detection of Airborne Antibiotic Resistance Genes from a Wastewater Treatment Plant. *Atmospheric Environment* 124 (Jan): 404–412. https://doi.org/10.1016/J.ATMOSENV.2015.06.030.

Lindmark, J., E. Thorin, R. B. Fdhila, and E. Dahlquist. 2014. Effects of Mixing on the Result of Anaerobic Digestion: Review. *Renewable and Sustainable Energy Reviews* 40 (Dec): 1030–1047. https://doi.org/10.1016/J.RSER.2014.07.182.

López-Causapé, C., E. Rojo-Molinero, X. Mulet, G. Cabot, B. Moyà, J. Figuerola, B. Togores, J. L. Pérez, and A. Oliver. 2013. Clonal Dissemination, Emergence of Mutator Lineages and Antibiotic Resistance Evolution in Pseudomonas Aeruginosa Cystic Fibrosis Chronic Lung Infection. *PLoS One* 8 (8): e71001. https://doi.org/10.1371/JOURNAL.PONE.0071001.

Mendoza-Espinosa, L., and T. Stephenson. 2009. A Review of Biological Aerated Filters (BAFs) for Wastewater Treatment. *Environmental Engineering Science* 16 (3): 201–216. https://doi.org/10.1089/EES.1999.16.201.

Minden, V., A. Deloy, A. M. Volkert, S. D. Leonhardt, and G. Pufal. 2017. Antibiotics Impact Plant Traits, Even at Small Concentrations. *AoB PLANTS* 9 (2): plx010. https://doi.org/10.1093/AOBPLA/PLX010.

Murray, C. J. L., K. S. Ikuta, F. Sharara, L. Swetschinski, G. R. Aguilar, A. Gray, C. Han, et al. 2022. Global Burden of Bacterial Antimicrobial Resistance in 2019: A Systematic Analysis. *The Lancet* 399 (10325): 629–655. https://doi.org/10.1016/S0140-6736(21)02724-0.

Naylor, N. R., R. Atun, N. Zhu, K. Kulasabanathan, S. Silva, A. Chatterjee, G. M. Knight, and J. V. Robotham. 2018. Estimating the Burden of Antimicrobial Resistance: A Systematic Literature Review. *Antimicrobial Resistance & Infection Control* 7 (1): 1–17. https://doi.org/10.1186/S13756-018-0336-Y.

O'Neill, J. 2016. Tackling Drug-Resistant Infections Globally: Final Report and Recommendations. The Review on Antimicrobial Resistance, Government of the United Kingdom, 2016. https://amr-review.org.

Papageorgiou, M., C. Kosma, and D. Lambropoulou. 2016. Seasonal Occurrence, Removal, Mass Loading and Environmental Risk Assessment of 55 Pharmaceuticals and Personal Care Products in a Municipal Wastewater Treatment Plant in Central Greece. *Science of The Total Environment* 543 (Feb): 547–569. https://doi.org/10.1016/J.SCITOTENV.2015.11.047.

Pei, R., S. C. Kim, K. H. Carlson, and A. Pruden. 2006. Effect of River Landscape on the Sediment Concentrations of Antibiotics and Corresponding Antibiotic Resistance Genes (ARG). *Water Research* 40 (12): 2427–2435. https://doi.org/10.1016/J.WATRES.2006.04.017.

Polianciuc, S. I., A. E. Gurzău, B. Kiss, M. G. Ştefan, and F. Loghin. 2020. Antibiotics in the Environment: Causes and Consequences. *Medicine and Pharmacy Reports* 93 (3): 231. https://doi.org/10.15386/MPR-1742.

Price, L. B., J. P. Graham, L. G. Lackey, A. Roess, R. Vailes, and E. Silbergeld. 2007. Elevated Risk of Carrying Gentamicin-Resistant Escherichia Coli among U.S. Poultry Workers. *Environmental Health Perspectives* 115 (12): 1738–1742. https://doi.org/10.1289/EHP.10191.

Pruden, A., M. Arabi, and H. N. Storteboom. 2012. Correlation between Upstream Human Activities and Riverine Antibiotic Resistance Genes. *Environmental Science and Technology* 46 (21): 11541–11549. https://doi.org/10.1021/ES302657R/SUPPL_FILE/ES302657R_SI_001.PDF.

Rafraf, I. D., I. Lekunberri, A. Sànchez-Melsió, M. Aouni, C. M. Borrego, and J. L. Balcázar. 2016. Abundance of Antibiotic Resistance Genes in Five Municipal Wastewater Treatment Plants in the Monastir Governorate, Tunisia. *Environmental Pollution* 219 (Dec): 353–358. https://doi.org/10.1016/J.ENVPOL.2016.10.062.

Rivera-Jaimes, J. A., C. Postigo, R. M. Melgoza-Alemán, J. Aceña, D. Barceló, and M. L. de Alda. 2018. Study of Pharmaceuticals in Surface and Wastewater from Cuernavaca, Morelos, Mexico: Occurrence and Environmental Risk Assessment. *Science of The Total Environment* 613–614 (Feb): 1263–1274. https://doi.org/10.1016/J.SCITOTENV.2017.09.134.

Sabri, N. A., H. Schmitt, B. Van Der Zaan, H. W. Gerritsen, T. Zuidema, H. H. M. Rijnaarts, and A. A. M. Langenhoff. 2020. Prevalence of Antibiotics and Antibiotic Resistance Genes in a Wastewater Effluent-Receiving River in the Netherlands. *Journal of Environmental Chemical Engineering* 8 (1): 102245. https://doi.org/10.1016/J.JECE.2018.03.004.

Serra-Compte, A., M. G. Pikkemaat, A. Elferink, D. Almeida, J. Diogène, J. A. Campillo, M. Llorca, D. Álvarez-Muñoz, D. Barceló, and S. Rodríguez-Mozaz. 2021. Combining an Effect-Based Methodology with Chemical Analysis for Antibiotics Determination in Wastewater and Receiving Freshwater and Marine Environment. *Environmental Pollution* 271 (Feb): 116313. https://doi.org/10.1016/J.ENVPOL.2020.116313.

Suarez, S., J. M. Lema, and F. Omil. 2009. Pre-Treatment of Hospital Wastewater by Coagulation–Flocculation and Flotation. *Bioresource Technology* 100 (7): 2138–2146. https://doi.org/10.1016/J.BIORTECH.2008.11.015.

Subedi, B., K. Balakrishna, R. K. Sinha, N. Yamashita, V. G. Balasubramanian, and K. Kannan. 2015. Mass Loading and Removal of Pharmaceuticals and Personal Care Products, Including Psychoactive and Illicit Drugs and Artificial Sweeteners, in Five Sewage Treatment Plants in India. *Journal of Environmental Chemical Engineering* 3 (4): 2882–2891. https://doi.org/10.1016/J.JECE.2015.09.031.

Tolmacheva, V. V., V. V. Apyari, A. A. Furletov, S. G. Dmitrienko, and Y. A. Zolotov. 2016. Facile Synthesis of Magnetic Hypercrosslinked Polystyrene and Its Application in the Magnetic Solid-Phase Extraction of Sulfonamides from Water and Milk Samples before Their HPLC Determination. *Talanta* 152 (May): 203–210. https://doi.org/10.1016/J.TALANTA.2016.02.010.

Ventola, C. L. 2015. The Antibiotic Resistance Crisis: Part 1: Causes and Threats. *Pharmacy and Therapeutics* 40 (4): 277. https://doi.org/Article.

Wang, M., W. Shen, L. Yan, X. H. Wang, and H. Xu, Environmental Pollution, and Undefined 2017. n.d. *Stepwise Impact of Urban Wastewater Treatment on the Bacterial Community Structure, Antibiotic Contents, and Prevalence of Antimicrobial Resistance*. Elsevier, Amsterdam, Netherlands. Accessed August 23, 2022. https://www.sciencedirect.com/science/article/pii/S0269749117327252.

Wohde, M., S. Berkner, T. Junker, S. Konradi, L. Schwarz, and R. A. Düring. 2016. Occurrence and Transformation of Veterinary Pharmaceuticals and Biocides in Manure: A Literature Review. *Environmental Sciences Europe* 28 (1): 23. https://doi.org/10.1186/S12302-016-0091-8.

Woolhouse, M., M. Ward, B. Van Bunnik, and J. Farrar. 2015. Antimicrobial Resistance in Humans, Livestock and the Wider Environment. *Philosophical Transactions of the Royal Society B: Biological Sciences* 370 (1670): 20140083. https://doi.org/10.1098/RSTB.2014.0083.

Yu, H., C. Du, F. Qu, J. He, and H. Rong. 2022. Efficient Biostimulants for Bacterial Quorum Quenching to Control Fouling in MBR. *Chemosphere* 286 (Jan): 131689. https://doi.org/10.1016/J.CHEMOSPHERE.2021.131689.

5 Aquaculture Antiviral Short Peptides

To Protect Aquaculture, Humans, and Animals and the Marine Water Quality

Nawal Abd El-Baky, Amro Abd Al Fattah Amara, and Arti Gupta

LIST OF ABBREVIATIONS

AR	Antibiotic resistance
ARG	Antibiotic resistance genes
AVPs	Antiviral peptides
IAV	Influenza A viruses
IHNV	Infectious hematopoietic necrosis virus
IPNV	Infectious pancreatic necrosis virus
NNV	Nervous necrosis virus
SHRV	Snakehead rhabdovirus
VHSV	Viral hemorrhagic septicemia virus
WHO	World Health Organization
WSSV	White spot syndrome virus

5.1 INTRODUCTION

Natural products and microbes are diverse (Amara, El-Masry, and Bogdady 2008, Hussain and Amara 2006, El-Baky and Al-Fattah Amara 2023)). Natural resources could cover the demand of many applications that need specific compounds such as antimicrobial (Amara, Salem-Bekhit, and Alanazi 2013), antiviral, anticancer (Amara, El-Masry, and Bogdady 2008), and anticlot (El-Baky, Amara, and Redwan 2023a). Viral infections lead to serious problems for the fish aquaculture industry. Cationic antivirus peptides are being used against several aquacultural viruses. For example, rhabdoviruses viral hemorrhagic septicemia virus (VHSV). Another example is infectious hematopoietic necrosis virus (IHNV). They are responsible for the losses in aquaculture production. Cecropine B and its synthetic analog CF17 have activity against major pathogenic viral infections in fish. They are active against hematopoietic necrosis virus (IHNV), VHSV, snakehead rhabdovirus (SHRV), and infectious pancreatic necrosis virus (IPNV). The mechanism of action is related to disrupting the viral envelop and disintegrating the viral capsids (Chiou et al. 2002, Mulder et al. 2013). Tilapia hepcidin 1–5 (TH 1–5) and cyclic shrimp anti-lipopolysaccharide factor (cSALF) exhibited noticeable antiviral activity *in vitro* against nervous necrosis virus (NNV). NNV caused mortality of several marine fish species at their larval stage. Both peptides cause agglutination of NNV virions into clumps and prevent viral cell entry (Chia et al. 2010, Mulder et al. 2013).

DOI: 10.1201/9781003269823-5

Together with hepcidin 1–5, the peptide hapcidin 1 has also shown antiviral activity *in vivo* against NNV in the Japanese rice fish medaka (*Oryzias latipes*). Pretreatment, cotreatment, or post-treatment with epinecidin-1 or hepcidin 1–5 has been shown to be effective in promoting an increase in medaka survival infected with NNV (Wang, Kung, and Chen 2010), Pleurocidin, and the HNP1. Pleurocidin MDPle, isolated from the Mud dab fish (*Limanda limanda*), has antiviral properties against VHSV. It is suggested that the peptide mechanism leads to disrupt the viral membrane (Falco et al. 2009). The HNP1 is a defensin demonstrated antiviral effects against MDPle. HNP1 can deactivate the enveloped virus by interacting with GAGs present on the viral surface. It alters the ability of these glycoproteins to bind to their receptors at the target cells and stimulate the immune modulatory system of infected fish cells by modulating IFN-related mechanisms (Falco et al. 2009, 2007).

Infection with viral diseases has exposed humanity to great dangers, as it is not far from us what happened in the past 5 years because of SARS-CoV-2, which force fast treatment medication (El-Baky and Amara 2022, El-Baky, Amara, and Redwan 2023b). Viruses are absolutely different from other disease inducers. They have no cell, can live for so long as genetic material, they completely control the cell they occupy, constantly mutate and undergo genetic evolution, mutate to new resistant structures, and sometimes fatal. Viruses are characterized by their successful evolution *via* genetic distinction, effective replication within cells of the host, transmission means variation, and their capability to persist (as genetic material or viral proteins) in the host (Wagner et al. 1999, El-Baky et al. 2023). All of these characteristics have made the treatment of viral diseases a hard and disturbing task. Thus, to eradicate them, there must be unique approaches with different tasks that will hinder the virus from invading, replicating, controlling, and killing the host cell and then transmit to another. The repeated outbreaks and epidemics make the situation worse. Till now, after the huge progress in drug industries, we do not have a reliable fast solution to face a new virus.

With the substantial increase in studies and reports revealing resistance of viruses and the occurrence as well as reoccurrence of viral epidemics, communities are under pressure to continuously discover innovative antiviral molecules. Among these molecules, antiviral peptides (AVPs) have presented themselves as promising therapeutic drugs. While their number is still low, AVPs have already shown enormous possibilities to be developed into pharmaceutically available and proficient antiviral medication (Vilas Boas et al. 2019).

The oceans cover roughly 70% of the surface of the globe and marine species represent about 50% of the estimates of entire universal biodiversity. Oceans are a rich mine of various bioactive molecules derived from marine organisms, for instance, sponges, mollusks, crustaceans, fish, and shellfish; therefore, they were comprehensively explored in the last decades for possible natural sources of new bioactive compounds to design novel drug leads (Hamed et al. 2015, Destoumieux-Garzón et al. 2016). Marine species are living in various complex marine habitats and are exposed to extremes of illumination, salinity, temperature, and pressure. The exposure to these extremes has led them to produce several unique secondary metabolites (Hamed et al. 2015). Marine bioactive molecules accomplish various biological roles with interest in therapeutics such as antimicrobial and anticancer activities, antioxidant effect, and anti-inflammatory properties (Lordan, Ross, and Stanton 2011). Marine inhabitants are taking the extremes as their home. Thus, it is expected that they will have special macromolecules that enable them to live according to highly standard measures and beat the hardest conditions, especially if unwelcomed foreign visitors are there in the surrounding (Amara 2019). We are getting used of the weapons marine inhabitants used to defend themselves against unwelcomed viruses and mimic them and develop them into weapons to defend ourselves, animals, and other aquaculture farming products.

In general, aquatic organisms are a rich source of AVPs, even though the clavanins (α-helical AVPs isolated from the tunicate *Styelaclava*) are the most profoundly studied marine AVPs. Yasin et al. (2000) showed that clavanin AK inhibited herpes simplex virus 1 (HSV-1) infection by about 70.4%. Another study revealed that clavanin A caused *in vitro* viral inactivation of nonenveloped adenovirus and rotavirus in the pretreatment (cells were pretreated with the peptide then inoculated with viruses) and simultaneous (cells were incubated with both the peptide and viruses simultaneously) inspections (Carriel-Gomes et al. 2007). Clavanin B was also proved to act against HIV (Ireland et al. 2008).

Plaza et al. isolated mirabamides A–D and three cyclic depsipeptides (A, C, and D) from the marine sponge *Siliquariaspongia mirabilis* (Plaza et al. 2007). These peptides could suppress HIV-1 fusion by obstructing the entry of virion to TZM-bl cells, consequently neutralizing the fusion of viral glycoprotein for CD4 and CCR5 (cell receptors of HIV). Furthermore, Lu et al. isolated mirabamides E–H (four cyclic depsipeptides) from the marine sponge *Stelletta clavosa* and reported them to have a noteworthy HIV-inhibitory effect (Lu et al. 2011). They demonstrated that mirabamide H displayed the best *in vitro* antiviral efficiency followed by F and G, then E. More recently, Stellettapeptins A and B (another depsipeptides) were also isolated from *Stelletta* genus and found to act against HIV in human T-cell line CEM-SS (Shin et al. 2015). The multifunctional AVP Pa-MAP 1 derived from the polar fish *Pleuronectes americanus* was reported to exert antibacterial, antifungal, and anticancer effects besides its antiviral potency against HSV-1 and HSV-2 (Migliolo et al. 2012). Later, this peptide was shown to have a virucidal mechanism of action as it interacts directly with viral envelope glycoproteins (Vilas Boas et al. 2017).

In a review article that was published in 2021, the authors stated that till the date they wrote their article, the greatest number of AVPs was identified from amphibians (1,057 peptides), followed by those obtained from mammals (222 peptides), while AVPs from human (123 peptides) were comparable to peptides from fish and shellfish (122 peptides) (Sumon et al. 2021). In most cases, marine peptides can be either linear or cyclic (e.g., depsipeptides in which amide bonds are either one or more replaced with ester ones), and their structure is complex that can also comprise secondary structures and varies depending on the sources of peptide isolation (Pelay-Gimeno, Tulla-Puche, and Albericio 2013). Some of them display the presence of branched amino acids and elevated percentage of proline (Dahiya et al. 2016), the existence of uncommon amino acids, for example, brominated tryptophan (Jimenez 2019), and finally the formation of secondary structures (α helix in case of short marine peptides and β sheets in case of longer ones) (Ravichandr et al. 2010).

Those AVPs from marine biomass vary in their approaches for constraining various viral infections and of course the virus they target. They can be new drug leads and expectancy for virus control. This review outlines many examples of natural AVPs from marine sources, starting from Didemnins (one of the first peptide groups with antiviral efficacy, isolated from marine source), passing with clavanins (the most-studied marine peptides of antiviral bio-effectiveness), till recently discovered marine peptides (e.g., Peptide A-3302-B), active against vital, immune-escaping, resistant, widespread, and fatal viruses.

Nevertheless, it should be taken into consideration that not all marine organisms can produce peptides which have these characteristics altogether, but rather, some characteristics are more common in definite groups. As an example, marine mollusks and sponges often produce cyclic peptides (Phyo et al. 2018), whereas marine bacteria produce cyclic depsipeptides (Ding et al. 2018).

5.2 PROTECTING THE NATURE FROM ANTIBIOTIC RESISTANCE

Fish farming is growing everywhere. It is a process that guarantees acceptable incomes and adds many values to the economy of any country. The health and the quality of the marine water during the aquaculture farming are important issues that affect the quality of the final products concerning any farming process. Meanwhile, fish farming outside of the natural habitat is not fully representing the natural habitat of the fish as well as the other aquaculture products. For that, additional protection like the use of antibiotics is used. Specific treatment after each cycle using chemical compounds is used to remove the unwanted by-products. Due to the accumulation of antibiotic residues, ammonia, and chemical residues, a closed aquaculture farming system is a source of microbial resistance strains to antibiotics. Over time, harmful antibiotic-resistant microbial strains are dominating the system. Those microbes might be harmful to other creatures like the human and some mammalian that based on their feed on fish' products (Lozano et al. 2018). It is wiser to use antimicrobial agents other than antibiotics such as the protein-based antimicrobial agents that could be degraded in a more perfect way than the antibiotic in the nature.

5.2.1 Example of an Antibiotic Resistance Problem in Marine Aquaculture

In Chile (1973 and 1976), the fish farming of salmonids was grown (Lozano et al. 2018). Practically, four species of salmonids were cultivated: rainbow trout (*Oncorhynchus mykiss*), Chinook (*Oncorhynchus tshawytscha*), coho (*Oncorhynchus kisutch*), and Atlantic salmon (*Salmo salar*). Between 1987 and 2010, diseases appeared in salmon farming. The diseases were caused by the ectoparasite Caligus and bacterial kidney disease in 1987 (Cabello 2004, Ibieta et al. 2011); the presence of *Piscirickettsia salmonis* in 1989 (Lozano et al. 2018); presenting later high mortality due to Infectious Pancreatic Necrosis (IPN) in 1997; and by the outbreaks of infectious salmon anemia disease in the years 2002, 2007, and 2008 (Cabello 2004). Some authors reclaimed that due to the use of antibiotics in 1989 (Cabello 2004, Ibieta et al. 2011).

5.2.2 Analysis of the Problem of Antibiotic Resistance

Why do diseases appear in a more frequent pattern in crowded farming? Microbes are diverse and even similar strains might be different due to different exposure to nutrients and chemicals around (Hussain and Amara 2006, Amara and Shib 2015, Amara 2022a). One could extract some facts from other farming styles like animal farming or poultry farming. Comparing the wild nature habitat and that designed by humans could help to understand from where the problems come. At least at the first, one should stop the source of the problem, which might be simply causing antibiotic residues to accumulate over time. Before the antibiotic, the nature was in balance. And nature has all the keys to retrieve its balance again (in case of sudden change). The effects of the antibiotics on human and farm animals' health are ranged from acceptable, tolerant to harmful. Their role in spreading antibiotic resistance genes (ARGs) and the possibility of correcting the global problem decreased unless we are able to control the antibiotic resistance (AR). Fastening the research and development of new antimicrobial therapies, including improved biosecurity measures only will lead to nothing if we do not understand what happened in the nature because our activity has led to such a level of accumulated antimicrobial and AR.

There remain too many shortages regarding the understanding of the patterns of antimicrobial use and release during manufacturing and use, and what this means for resistance and, ultimately, human health. Some governments are adopting a broad "One Health" approach to tackling the issue of resistance as a part of the AR controlling strategy. Full success can be achieved by considering a complete controlling strategy including many tactics. For example, in animal farming including aquaculture closed system farming, different successive strategies are used such as applying antibiotics and some other antimicrobial agents such as therapy, prophylaxis (prevention), or growth promotion.

The art of the antimicrobial agents was well studied concerning their mechanisms, the microbial death rate, the structure of the agents, their toxicity and side effects, etc. If the antibiotic producer is a prokaryote (e.g., bacteria), that is a sign that these prokaryotes have one/more than one resistance gene or one/more than one resistance mechanism for its own product (the antibiotic). A fact even logic but was not observed early by the scientists. Some resistance mechanisms not only protect the mother strain but also could protect sensitive microbes around like the microbial biofilm (Amara 2011). The biofilm is of great importance while it might give sensitive microbes enough time to protect themselves till acquiring a resistance gene or building/activating a resistance mechanism (Amara 2011). No place will be better than the crowded animals/poultry fast farming for forming different types of biofilms. The World Health Organization (WHO) defined AMR as "antimicrobial resistance is the ability of a microorganism (like bacteria, viruses, and some parasites) to stop an antimicrobial (such as antibiotics, antivirals, and antimalarials) from working against it. As a result, standard treatments become ineffective; infections persist and may spread to others" (Furtado et al. 2015, Medina-Pizzali et al. 2021).

Antibiotics were topically used to combat disease in humans, animals, and plants. The British bacteriologist Alexander Fleming discovered penicillin in 1928. For more details, refer to Amara et al. (2024). After their first purification, antibiotics start to dominate the image, and other common and public, even traditional and folk-based alternative start to disappear. Moreover, processing assistance and alternatives were blocked (e.g., bacteriophages). The overall medicinal system negatively complements in omitting many processing alternatives and install its own curriculum for medication. Perhaps only nonmodernized countries and a minute number of other countries keep other alternatives running beside. For more details, revise the history of the probiotic (Amara 2023), bacteriophage (Amara 2024), and the aroma therapy and where they are practiced.

Perhaps we, our families, our generation are almost antibiotic users. Seldom who tries to treat against/protect from pathogens using, for example, volatile oil, bacteriophages, etc.? Perhaps, the only successfully running alternatives to the antibiotic were the vaccines and to some extent the different types of disinfectants. Of course, besides some miscellaneous other types in a narrow range of activity are usually in the form of soap, liquid disinfectants, and the like. Within the world of antibiotics, there are many leakages such as (1) the excessive use, (2) the nonspecific use, (3) synthetic antibiotic that contain dextrorotatory structures, and (4) antibiotic residues that contaminate different ecosystems and our food, water, marine aquaculture, and many others. But perhaps the most interesting one is the AR to the antibiotic itself.

Nowadays, progress in different scientific sectors enables different kinds of inspecting new natural product molecules with specific activities as well as designing *de novo* molecules (Amara 2013, 2015). Some molecules could have more than one positive beneficial effect and could protect from additional illness if used in an adequate amount (Amara, Salem, and Shabeb 2009).

The role of the nutrient in elevating the mutants was proposed by Amara Shibl, and Tawfik (2012). It is based on screening the similarity and differences between the four different classes of β-lactamases based on their protein sequences. The investigation shows that a single strain (*E. coli* ATCC 8739 found in the protein database) has different β-lactamases. By investigating the Blast protein database for one protein of different *E. coli* β-lactamases, the result shows that this protein can be found in hundreds of different microbes with 100% identity (Amara, Shibl, and Tawfik 2012). Some basic evidence about the stability of the biological system and the parts involved also a discussion about why similar or different mutants exist in the same gene and the factors involved in their elevation. Also, the role of the nutrient in the stability of the gene structure in a particular strain and in its role in changing other strains was discussed and proposed. The question is, did the differences and the similarities between the β-lactamase is because of mutations, host adaptation, its mobility, all that, or something else. The investigated strain *E. coli* ATCC 8739 that published separately in different places shows different β-lactamase protein sequences from the same classes. That strongly supports the hypothesis of the environmental effect on a specific gene(s), such as the presence of the alginate (enabling sublethal dosage) (Amara 2022b) and nutritional factors (epigenetic-like mechanism), that could support inducing mutation in the β-lactamase genes.

Before purifying the fungal antibiotic, penicillin, there were many types of antimicrobial agents that were historically used by many. For example, the ethanol by the ancient Egyptian; the vinegar by the Arab; the salt by many; the volatile oil and the river water (containing bacteriophages) by the Indian and the middle and Far East; the silver by the Nobel (by chance) and the like. But apparently, none of the old used antimicrobial agents has the power of the antibiotics. Because it is selective to the prokaryotic (could have toxicity to eukaryotic in high dosage) and nearly can reach any part of the body.

Synthetic antibiotics are less degradable in nature because they are less compatible when compared with the naturally produced antibiotics.

They are designed specifically (rational design) or randomly (randomly synthesized) to be less degradable. Synthetic compounds are not only different from the naturally produced but also could have incomplete structure and dextrorotatory isomers. Upon being discharged to the nature, no one could guarantee their mode of action or their side effect. Generally, the synthetic compound if does not resemble a natural structure, they are mostly incompatible with the nature. They will kill microbes around (if the correct criteria exist) with a longer life span than that belonging to the natural antimicrobial agents (Figure 5.1). Only resistance microbes (by a gene(s) or by a mechanism) will be able to survive. Again, this will let resistance microbes if succeeded, by any means, to build or to acquire resistance to dominate the image! In fact, we built a perfect protocol for changing the nature biodiversity, dominating resistance microbes and keep a guard who guarantees the success of this process to lasting for a long time (Figure 5.1); it is the antibiotic residues in the nature (Figure 5.1) (Amara et al. 2024). For more details, refer to Amara et al. (2024).

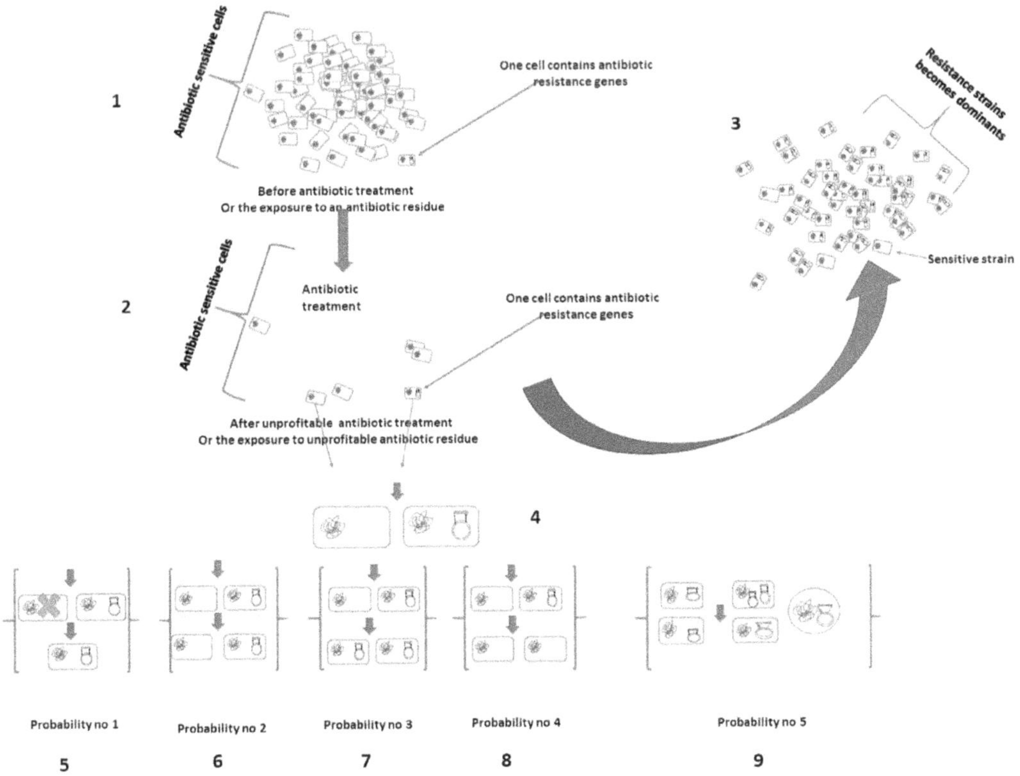

FIGURE 5.1 (1) Before antibiotic treatment, microbes in normal biodiversity (many sensitive cells and few resistant cells); (2) after antibiotic treatment; most sensitive cells are killed, and resistance cells are survive; (3) AR cells become dominant after time due to the presence of the antibiotic residue(s) in the nature; (4) two types of cells are existed, sensitive cell and resistance cell with AR gene carried on plasmid; (5) probability number 1 all sensitive cells are dead and resistance cells are survive; (6) probability number 2, both types of cells are survive; (7) probability number 3, sensitive cells are acquired AR gene(s) and all cells become resistance; (8) probability number 4, resistance cells loss its resistance gene on the mobile element (plasmid (plasmid curing)) and they all become sensitive again (due to the absence of the antibiotic); (9) probability no 5, some resistance cell gain additional AR gene(s) and become multiple resistance while some stay with one resistance gene, later the two acquired resistance genes are located in one plasmid. (The figure has been generated in this study.)

5.2.3 Microbial Biodiversity

Generally and collectively microbes are not all evil. Pathogenic microbes are few in their number and existence. Apparently, the microbes' major duty in nature is to clean nature from different wastes including the antibiotics. Most other important roles might be produced as a result of the waste (and any available nutrient source) degradation. They do this activity in the desert, cities, farms, forests, marine ecosystems, and nearly exist in every place, and inside our body and on our skin as well. The general role for the microbes is "feed and feed anything can be considered as a feed." Microbes that are resistant to antimicrobial agents either having resistance genes for one or more of the antimicrobial agents or can acquire resistance genes (or additional resistance gene) from others that have gained through different transformations/other acquiring mechanisms. Also, they either have built-in structures made them resistance or could readjust the function/structure of some of their macromolecules and cells' structures to bypass the activity of some of the antimicrobial drugs. But those microbes live in nature with many other microbes, those which fix nitrogen, degrade cellulose, other hydrocarbons degradable, degrade protein, degrade the corpus of the dead animals and human, clean the nature of many contaminants, etc. They all have the power to survive. But not all of them have ARG genes. Exposing the nature to antibiotic residue will kill them if they do not acquire resistance genes or build a successful resistance mechanism. We enrich nature with the factors that shift its balance toward favoring the condition for the AMRs (Figure 5.1).

5.3 ANTIVIRAL MARINE PEPTIDES AS A PRODUCT COULD ASSIST THE AQUACULTURE FARMING

Many of the virus infections if prevented could block the secondary infection with pathogenic microns. There is always a link between the virus infection and the microbial infections. This link can affect positively or negatively the power of the immune system. The microbial infections can become first and then the virus infection, or *vice versa*. The immune system might be exhausted or impaired due to one or more than one type of any of the viral or the microbial infection. Our attention should be expanded to other factors that could include: environmental factors such as the aquaculture water quality, crowding, waste removing, temperature, feed components, and antimicrobial agents.

Active peptides are one of the proposed alternatives to antibiotics and should get more attention.

5.3.1 History of Marine Peptides with Antiviral Efficacy

Pardaxin was the first characterized marine peptide with antimicrobial efficiency, which was reported in 1980 derived from the Red Sea flatfish *Pardachirus marmoratus* (Primor and Tu 1980). Later, peptides that have *in vitro* or *in vivo* (sometimes both of them) antiviral efficacy were obtained from numerous sponges, shellfish, fish, etc. Their examples include β-defensin from *Epinephelus coioides* (Jin et al. 2010), pro-hepcidin from *Oryzias melastigmus* (Cai et al. 2012), microspinosamide from *Sidonops microspinosa* (Rashid et al. 2000), hepcidins from *Scophthalmus maximus* (Pereiro, Figueras, and Novoa 2012), and Scygonadin from *Scylla paramamosain* (Peng et al. 2012).

Although particular drugs for viral infections were available since 1946, AVPs remained an encouraging substitute in designing viral diseases therapeutics (Field and De Clercq 2004). Didemnins were the earliest of these promising marine AVPs (Rinehart et al. 1981). The following decade witnessed the isolation of various marine AVPs like polyphemusin II from horseshoe crab *Tachypleus tridentatus* and its synthetic peptide T22 (Nakashima et al. 1992), tachyplesin I (Morimoto et al. 1991), magainin from *Xenopus laevis* (Aboudy et al. 1994), and pokeweed antiviral protein from *Phytolacca americana* that affects plant viruses (Chen et al. 1991).

On the other hand, fishes and shellfishes were not explored for most of their AVPs till the 21st century. The most famous of them was Pa-MAP (Migliolo et al. 2012). In 2010, Chia et al. announced that they found tilapia hepcidin (TH) 1–5 and Epinecidin-1 in *Oreochromis mossambicus* (fish) in addition to cSALF in *Penaeus monodon* (shrimp), which could act against nodavirus in fish (Chia et al. 2010).

After the catastrophic situation occurred with COVID-19, attention was drawn to the possibility of exploiting these peptides to control SARS-CoV-2. In 2020, Chowdhury et al. employed molecular dynamics simulation to find S2P25 and S2P26 peptides; two previously reported anti-SARS-CoV-1 peptides, which hold promise to block SARS-CoV-2 entry to cells (Chowdhury et al. 2020). Moreover, computational approaches were involved to anticipate AVPs efficiency against SARS-CoV-2 (Fakih 2020, Ling et al. 2020).

5.3.2 Isolation of Marine Peptides with Bio-Effectiveness (on General Regard)

Marine peptides with bio-effectiveness were extracted from tissues of various species (mostly evolutionarily distant) of marine vertebrates or invertebrates besides marine microorganisms such as algae, bacteria, and fungi. The typical protocol to follow for the discovery, isolation, then structural and functional characterization of marine peptides with bio-effectiveness (either linear or cyclic peptides) involves four key steps: (1) selection of the sources from which peptides with bioactivity are first isolated, (2) extraction and/or hydrolysis, screening for a particular bioactivity, and fractionation with a bioassay-guided fractionation method, (3) separation and isolation of a single bioactive peptide, and (4) identification of its structure and function (Lazcano-Perez et al. 2012).

Peptides with bioactivity extracted from tissues of marine animals require efficient solubilization to retain their bioactivity and evade their rapid degradation. Target marine peptides or proteins are usually isolated from their natural sources using water or aqueous buffers to achieve high yield recovery and keep their functionality. The fractionation process is usually carried out *via* salting out by ammonium sulfate or using acetone fractionation. The chromatographic techniques for purification of peptides with bioactivity generally include ionic exchange, gel filtration, and reversed-phase high-performance liquid chromatography (RP-HPLC) (Lazcano-Perez et al. 2012). Lastly, identification of peptide primary sequence is performed using protease digestion or Edman degradation and then mass spectrometry follows. Circular dichroism identifies the secondary structure of peptides whereas nuclear magnetic resonance offers information concerning their tertiary structure.

The first step in the marine cyclic peptides isolation protocol is usually performed using organic solvents such as ethyl acetate or methanol. Subsequently, the produced extract is concentrated and then partitioned by other solvents (e.g., carbon tetrachloride, hexane, or dichloromethane). The extract is subjected afterward to size exclusion chromatography (gel filtration) or silica gel and eluted using solvents that have increasing polarity. The final step is carried out by using RP-C18 HPLC (Du et al. 2014).

On the other hand, the first step in the marine linear peptides extraction procedure includes size fractionation of their aqueous extract with gel filtration. After concentration of the target fraction, it is applied to anion/cation-exchange resin column. The elution process is carried out according to the target peptide basicity. The final purification step includes RP-C18 HPLC with a further step of repeated chromatography on the same column for eliminating impurities (Lebbe et al. 2014).

Furthermore, marine peptides with bio-effectiveness might be produced by enzymatic digestion (frequently trypsin, pepsin, α-chymotrypsin, commercial protease combinations, or papain is used) of proteins derived from marine organisms (Qian, Je, and Kim 2007). The obtained hydrolysates are thoroughly screened for numerous bioactivities after their fractionation by ultrafiltration based on size (Pavlicevic, Maestri, and Marmiroli 2020). The fraction displaying the superior bioactivity is subjected to either RP-HPLC or gel filtration to resolve individual peptides based on their hydrophobicity and size, respectively. These fractions of individual peptides are characterized *via* mass spectrometry in addition to protein sequencing (Iijima et al. 2003).

For industrial-scale production, marine peptides with bio-effectiveness can be isolated from proteins derived from marine species by several approaches. Previously, organic solvent protein extraction technique was usually used, but it was found to be expensive, time consuming, and environmentally unfriendly method. Currently, alternative and efficient protein isolation techniques are used such as ultrasound-assisted extraction, microwave-assisted extraction, pressurized solvent extraction, supercritical fluid extraction, and enzyme-assisted extraction (Grosso et al. 2015).

After extraction, the proteins are hydrolyzed into functionally active peptides. For pharmaceutical and nutraceutical industries, enzyme-assisted protein extraction and hydrolysis is commonly desired (Zhao et al. 2009).

5.3.3 Marine AVP Structural Characteristics

AVPs derived from marine sources are short; their length ranges from 12 to 50 amino acids and their charge is mostly positive (may reach +9). These amphipathic peptides are gene encoded (Mahlapuu et al. 2016, Cappello and Nieri 2021). All of these characteristics empower their antimicrobial efficiency but definitely hydrophobicity is crucial for targeting of enveloped viruses by these peptides (Badani, Garry, and Wimley 2014, Wang, Shih, and Chang 2017). It must be said in this context that the amino acid composition studied by statistical exploration on antimicrobial peptide database revealed the abundance of hydrophobic cysteine in peptides with antiviral bio-effectiveness (Mishra and Wang 2012).

The fascinating thing is that aquatic organisms like fishes produce these peptides in nature as a primary weapon to defend themselves against pathogens as a part of their innate immunity (Smith, Desbois, and Dyrynda 2010). As mentioned above, there are 122 reported marine peptides with antiviral bio-effectiveness in fishes and shellfishes, which were classified into β-defensins, hepcidins, histone-derived peptides, fish-specific piscidins, and cathelicidins according to structural characteristics (Wang, Li, and Wang 2016).

Defensins are rich in cysteine and have a molecular weight that ranges between 3 and 6 kDa. They defend their producer (vertebrates and invertebrates) against enveloped as well as nonenveloped viruses (Bulet, Stöcklin, and Menin 2004). These peptides are stabilized by disulfide bridges that link between the structures of α-helix and β-pleated sheet. EcDefensin (β-defensin) is one of this class of peptides derived from orange-spotted grouper liver and acts against nonenveloped RNA and enveloped DNA viruses (Guo et al. 2012). In 2008, Falco et al. stated that β-defensin could be isolated from *Oncorhynchus mykiss* (rainbow trout) that can act against rhabdovirus; VHSV (Falco et al. 2008). Piscidins are α-helical peptides obtained from hybrid striped bass mast cells and act against the enveloped channel catfish virus (Silphaduang and Noga 2001).

Interestingly, nine conserved regions were found in multiple sequence alignment performed *via* Clustal Omega software among four α-defensins of human origin and five ones from aquatic organisms (*Salmo salar*, *Gadus morhua*, *Tetraodon nigroviridis*, *Fundulus heteroclitus*, and *Pseudochaenichthys georgianus*). Another time, 19 conserved regions were found in multiple sequence alignment performed *via* Clustal Omega among four hepcidins of aquatic origin (*Nematolebias whitei*, *Morone saxatilis*, *Sorsogona tuberculata*, and *Dicentrarchus labrax*) and two hepcidins from human. With nucleotide sequence, "CCNCC" being found, among all hepcidins (Sumon et al. 2021).

5.3.4 Marine AVP Action Mechanisms

Generally speaking, any efficient antiviral drug should act upon particular virus/s and interfere at specific phase/s of the life cycle of the targeted virus. Marine peptides act against enveloped viruses *via* interfering with entry, synthesis, or assemblage (fundamental life cycle stages). They can inhibit virion (which is a direct virucidal action), viral penetration, cellular association, uncoating of the virus, replication of its genome, besides viral packaging, assemblage, and discharge (Vilas Boas et al. 2019, Lou, Sun, and Rao 2014). The following are marine AVP action mechanisms: first, direct binding or hindering viral pathogenesis (direct action mechanism); second, viral gene expression suppression, along with preventing viral enzymes from doing their jobs in replication of the virus inside the host (indirect action mechanism); third, entry inhibition (Sukmarini 2022). Some of the explicit antiviral modes of action of marine peptides are summarized in Table 5.1.

TABLE 5.1
Some of the Explicit Antiviral Modes of Action for Marine Peptides (Sukmarini 2022)

Modes of Action	Peptide	Virus	Peptide Derived From
Virucidal action	Pa-MAP	HSV-1/HSV-2	*Pleuronectes americanus*
Entry inhibition	Stellettapeptins A and B	HIV-1	*Stelletta* sp
	Mirabamides E–H	HIV-1	*Stelletta clavosa*
	Aspergillipeptide D	HSV-1	*Aspergillus* sp. SCSIO 41501
	Divamide A	HIV-1	*Didemnum molle* E11-036
NS3-4A protease inhibition	Cyclo(L-Tyr-L-Pro) diketopiperazine	HCV	*Aspergillus versicolor*
Integrase inhibition	Mollamide F	HIV-1	*Didemnum molle* PNG07-2-050
Replication inhibition	Al(III)-acremonpeptide D	-	*Acremonium persicinum* SCSIO 115
	Acremonpeptides A and B	HSV-1	*Acremonium persicinum* SCSIO 115
	Plitidepsin	SARS-CoV-2	*Aplidium albicans*
	Myticin C	HSV-1/HSV-2	*Mytilus galloprovincialis*

5.4 MARINE AVP TREATMENTS ADOPTED FOR HUMAN AND ANIMAL WELL-BEING

Numerous studies have confirmed the effectiveness of marine AVPs functional against viruses infecting humans as drugs. These peptides were generated from several marine sources and act by different modes against numerous viruses (Table 5.2).

TABLE 5.2
Marine Peptides with Antiviral Bio-Effectiveness from Diverse Sources, Their Function, and Precise Modes of Action Against Viruses Infecting Humans

Peptide Name	Peptide Isolated from	Classes	Modes of Action	Functional Against	References
Didemnins A and B	*Trididemnum*	Tunicate	Constrains DNA/RNA (viral) duplication	Parainfluenza HSV Dengue virus	Rinehart et al. (1981), Aneiros and Garateix (2004), Xu et al. (2012)
Pa-MAP	*Pleuronectes americanus*	Fish	Interferes with the envelope of HSV	HSV	Migliolo et al. (2012)
Microspinosamide	*Sidonops microspinosa*	Sponge	Hinders cytopathic effectiveness	HIV	Rashid et al. (2000), Uzair, Mahmood, and Tabassum (2011)
Polyphemusin	*Tachypleus tridentatus*	Horseshoe crab	Interacts with CD4 and gp120	HIV	Nakashima et al. (1992), Tamamura et al. (1996)
Tachyplesin	*Tachypleus tridentatus*	Horseshoe crab	Interferes with envelope Inactivation of virion Stops fusion between virus and cell	Influenza A viruses (IAV) Vesicular stomatitis virus (VSV) HSV HIV	Yasin et al. (2000), Morimoto et al. (1991), Murakami et al. (1991)

TABLE 5.3
Marine Peptides with Antiviral Bio-Effectiveness from Diverse Sources, Their Function, and Precise Modes of Action Against Pathogenic Viruses Infecting Fishes

Peptide Name	Peptide Isolated From	Classes	Modes of Action	Functional Against	References
TH 1-5	*Oreochromis mossambicus*	Fish	Virions clumping	NNV	Chia et al. (2010)
Epinecidin-1	*Epinephelus coioides*	Fish			
cSALF	*Penaeus Monodon*	Shrimp			
Cecropin B	*Hyalophora cecropia*	Giant silkmoth (synthesized)	Replication inhibition	VHSV IHNV SHRV	Sumon et al. (2021), Chiou et al. (2002)
CF17 (synthetic analog of cecropin B)					
Pleurocidin MDPle	*Limanda limanda*	Fish	Disrupting membrane of the virus by forming toroidal pore	VHSV	Falco et al. (2009)

NNV, nervous necrosis virus; IHNV, infectious hematopoietic necrosis virus; SHRV, snakehead rhabdovirus.

Didemnin B, a depsipeptide reported in Caribbean Sea-squirt *Trididemnum solidum* was the first marine peptide drug proven clinically in humans (Xu et al. 2012). This marine peptide has mutually antiviral and significant immunosuppressive efficacy.

The cyclic depsipeptide microspinosamide was generated from marine sponge *Sidonops microspinosa* and hindered HIV-1 cytopathogenicity (Rashid et al. 2000). The first time it was announced to be of pharmacological (antiviral) potential was in 2001 in United States (Uzair, Mahmood, and Tabassum 2011).

Plitidepsin is an analog of didemnin since it comprises a pyruvic acid in place of didemnin B lactic acid. Although it has lower toxicity, it was approved only for drug grade as multiple myeloma and acute lymphoblastic leukemia therapy in 2004 by FDA (Falco et al. 2022). The commercial name of the drug is Aplidin®. It causes eEF1A inhibition, eukaryotic-translation-elongation-factor 1 α1, a mechanism by which plitidepsin could perform effective antiviral action against SARS-CoV-2; thus, it was introduced to phase 3 clinical assessment to treat COVID-19 (Muñoz-Alonso et al. 2009, White et al. 2021).

As a general rule, the antiviral action mechanisms of marine AVPs against viruses infecting humans include: (1) inhibiting attachment of virus to host cell *via* peptide binding to targets of virus on surface of host cell, (2) hindering viral fusion, (3) inhibiting virus spread intracellularly by suppressing the expression of viral genes, and (4) immune modulatory actions (Mulder et al. 2013).

Few reports stated the impending application of marine AVPs against viruses infecting animals. The most important use of these peptides is in aquaculture. Diseases caused by viruses are of major concern to the fish aquaculture industry, and research has already found capable marine peptides active against various pathogenic viruses initiating diseases in fishes such as rhabdoviruses (e.g., VHSV) and NNV (Nodaviridae) (Table 5.3). Rhabdoviruses provoked serious illnesses in farmed fish leading to utmost fatalities in aquaculture production.

5.5 AVPs DERIVED FROM MARINE FUNGI AND BACTERIA

Peptides derived from marine fungi exhibited remarkable antiviral activities, with less side effects and minimal human toxicity compared to synthetic drugs (Youssef et al. 2019). For example, Asperterrestide A (cyclic tetrapeptide) derived from the marine *Aspergillus terreus* SCSGAF0162 accomplished antiviral action on H1N1 and H3N2 (Influenza A virus) at IC_{50} of 15 and 8.1 µM, in that order (He et al. 2013, Abdelmohsen et al. 2017).

Furthermore, four peptides with antiviral bio-effectiveness (a cyclic pentapeptide in addition to three linear peptides), named aspergillipeptides D–G were extracted from a culture broth of *Aspergillus* sp. SCSIO 41501 (aquatic) and presented antiviral action against HSV-1 (Ma et al. 2017). Additionally, Simplicilliumtide J (a novel antiviral cyclic hexapeptide) and analogs Verlamelins A and B (cyclic lipodepsipeptides) were attained from *Simplicillium obclavatum* EIODSF 020 (a fungus found in deepsea) and performed powerful anti-HSV-1 efficacy (Liang et al. 2017).

In 2022, Sureram et al. announced the extraction of peptide A-3302-B from marine *Micromonospora* sp. MAG 9-7 (bacterial strain) and declared its efficiency against human cytomegalovirus and HSV-1/HSV-2 (Sureram et al. 2022). The team exhaustively inspected A-3302-B mode of action. The peptide actually has a unique antiviral action instead of affecting viral proteins expression as usual; it repressed the replicative cycle of HSV-2 at its late events *via* stopping HSV-2 egress from infected host cells (Sureram et al. 2022). They suggested its development as a therapy to diminish HSV-2 transmission.

5.6 AVPs DERIVED FROM MARINE INVERTEBRATES

The most promising AVPs from sponges were mirabamides A–D and mirabamides E–H derived from the marine sponges *Siliquariaspongia mirabilis* and *Stelletta clavosa*, respectively (Plaza et al. 2007, Lu et al. 2011).

Isolation of AVPs was also described from mollusks. For example, Myticin class C (Myt C) derived from the mussel *Mytilus galloprovincialis* was stated to have antiviral bio-efficiency against VHSV infection in CHSE-214 (Chinook salmon embryo) cells (Balseiro et al. 2011).

In case of crustaceans, most research focused on the isolation of AVPs against White spot syndrome virus (WSSV). This is because of the fact that WSSV causes a severe epidemic disease in farmed animals (Zheng, Xu, and Liu 2019). It was found that AVPs involved in crustacean immune response have an anti-WSSV effect (Zheng, Xu, and Liu 2019). The antiviral potential of Scygonadin (an AVP from the crab *Scylla paramamosain*) on WSSV was investigated and confirmed (Peng et al. 2012). Additionally, crustins, a family of crustacean peptides with antimicrobial bio-efficiency, caused a decrease in WSSV virion load in *in vitro* assessment (Du et al. 2019). LvHcL48, a shrimp hemocyanin-generated AVP, was also reported to exhibit anti-WSSV activity *via* attenuating wsv069 and wsv421 (WSSV genes) transcription both in *in vitro* and *in vivo* experiments (Zhan et al. 2019). Moreover, LvHcL48 has the ability to directly interfere with the envelope protein VP28 of WSSV (Zhan et al. 2019).

Tachyplesin I derivative from the hemocytes of the horseshoe crab (marine arthropod) *Tachypleus tridentatus* was stated to have an antiviral action against nodavirus and iridovirus (Xie, Wei, and Qin 2016).

Mollamides E and F (two novel cyclopeptides) were generated from the methanol extract of the colonial tunicate *Didemnum molle* and found to display anti-HIV-1 function (Lu et al. 2012). Mollamide F was able to reduce HIV-1 viral replication while mollamide E was active on HIV-1 integrase (Riccio et al. 2020).

A group of anti-HIV peptides were identified from marine sponges in the time period between 1990 and 2010 as follows (Falco et al. 2022): Callipeltin A from *Callipelta* sp. (Zampella et al. 1996), Neamphamide A from *Neamphius huxleyi* (Oku et al. 2004), Theopapuamides A–D from *T. swinhoei* and *S. mirabilis* (Ratnayake et al. 2006), Homophymine A from *Homophymia* sp. (Zampella et al. 2008), Papuamide A–D from *T. swinhoei* and *S. mirabilis* (Andjelic, Planelles, and Barrows 2008), Celebeside A from *Siliquariaspongia mirabilis* (Plaza et al. 2009), and Koshikamides F–H from *T. swinhoei* and *Theonella cupola* (Plaza et al. 2010).

5.7 AVPs DERIVED FROM MARINE VERTEBRATES

Research in the last decade described a few investigations addressing antiviral effects performed by marine vertebrates. The main focus of this research was on AVPs derived from fishes (Riccio et al. 2020). The first encounter between virus particles and host cells initiates in the fish mucosal tissue and then numerous defense machineries against virus evolve at the mucus layer. These defense machineries include mucosal antibodies, mucus shedding and reproduction, and our focus AVPs (Rombout, Yang, and Kiron 2014, Collet 2014).

The most imperative example of these NNV from fishes is the antifreeze peptide Pa-MAP generated from the marine polar fish *Pleuronectes americanus*, which acts against HSV-1 and HSV-2 (Migliolo et al. 2012, Vilas Boas et al. 2017). Another illustration is EcDefensin (β-defensin) resultant from the liver of *Epinephelus coioides* (orange-spotted grouper), which was described to constrain infection as well as replication of Viral NNV and Singapore Grouper Iridovirus both of which are marine fish viruses (Guo et al. 2012). This suggests the would-be application of this peptide as an antiviral mediator in the fisheries industry. Providentially, most microbial pathogens are still susceptible to defensins as antimicrobial means (Cheung et al. 2014).

Zhang et al. acknowledged two hepcidins (cysteine-rich antimicrobial peptides) in *Scophthalmus maximus* (flatfish belongs to the family Scophthalmidae). In addition to their bactericidal bio-effectiveness on Gram-negative/positive bacteria, both peptides presented significantly enhanced resistance against viral infection in *in vivo* assessment (Zhang et al. 2014). Another two hepcidins (SA-hepcidin1 and SA-hepcidin2) were derived from *Scatophagus argus* (spotted scat belongs to the family Scatophagidae), and SA-hepcidin2 was stated to be active against Largemouth bass *Micropterus salmoides* reovirus (MsReV) and rhabdovirus *Siniperca chuatsi* rhabdovirus, but with no such action by SA-hepcidin1 (Gui et al. 2016).

Whether all of the above-mentioned and other marine peptides with antiviral bio-effectiveness are a new hope for antiviral therapy or not will be answered in the following points from 10 to 14.

5.8 IMPORTANT APPLICATIONS FOR MARINE PEPTIDES

5.8.1 Marine Peptides Acting Against Influenza Virus

Influenza viruses are one of the most important resistant viruses that have concerned scientists. By nature, they create wide-ranging changeability of strains and have speedy evolution; predominantly, the IAV is the best example of these behaviors. Epidemics caused by these viruses have occurred over many years starting from Spanish flu (in 1918 caused by H1N1), Asian flu (in 1957 caused by H2N2), Hong Kong flu (in 1968 caused by H3N2), South-East Asia influenza (in 1998 caused by A/H5N1), and the Belgium and Netherlands influenza (in 2003 caused by A/H7N7).

Till now, influenza drugs such as oseltamivir and zanamivir that hinder the virulence factor neuraminidase or adamantane medicines do not have the capability to diminish adverse effects (Gogineni, Schinazi, and Hamann 2015). Thus, asperterrestide A represents a modern anticipative less toxic inhibitor against influenza viruses (He et al. 2013, Abdelmohsen et al. 2017).

As declared previously in the context of this review, asperterrestide A was obtained from marine fungus. It has a rare moiety of 3-hydroxy-N-methyl-Phe. Cytopathic check of this peptide revealed its action against H1N1 (M2-resistant strain A/W/33) in addition to H3N2 (M2-sensitive strain A/Hong Kong/8/68) (He et al. 2013). Most likely, the reason for this action was because of 3-hydroxy-N-methyl-Phe presence in the peptide (He et al. 2013), but still this is a rather unsatisfactory explanation for the mechanism by which asperterrestide A acts on influenza viruses.

5.8.2 Marine Peptides Acting Against HIV

HIV is a possibly fatal virus that directly hits the immune system. Anti-HIV drugs already in the market face various restraints, counting side effects, resistance, and toxicity (Singh and Bodiwala 2010). Yet, the generation of one (a lectin; griffithsin from red alga) of anti-HIV molecules from marine biomass introduced to clinical estimation unveils the exceptional possibility of aquatic natural compounds, emphasizing the unceasing necessity of advancement in anti-HIV medications derivative from living beings (Lee 2019).

Three examples for anti-HIV peptides from marine sponges are already mentioned above along with their mode of action; mollamides F, stellettapeptins A and B, and mirabamides E–H. Interestingly, mirabamides were obtained from sponges that have diverse phylogeny, which proposes that mirabamides may be derivatives of microbes associated with the sponges (Lu et al. 2011). Papuamides are another model for HIV-1 entry inhibitor (Andjelic, Planelles, and Barrows 2008).

Mirabamides F and G exhibited dual-fold boost in their anti-HIV action because of having an unmatched moiety (2-amino-2-butenoic acid) in place of threonine (Plaza et al. 2007, Lu et al. 2011). Hydrophobicity is also an effective factor in this anti-HIV action potency of mirabamides (Lu et al. 2011).

On the other hand, Stellettapeptins enclose noncanonical amino acid; 3-hydroxy-6, 8-dimethylnon-4-(Z)-enoic acid that has a significant impact on its antiviral action (Shin et al. 2015). As suggested in the case of mirabamides, these peptides are thought to be derived from bacteria, as they resemble nonribosomal peptides features and differ phylogenetically from papuamides and callipeptines came from sponges (Andjelic, Planelles, and Barrows 2008, Plaza et al. 2010).

A conventional campaign of anti-HIV drug screening and discovery guided by bioactivity that was organized by United States and Papua New Guinea witnessed the discovery of mollamides E and F and molleurea. The former is thiazoline cyclopeptides (Lu et al. 2012, Riccio et al. 2020), while the last is Tris-phenethyl urea (Lu et al. 2012). This was the first time to entirely portray these peptides as isolated marine bioactive molecules based on natural products.

In 2000, Mitchell et al. found that dolastatin 3 (that resembles mollamides structure) had anti-HIV integrase efficacy (Mitchell et al. 2000). Malformin C (cyclopentapeptide) was rediscovered as an anti-HIV peptide (robustly hindering receptor CCRC5 in TMZ-bl cells infected by HIV-1 SF162) obtained from marine *Aspergillus niger* SCSIO Jscw6F30 (Zhou et al. 2016). Remarkably, this activity of malformin C was comparable to abacavir (HIV drug).

5.8.3 Marine Peptides Acting Against HCV

In the last three decades, HCV carried on to be a severe load on global health, resulting in cirrhosis, cancer, and liver collapse along with the unavailability of vaccine. What is more, HCV remedy even advanced ones are still expensive and with critical adverse events. Consequently, innovative drugs against HCV are totally welcomed.

Lately, marine anti-HCV peptides, comprising valinomycin, cyclo (L-Tyr-L-Pro) diketopiperazine (protease inhibitor from *Aspergillus versicolor* associated with *Spongia officinalis*), streptodepsi peptide SV21, and streptodepsi peptide P11A (from *Streptomyces* sp. SV21 living with sea cucumber) were found (Ahmed et al. 2017, Wibowo et al. 2021).

Streptodepsipeptide SV21, streptodepsipeptide P11A, and valinomycin proved to have comparable efficacy against HCV to that of epigallocatechin gallate (Wibowo et al. 2021). Knowing that epigallocatechin gallate in the first place is a robust inhibitor for HCV entry/cell surface attachment (Calland et al. 2012). Wibowo et al. stated that streptodepsipeptide SV21, streptodepsipeptide P11A, and valinomycin efficiency are dependent on the ring system symmetry (Wibowo et al. 2021).

5.8.4 Marine Peptides Acting Against HSV

On January 2024, Hao et al. published an article, in which they reviewed compounds from marine sources and their action against HSV (Hao et al. 2024). The most noteworthy example of these peptides they cited was peptide A-3302-B, which is already discussed above with its explicit mode of action (Sureram et al. 2022).

As previously cited in the context of this review, Didemnins were the earliest of these marine anti-HSV peptides (Rinehart et al. 1981). Simpliciliumtides J is another model for marine anti-HSV peptides (active against HSV-1/106 ACV resistant, HSV-1/F, and HSV-1/153 ACV resistant) that was obtained from marine fungi (Liang et al. 2017, Wang et al. 2020). Halovirs A generated from marine Scytalidium straightly disable HSV-1 virions (Rowley 2003).

5.8.5 Marine Peptides against SARS-CoV-2

At a time when scientists were searching for a way to confront the catastrophic situation of the COVID-19 pandemic that surprised the entire world, AVPs were not absent from the scene. Research using AVPs has been spread aiming to find a solution for COVID-19 pandemic (Mahendran et al. 2020, Raihan et al. 2021, Tonk, Růžek, and Vilcinskas 2021, Heydari et al. 2021, Mousavi Maleki, Rostamian, and Madanchi 2021, Mahmud et al. 2021). This research resulted in finding plitidepsin (principally multiple myeloma drugs from *Aplidium albicans*, a marine tunicate).

This cyclodepsipeptide was totally synthesized (Alonso-Álvarez et al. 2017). *In silico* approach could anticipate binding of this peptide in a more efficient way with SARS-CoV-2 key protease than remdesivir, hydroxychloroquine, and lopinavir (Kalhotra et al. 2021, Vishvakarma et al. 2022). Its *in vitro* anti-SARS-CoV-2 efficiency was confirmed by White et al. (2021). They stated that *in vitro* activity of plitidepsin was superior to that of remdesivir along with less toxicity (White et al. 2021). They found another mode of anti-SARS-CoV-2 action for plitidepsin; constraining eEF1A, which is a crucial factor of host involved in the replication of the virus.

In vivo murine investigation revealed plitidepsin ability to decrease SARS-CoV-2 replication and weaken the inflammation of lungs. Therefore, PharmaMar tried plitidepsin clinically at phase I/II, then it was introduced to phase II/III trials to treat COVID-19 (White et al. 2021, Papapanou et al. 2021). The most significant investigations were those stated that plitidepsin can act proficiently against all SARS-CoV-2 variants in animals and can treat persistent SARS-CoV-2 replication in a COVID-19 patient who has chronic lymphocytic leukemia (Varona et al. 2022, Guisado-Vasco et al. 2022).

5.9 FUTURE PERCEPTION

Overall, every organism living in marine habitats or microorganism (associated or free) may be a producer for AVPs active (moderate-to-significant efficacy) against viruses particularly the aquaculture humans' and animals' viruses. Between 2011 and 2021, twenty-five compounds were reported, mostly from fungi, and are generated chiefly from nonribosomal peptides. Contrariwise, myticin C, divamide A, and Pa-MAP are ribosomal (semi-)synthetic deviations underwent alterations in their structure.

Captivatingly, practically all of them are cyclic (with enhanced permeability into cells and improved efficacy and resistance to proteases compared to linear ones), extending from cyclic dipeptide diketopiperazine to cyclodepsipeptides (Zorzi, Deyle, and Heinis 2017, Abdalla and McGaw 2018).

All of these favorable characteristics make them an appealing applicant for therapeutics. Additionally, there are chances to even enhance their efficacy *via* alterations in structure and chemistry, and stable D-amino acids application. Yet, we should take into consideration their shortcomings like cell permeability and bioavailability. Nanoencapsulation and cell-penetrating peptides are among solutions (Bechara and Sagan 2013).

While more than a few marine AVPs are presently under consideration preclinically and clinically, or have even reached the market (Macedo et al. 2021), not a single peptide is an antiviral top-drug up till now. And worse, most peptides discussed here did not reach the stages of evaluation *in vivo* in animals or human trials. However, the change of purpose occurs in case of plitidepsin from anticancer to anti-SARS-CoV-2 drug under phase II/III clinical trial is encouraging.

Acceleration can also occur through synthetic biology and (meta) genomics. Fermentation may substitute chemical synthesis if marine microbes are the source. Their adverse effects and less toxicity have to be additionally studied. Seemingly, marine AVPs can lead in antiviral therapy in the nearby future.

There is a promising AVP for treating serious aquaculture viruses like hematopoietic necrosis virus (IHNV). VHSV, SHRV, IPNV, and NNV were mentioned in this chapter.

5.10 CONCLUSION

There is a real need for a safe alternative to antibiotics. That is mainly because of the elevation of the antimicrobial and the AR and the accumulation of their related structures in the nature. Marine AVP is a treasure we should mine for to have a better control over the several viral outbursts we are having since many years. They have both explicit structures and mode of action that deserve a real chance for developing into leading drugs. The pharmaceutical sector besides public funding has lesser interest in antiviral drugs than in antibacterial ones, and the evidence is the fact that there are just 118 antivirals until 2020 permitted by FDA and 52 of them directed against HIV1. Another evidence of this is what happened recently with the COVID-19 pandemic and the confusion and lack of clarity of vision in how to deal with this epidemic, which proves indisputably that we have not yet matured in dealing with new viral diseases.

Existing treatments are consequently commonly costly, infrequent, and necessitate diagnosis to be early to function well. On the other hand, antivirals for chronic infections are recommended long term and sometimes for life, a problem that results in the patient not sustaining its use. Thus, we have to deal better and give more interest for viral infections than we are now. Sustainable research to find and develop antivirals is the key. Marine AVPs that proved more efficient than already-existing drugs are highlighting the way we should walk into. Peptide from marine resources can play a bigger role in marine aquaculture.

REFERENCES

Abdalla, M. A., and L. J. McGaw. 2018. Natural cyclic peptides as an attractive modality for therapeutics: A mini review. *Molecules* 23 (8):2080. doi: 10.3390/molecules23082080.

Abdelmohsen, U. R., S. Balasubramanian, T. A. Oelschlaeger, T. Grkovic, N. B. Pham, R. J. Quinn, and U. Hentschel. 2017. Potential of marine natural products against drug-resistant fungal, viral, and parasitic infections. *The Lancet Infectious Diseases* 17 (2):e30–e41. doi: 10.1016/s1473-3099(16)30323-1.

Aboudy, Y., E. Mendelson, I. Shalit, R. Bessalle, and M. Fridkin. 1994. Activity of two synthetic amphiphilic peptides and magainin-2 against herpes simplex virus types 1 and 2. *International Journal of Peptide and Protein Research* 43 (6):573–582. doi: 10.1111/j.1399-3011.1994.tb00559.x.

Ahmed, E. F., M. E. Rateb, L. T. A. El-Kassem, and U. W. Hawas. 2017. Anti-HCV protease of diketopiperazines produced by the Red Sea sponge-associated fungus *Aspergillus versicolor*. *Applied Biochemistry and Microbiology* 53 (1):101–106. doi: 10.1134/s0003683817010021.

Alonso-Álvarez, S., E. Pardal, D. Sánchez-Nieto, M. Navarro, M. D. Caballero, M. V. Mateos, and A. Martín. 2017. Plitidepsin: Design, development, and potential place in therapy. *Drug Design, Development and Therapy* 11:253–264. doi: 10.2147/DDDT.S94165.

Amara, A. A. 2011. Opportunistic pathogens and their biofilm food for thought in science against microbial pathogens: Communicating current research and technology advances: Edit by Mendez-Vilas. *FORMATEX Microbiology Series* 3 (2):813–825.

Amara, A. A. 2013. Pharmaceutical and industrial protein engineering: Where we are? *Pakistan Journal of Pharmaceutical Sciences* 26 (1):217–232.

Amara, A. A. 2015. An overview of the molecular and cellular interactions of some bioactive compounds. In Vijai, K. G., G. T. Maria, A. O'Donovan, and M. Lohani (Eds) *Biotechnology of Bioactive Compounds: Sources and Applications*, pp. 525–554. Wiley Blackwell, Hoboken, NJ.

Amara, A. A. 2019. An overview about our understanding of protein activity, specificity, structure and function. In Gupta, V. K., and M. G. Tuohy (Eds) *Microbial Catalysts*, pp. 20–34, Vol. 1. Nova Publishers, Enfield, NH.

Amara, A. A. 2022a. The role of divalent cations in antibiotic sensitivity: A molecular aspect. In Gupta, V. K. (Eds) *Biomolecules from Natural Sources: Advances and Applications*, pp. 252–277. Wiley, Hoboken, NJ.

Amara, A. A. 2023. Fermented food microbes and their preservative byproducts. In Gupta, A., and R. Prasad (Eds) *Antimicrobials in Food Science and Technology*, pp. 30–71. CRC Press, Boca Raton, FL.
Amara, A. A. 2024. Phags and their role in natural and directed microbial control. In Varjani, S., I. Haq, A. Pandey, V. K. Gupta, and X. T. Bui (Eds) *Waste Management in Climate Change and Sustainability Perspectives*, p. 178. CRC Press, Boca Raton, FL.
Amara, A. A., N. A. El-Baky, M. M. Salem-Bekhit, K. M. A. Zoheir, and M. M. M. Ahmed. 2024. Antibiotic resistance from theory to facts. In Gupta, A., R. P. Gupta, and A. R. Prasad (Eds) *Antimicrobial Resistance in Agriculture and Its Consequences*, p. 204. CRC Press, Boca Raton, FL.
Amara, A. A., M. H. El-Masry, and H. H. Bogdady. 2008. Plant crude extracts could be the solution: Extracts showing in vivo antitumorigenic activity. *Pakistan Journal of Pharmaceutical Sciences* 21 (2):159–171.
Amara, A. A., M. M. Salem-Bekhit, and F. K. Alanazi. 2013. Sponge-like: A new protocol for preparing bacterial ghosts. *The Scientific World Journal* 2013:545741. doi: 10.1155/2013/545741.
Amara, A. A., S. R. Salem, and M. S. A. Shabeb. 2009. The possibility to use bacterial protease and lipase as biodetergent. *Global Journal of Biotechnology & Biochemistry* 4 (2):104–114.
Amara, A. A., and A. Shib. 2015. A hypothesis about the variation within β-lactamases: An epigenetic-like mechanism. *IIOABJ* 6 (1):1–11.
Amara, A. A., A. M. Shibl, and A. F. Tawfik. 2012. Deeper in the antibiotics paradigm: Microbes react individually and collectively for spreading the resistance. *African Journal of Microbiology Research* 6 (9):1224. doi: 10.5897/ajmr11.1224.
Amara, A. A. A. F. 2022b. Natural polymer types and applications. In Gupta, V. K., S. D. Sarker, M. Sharma, M. E. Pirovani, Z. Usmani, and C. Jayabaskaran (Eds) *Biomolecules from Natural Sources*, pp. 31–81. Wiley Blackwell, Hoboken, NJ.
Andjelic, C. D., V. Planelles, and L. R. Barrows. 2008. Characterizing the anti-HIV activity of papuamide A. *Marine Drugs* 6 (4):528–549. doi: 10.3390/md20080027.
Aneiros, A., and A. Garateix. 2004. Bioactive peptides from marine sources: Pharmacological properties and isolation procedures. *Journal of Chromatography B* 803 (1):41–53. doi: 10.1016/j.jchromb.2003.11.005.
Badani, H., R. F. Garry, and W. C. Wimley. 2014. Peptide entry inhibitors of enveloped viruses: The importance of interfacial hydrophobicity. *Biochimica et Biophysica Acta* 1838 (9):2180–2197. doi: 10.1016/j.bbamem.2014.04.015.
Balseiro, P., A. Falcó, A. Romero, S. Dios, A. Martínez-López, A. Figueras, A. Estepa, and B. Novoa. 2011. Mytilus galloprovincialis myticin C: A chemotactic molecule with antiviral activity and immunoregulatory properties. *PLoS One* 6 (8):e23140. doi: 10.1371/journal.pone.0023140.
Bechara, C., and S. Sagan. 2013. Cell-penetrating peptides: 20 years later, where do we stand? *FEBS Letters* 587 (12):1693–1702. doi: 10.1016/j.febslet.2013.04.031.
Bulet, P., R. Stöcklin, and L. Menin. 2004. Anti-microbial peptides: From invertebrates to vertebrates. *Immunological Reviews* 198 (1):169–184. doi: 10.1111/j.0105-2896.2004.0124.x.
Cabello, F. C. 2004. Antibióticos y acuicultura en Chile: Consecuencias para la salud humana y animal. *Revista médica de Chile* 132 (8):1001–1006. doi: 10.4067/s0034-98872004000800014.
Cai, L., J.-J. Cai, H.-P. Liu, D.-Q. Fan, H. Peng, and K.-J. Wang. 2012. Recombinant medaka (*Oryzias melastigmus*) pro-hepcidin: Multifunctional characterization. *Comparative Biochemistry and Physiology Part B: Biochemistry and Molecular Biology* 161 (2):140–147. doi: 10.1016/j.cbpb.2011.10.006.
Calland, N., A. Albecka, S. Belouzard, C. Wychowski, G. Duverlie, V. Descamps, D. Hober, J. Dubuisson, Y. Rouillé, and K. Séron. 2012. (−)-Epigallocatechin- 3 -gallate is a new inhibitor of hepatitis C virus entry. *Hepatology* 55 (3):720–729. doi: 10.1002/hep.24803.
Cappello, E., and P. Nieri. 2021. From life in the sea to the clinic: The marine drugs approved and under clinical trial. *Life* 11 (12):1390. doi: 10.3390/life11121390.
Carriel-Gomes, M. C., J. M. Kratz, M. A. Barracco, E. Bachére, C. R. M. Barardi, and C. M. O. Simões. 2007. *In vitro* antiviral activity of antimicrobial peptides against herpes simplex virus 1, adenovirus, and rotavirus. *Memórias do Instituto Oswaldo Cruz* 102 (4):469–472. doi: 10.1590/s0074-02762007005000028.
Chen, Z. C., R. F. White, J. F. Antoniw, and Q. Lin. 1991. Effect of pokeweed antiviral protein (PAP) on the infection of plant viruses. *Plant Pathology* 40 (4):612–620. doi: 10.1111/j.1365-3059.1991.tb02426.x.
Cheung, R. C. F., J. H. Wong, W. L. Pan, Y. S. Chan, C. M. Yin, X. L. Dan, H. X. Wang, E. F. Fang, S. K. Lam, P. H. K. Ngai, L. X. Xia, F. Liu, X. Y. Ye, G. Q. Zhang, Q. H. Liu, O. Sha, P. Lin, C. Ki, A. A. Bekhit, A. E. D. Bekhit, D. C. C. Wan, X. J. Ye, J. Xia, and T. B. Ng. 2014. Antifungal and antiviral products of marine organisms. *Applied Microbiology and Biotechnology* 98 (8):3475–3494. doi: 10.1007/s00253-014-5575-0.
Chia, T.-J., Y.-C. Wu, J.-Y. Chen, and S.-C. Chi. 2010. Antimicrobial peptides (AMP) with antiviral activity against fish nodavirus. *Fish & Shellfish Immunology* 28 (3):434–439. doi: 10.1016/j.fsi.2009.11.020.

Chiou, P. P., C.-M. Lin, L. Perez, and T. T. Chen. 2002. Effect of cecropin B and a Synthetic analogue on propagation of fish viruses *in vitro*. *Marine Biotechnology* 4 (3):294–302. doi: 10.1007/s10126-002-0021-1.

Chowdhury, S. M., S. A. Talukder, A. M. Khan, N. Afrin, M. A. Ali, R. Islam, R. Parves, A. A. Mamun, M. A. Sufian, M. N. Hossain, M. A. Hossain, and M. A. Halim. 2020. Antiviral peptides as promising therapeutics against SARS-CoV-2. *The Journal of Physical Chemistry B* 124 (44):9785–9792. doi: 10.1021/acs.jpcb.0c05621.

Collet, B. 2014. Innate immune responses of salmonid fish to viral infections. *Developmental & Comparative Immunology* 43 (2):160–173. doi: 10.1016/j.dci.2013.08.017.

Dahiya, R., S. Singh, A. Sharma, S. V. Chennupati, and S. Maharaj. 2016. First total synthesis and biological screening of a proline-rich cyclopeptide from a Caribbean marine sponge. *Marine Drugs* 14 (12):228. doi: 10.3390/md14120228.

Destoumieux-Garzón, D., R. D. Rosa, P. Schmitt, C. Barreto, J. Vidal-Dupiol, G. Mitta, Y. Gueguen, and E. Bachère. 2016. Antimicrobial peptides in marine invertebrate health and disease. *Philosophical Transactions of the Royal Society of London. Series B, Biological Sciences* 371 (1695):20150300. doi: 10.1098/rstb.2015.0300.

Ding, C. Y. G., J. F. M. Ong, H. C. Goh, C. R. Coffill, and L. T. Tan. 2018. Benderamide A, a cyclic depsipeptide from a Singapore collection of marine cyanobacterium cf. Lyngbya sp. *Marine Drugs* 16 (11):409. doi: 10.3390/md16110409.

Du, F.-Y., X.-M. Li, P. Zhang, C.-S. Li, and B.-G. Wang. 2014. Cyclodepsipeptides and other O-containing heterocyclic metabolites from Beauveria felina EN-135, a marine-derived entomopathogenic fungus. *Marine Drugs* 12 (5):2816–2826. doi: 10.3390/md12052816.

Du, Z.-Q., Y. Wang, H.-Y. Ma, X.-L. Shen, K. Wang, J. Du, X.-D. Yu, W.-H. Fang, and X.-C. Li. 2019. A new crustin homologue (SpCrus6) involved in the antimicrobial and antiviral innate immunity in mud crab, *Scylla paramamosain*. *Fish & Shellfish Immunology* 84:733–743. doi: 10.1016/j.fsi.2018.10.072.

El-Baky, N., and A. A. Amara. 2022. Depending on epitope profile of COVID-19 mRNA vaccine recipients: Are they more efficient against the arising viral variants? An opinion article. *Frontiers in Medicine* 9:903876.

El-Baky, N. A., and A. A. A. F. Amara. 2023. Natural antimicrobial therapeutic peptides: Milk lactoferricin and *Spirulina platensis* peptides. In Gupta, A., and R. Prasad (Eds) *Antimicrobials in Pharmaceutical and Medicinal Research*, pp. 19–48. CRC Press, Boca Raton, FL.

El-Baky, N. A., A. A. A. F. Amara, and E. M. Redwan. 2023a. Nutraceutical and therapeutic importance of clots and their metabolites. In Inamuddin, M. A., T. Altalhi, and J. N. Cruz (Eds) *Nutraceuticals: Sources, Processing Methods, Properties, and Applications*, pp. 241–268. Elsevier, Amsterdam, Netherlands.

El-Baky, N. A., A. A. Amara, and E. M. Redwan. 2023b. HLA-I and HLA-II peptidomes of SARS-CoV-2: A review. *Vaccines* 11 (3):548. doi: 10.3390/vaccines11030548.

El-Baky, N. A., A. A. Amara, V. N. Uversky, and E. M. Redwan. 2023. Intrinsic factors behind long COVID: III. Persistence of SARS-CoV-2 and its components. *Journal of Cellular Biochemistry* 125 (1):30514. doi: 10.1002/jcb.30514.

Fakih, T. M. 2020. Dermaseptin-based antiviral peptides to prevent COVID-19 through *in silico* molecular docking studies against SARS-CoV-2 spike protein. *Pharmaceutical Sciences and Research* 7 (4):65–70. doi: 10.7454/psr.v7i4.1079.

Falco, A., V. Chico, L. Marroquí, L. Perez, J. M. Coll, and A. Estepa. 2008. Expression and antiviral activity of a *β*-defensin-like peptide identified in the rainbow trout (*Oncorhynchus mykiss*) EST sequences. *Molecular Immunology* 45 (3):757–765. doi: 10.1016/j.molimm.2007.06.358.

Falco, A., V. Mas, C. Tafalla, L. Perez, J. M. Coll, and A. Estepa. 2007. Dual antiviral activity of human alpha-defensin-1 against viral haemorrhagic septicaemia rhabdovirus (VHSV): Inactivation of virus particles and induction of a type I interferon-related response. *Antiviral Research* 76 (2):111–123. doi: 10.1016/j.antiviral.2007.06.006.

Falco, A., M. Ortega-Villaizan, V. Chico, I. Brocal, L. Perez, J. Coll, and A. Estepa. 2009. Antimicrobial peptides as model molecules for the development of novel antiviral agents in aquaculture. *Mini-Reviews in Medicinal Chemistry* 9 (10):1159–1164. doi: 10.2174/138955709789055171.

Falco, A., M. Adamek, P. Pereiro, D. Hoole, J. A. Encinar, B. Novoa, and R. Mallavia. 2022. The immune system of marine organisms as source for drugs against infectious diseases. *Marine Drugs* 20 (6):363. doi: 10.3390/md20060363.

Field, H. J., and E. De Clercq. 2004. Antiviral drugs-a short history of their discovery and development. *Microbiology Today* 31:58–61.

Furtado, D. N., S. D. Todorov, M. Landgraf, M. T. Destro, and B. D. Franco. 2015. Bacteriocinogenic *Lactococcus lactis* subsp. lactis DF04Mi isolated from goat milk: Application in the control of Listeria monocytogenes in fresh Minas-type goat cheese. *Brazilian Journal of Microbiology* 46 (1):201–206. doi: 10.1590/s1517-838246120130761.

Gogineni, V., R. F. Schinazi, and M. T. Hamann. 2015. Role of marine natural products in the genesis of antiviral agents. *Chemical Reviews* 115 (18):9655–9706. doi: 10.1021/cr4006318.

Grosso, C., P. Valentão, F. Ferreres, and P. B. Andrade. 2015. Alternative and efficient extraction methods for marine-derived compounds. *Marine Drugs* 13 (5):3182–3230. doi: 10.3390/md13053182.

Gui, L., P. Zhang, Q. Zhang, and J. Zhang. 2016. Two hepcidins from spotted scat (*Scatophagus argus*) possess antibacterial and antiviral functions in vitro. *Fish & Shellfish Immunology* 50:191–199. doi: 10.1016/j.fsi.2016.01.038.

Guisado-Vasco, P., M. M. Carralón-González, J. Aguareles-Gorines, E. M. Martí-Ballesteros, M. D. Sánchez-Manzano, D. Carnevali-Ruiz, M. García-Coca, R. Barrena-Puertas, R. García de Viedma, J. M. Luque-Pinilla, G. Sotres-Fernandez, J. M. Fernández-Sousa, X. E. Luepke-Estefan, J. A. López-Martín, and J. M. Jimeno. 2022. Plitidepsin as a successful rescue treatment for prolonged viral SARS-CoV-2 replication in a patient with previous anti-CD20 monoclonal antibody-mediated B cell depletion and chronic lymphocytic leukemia. *Journal of Hematology & Oncology* 15 (1):4. doi: 10.1186/s13045-021-01220-0.

Guo, M., J. Wei, X. Huang, Y. Huang, and Q. Qin. 2012. Antiviral effects of β-defensin derived from orange-spotted grouper (*Epinephelus coioides*). *Fish & Shellfish Immunology* 32 (5):828–838. doi: 10.1016/j.fsi.2012.02.005.

Hamed, I., F. Özogul, Y. Özogul, and J. M. Regenstein. 2015. Marine bioactive compounds and their health benefits: A review. *Comprehensive Reviews in Food Science and Food Safety* 14 (4):446–465. doi: 10.1111/1541-4337.12136.

Hao, C., Z. Xu, C. Xu, and R. Yao. 2024. Anti-herpes simplex virus activities and mechanisms of marine derived compounds. *Frontiers in Cellular and Infection Microbiology* 13:1302096. doi: 10.3389/fcimb.2023.1302096.

He, F., J. Bao, X.-Y. Zhang, Z.-C. Tu, Y.-M. Shi, and S.-H. Qi. 2013. Asperterrestide A, a cytotoxic cyclic tetrapeptide from the marine-derived fungus *Aspergillus terreus* SCSGAF0162. *Journal of Natural Products* 76 (6):1182–1186. doi: 10.1021/np300897v.

Heydari, H., R. Golmohammadi, R. Mirnejad, H. Tebyanian, M. Fasihi-Ramandi, and M. M. Moghaddam. 2021. Antiviral peptides against coronaviridae family: A review. *Peptides* 139:170526. doi: 10.1016/j.peptides.2021.170526.

Hussain, M. Z., and A. A. Amara. 2006. Case-by-case study using antibiotic-EDTA combination to control *Pseudomonas aeruginosa*. *Pakistan Journal of Pharmaceutical Sciences* 19 (3):236–243.

Ibieta, P., V. Tapia, C. Venegas, M. Hausdorf, and H. Takle. 2011. Chilean salmon farming on the horizon of sustainability: Review of the development of a highly intensive production, the ISA crisis and implemented actions to reconstruct a more sustainable aquaculture industry. In Barbara, S. (Eds) *Aquaculture and the Environment - A Shared Destiny*, pp. 215–246. InTech, London.

Iijima, N., N. Tanimoto, Y. Emoto, Y. Morita, K. Uematsu, T. Murakami, and T. Nakai. 2003. Purification and characterization of three isoforms of chrysophsin, a novel antimicrobial peptide in the gills of the red sea bream, *Chrysophrys major*. *European Journal of Biochemistry* 270 (4):675–686. doi: 10.1046/j.1432-1033.2003.03419.x.

Ireland, D. C., C. K. L. Wang, J. A. Wilson, K. R. Gustafson, and D. J. Craik. 2008. Cyclotides as natural anti-HIV agents. *Biopolymers* 90 (1):51–60. doi: 10.1002/bip.20886.

Jimenez, E. C. 2019. Bromotryptophan and its analogs in peptides from marine animals. *Protein Peptide Letters* 26 (4):251–260. doi: 10.2174/0929866526666190119170020.

Jin, J.-Y., L. Zhou, Y. Wang, Z. Li, J.-G. Zhao, Q.-Y. Zhang, and J.-F. Gui. 2010. Antibacterial and antiviral roles of a fish β-defensin expressed both in pituitary and testis. *PLoS One* 5 (12):e12883. doi: 10.1371/journal.pone.0012883.

Kalhotra, P., V. C. S. R. Chittepu, G. Osorio-Revilla, and T. Gallardo-Velazquez. 2021. Field-template, QSAR, ensemble molecular docking, and 3D-RISM solvation studies expose potential of FDA-approved marine drugs as SARS-CoVID-2 main protease inhibitors. *Molecules* 26 (4):936. doi: 10.3390/molecules26040936.

Lazcano-Perez, F., S. A. Roman-Gonzalez, N. Sanchez-Puig, and R. A. Espinosa. 2012. Bioactive peptides from marine organisms: A short overview. *Protein & Peptide Letters* 19 (7):700–707. doi: 10.2174/092986612800793208.

Lebbe, E. K. M., S. Peigneur, M. Maiti, P. Devi, S. Ravichandran, E. Lescrinier, C. Ulens, E. Waelkens, L. D'Souza, P. Herdewijn, and J. Tytgat. 2014. Structure-function elucidation of a new α-conotoxin, Lo1a, from *Conus longurionis*. *The Journal of Biological Chemistry* 289 (14):9573–9583. doi: 10.1074/jbc.M114.556175.

Lee, C. 2019. Griffithsin, a highly potent broad-spectrum antiviral lectin from red algae: From discovery to clinical application. *Marine Drugs* 17 (10):567. doi: 10.3390/md17100567.

Liang, X., X.-H. Nong, Z.-H. Huang, and S.-H. Qi. 2017. Antifungal and antiviral cyclic peptides from the deep-sea-derived fungus *Simplicillium obclavatum* EIODSF 020. *Journal of Agricultural and Food Chemistry* 65 (25):5114–5121. doi: 10.1021/acs.jafc.7b01238.

Ling, R., Y. Dai, B. Huang, W. Huang, J. Yu, X. Lu, and Y. Jiang. 2020. *In silico* design of antiviral peptides targeting the spike protein of SARS-CoV-2. *Peptides* 130:170328. doi: 10.1016/j.peptides.2020.170328.

Lordan, S., R. P. Ross, and C. Stanton. 2011. Marine bioactives as functional food ingredients: Potential to reduce the incidence of chronic diseases. *Marine Drugs* 9 (6):1056–1100. doi: 10.3390/md9061056.

Lou, Z., Y. Sun, and Z. Rao. 2014. Current progress in antiviral strategies. *Trends in Pharmacological Sciences* 35 (2):86–102. doi: 10.1016/j.tips.2013.11.006.

Lozano, I., N. F. Díaz, S. Muñoz, and C. Riquelme. 2018. Antibiotics in Chilean aquaculture: A review. In Sara, S. (Eds) *Antibiotic Use in Animals*, p. 71780. InTech, London.

Lu, Z., M. K. Harper, C. D. Pond, L. R. Barrows, C. M. Ireland, and R. M. Van Wagoner. 2012. Thiazoline peptides and a tris-phenethyl urea from Didemnum molle with anti-HIV activity. *Journal of Natural Products* 75 (8):1436–1440. doi: 10.1021/np300270p.

Lu, Z., R. M. Van Wagoner, M. K. Harper, H. L. Baker, J. N. A. Hooper, C. A. Bewley, and C. M. Ireland. 2011. Mirabamides E-H, HIV-inhibitory depsipeptides from the sponge *Stelletta clavosa*. *Journal of Natural Products* 74 (2):185–193. doi: 10.1021/np100613p.

Ma, X., X.-H. Nong, Z. Ren, J. Wang, X. Liang, L. Wang, and S.-H. Qi. 2017. Antiviral peptides from marine gorgonian-derived fungus *Aspergillus sp.* SCSIO 41501. *Tetrahedron Letters* 58 (12):1151–1155. doi: 10.1016/j.tetlet.2017.02.005.

Macedo, M. W. F. S., N. B. da Cunha, J. A. Carneiro, R. A. da Costa, S. A. de Alencar, M. H. Cardoso, O. L. Franco, and S. C. Dias. 2021. Marine organisms as a rich source of biologically active peptides. *Frontiers in Marine Science* 8:667764. doi: 10.3389/fmars.2021.667764.

Mahendran, A. S. K., Y. S. Lim, C.-M. Fang, H.-S. Loh, and C. F. Le. 2020. The potential of antiviral peptides as COVID-19 therapeutics. *Frontiers in Pharmacology* 11:575444. doi: 10.3389/fphar.2020.575444.

Mahlapuu, M., J. Håkansson, L. Ringstad, and C. Björn. 2016. Antimicrobial peptides: An emerging category of therapeutic agents. *Frontiers in Cellular and Infection Microbiology* 6:194. doi: 10.3389/fcimb.2016.00194.

Mahmud, S., S. Biswas, G. K. Paul, M. A. Mita, S. Afrose, M. R. Hasan, M. S. S. Shimu, M. A. R. Uddin, M. S. U. Shahriar Zaman, K. M. Kaderi Kibria, M. A. Khan, T. B. Emran, and M. A. Saleh. 2021. Antiviral peptides against the main protease of SARS-CoV-2: A molecular docking and dynamics study. *Arabian Journal of Chemistry* 14 (9):103315. doi: 10.1016/j.arabjc.2021.103315.

Medina-Pizzali, M. L., S. M. Hartinger, G. Salmon-Mulanovich, A. Larson, M. Riveros, and D. Mäusezahl. 2021. Antimicrobial resistance in rural settings in Latin America: A scoping review with a one health lens. *International Journal of Environmental Research and Public Health* 18 (18):9837. doi: 10.3390/ijerph18189837.

Migliolo, L., O. N. Silva, P. A. Silva, M. P. Costa, C. R. Costa, D. O. Nolasco, J. A. R. G. Barbosa, M. R. R. Silva, M. P. Bemquerer, L. M. P. Lima, M. T. V. Romanos, S. M. Freitas, B. S. Magalhães, and O. L. Franco. 2012. Structural and functional characterization of a multifunctional alanine-rich peptide analogue from *Pleuronectes americanus*. *PLoS One* 7 (10):e47047. doi: 10.1371/journal.pone.0047047.

Mishra, B., and G. Wang. 2012. The importance of amino acid composition in natural AMPs: An evolutional, structural, and functional perspective. *Frontiers in Immunology* 3:221. doi: 10.3389/fimmu.2012.00221.

Mitchell, S. S., D. J. Faulkner, K. Rubins, and F. D. Bushman. 2000. Dolastatin 3 and two novel cyclic peptides from a Palauan collection of *Lyngbya majuscula*. *Journal of Natural Products* 63 (2):279–282. doi: 10.1021/np990353f.

Morimoto, M., H. Mori, T. Otake, N. Ueba, N. Kunita, M. Niwa, T. Murakami, and S. Iwanaga. 1991. Inhibitory effect of tachyplesin I on the proliferation of human immunodeficiency virus *in vitro*. *Chemotherapy* 37 (3):206–211. doi: 10.1159/000238855.

Mousavi, M., M. Sadat, M. Rostamian, and H. Madanchi. 2021. Antimicrobial peptides and other peptide-like therapeutics as promising candidates to combat SARS-CoV-2. *Expert Review of Anti-Infective Therapy* 19 (10):1205–1217. doi: 10.1080/14787210.2021.1912593.

Mulder, K. C. L., L. A. Lima, V. J. Miranda, S. C. Dias, and O. L. Franco. 2013. Current scenario of peptide-based drugs: The key roles of cationic antitumor and antiviral peptides. *Frontiers in Microbiology* 4:321. doi: 10.3389/fmicb.2013.00321.

Muñoz-Alonso, M. J., L. González-Santiago, T. Martínez, A. Losada, C. M. Galmarini, and A. Muñoz. 2009. The mechanism of action of plitidepsin. *Current Opinion in Investigational Drugs* 10:536–542.

Murakami, T., M. Niwa, F. Tokunaga, T. Miyata, and S. Iwanaga. 1991. Direct virus inactivation of tachyplesin I and its isopeptides from horseshoe crab hemocytes. *Chemotherapy* 37 (5):327–334. doi: 10.1159/000238875.

Nakashima, H., M. Masuda, T. Murakami, Y. Koyanagi, A. Matsumoto, N. Fujii, and N. Yamamoto. 1992. Anti-human immunodeficiency virus activity of a novel synthetic peptide, T22 ([Tyr-5,12, Lys-7]polyphemusin II): A possible inhibitor of virus-cell fusion. *Antimicrobial Agents and Chemotherapy* 36 (6):1249–1255. doi: 10.1128/AAC.36.6.1249.

Oku, N., K. R. Gustafson, L. K. Cartner, J. A. Wilson, N. Shigematsu, S. Hess, L. K. Pannell, M. R. Boyd, and J. B. McMahon. 2004. Neamphamide A, a new HIV-inhibitory depsipeptide from the Papua New Guinea marine sponge *Neamphius huxleyi*. *Journal of Natural Products* 67 (8):1407–1411. doi: 10.1021/np040003f.

Papapanou, M., E. Papoutsi, T. Giannakas, and P. Katsaounou. 2021. Plitidepsin: Mechanisms and clinical profile of a promising antiviral agent against COVID-19. *Journal of Personalized Medicine* 11 (7):668. doi: 10.3390/jpm11070668.

Pavlicevic, M., E. Maestri, and M. Marmiroli. 2020. Marine bioactive peptides-an overview of generation, structure and application with a focus on food sources. *Marine Drugs* 18 (8):424. doi: 10.3390/md18080424.

Pelay-Gimeno, M., J. Tulla-Puche, and F. Albericio. 2013. Head-to-side-chain cyclodepsipeptides of marine origin. *Marine Drugs* 11 (5):1693–1717. doi: 10.3390/md11051693.

Peng, H., H.-P. Liu, B. Chen, H. Hao, and K.-J. Wang. 2012. Optimized production of scygonadin in *Pichia pastoris* and analysis of its antimicrobial and antiviral activities. *Protein Expression and Purification* 82 (1):37–44. doi: 10.1016/j.pep.2011.11.008.

Pereiro, P., A. Figueras, and B. Novoa. 2012. A novel hepcidin-like in turbot (*Scophthalmus maximus* L.) highly expressed after pathogen challenge but not after iron overload. *Fish & Shellfish Immunology* 32 (5):879–889. doi: 10.1016/j.fsi.2012.02.016.

Phyo, Y. Z., J. Ribeiro, C. Fernandes, A. Kijjoa, and M. M. M. Pinto. 2018. Marine natural peptides: Determination of absolute configuration using liquid chromatography methods and evaluation of bioactivities. *Molecules* 23 (2):306. doi: 10.3390/molecules23020306.

Plaza, A., G. Bifulco, J. L. Keffer, J. R. Lloyd, H. L. Baker, and C. A. Bewley. 2009. Celebesides A-C and theopapuamides B-D, depsipeptides from an Indonesian sponge that inhibit HIV-1 entry. *The Journal of Organic Chemistry* 74 (2):504–512. doi: 10.1021/jo802232u.

Plaza, A., G. Bifulco, M. Masullo, J. R. Lloyd, J. L. Keffer, P. L. Colin, J. N. A. Hooper, L. J. Bell, and C. A. Bewley. 2010. Mutremdamide A and koshikamides C-H, peptide inhibitors of HIV-1 entry from different Theonella species. *The Journal of Organic Chemistry* 75 (13):4344–4355. doi: 10.1021/jo100076g.

Plaza, A., E. Gustchina, H. L. Baker, M. Kelly, and C. A. Bewley. 2007. Mirabamides A–D, Depsipeptides from the Sponge *Siliquariaspongia mirabilis* That Inhibit HIV-1 Fusion. *Journal of Natural Products* 70 (11):1753–1760. doi: 10.1021/np070306k.

Primor, N., and A. T. Tu. 1980. Conformation of pardaxin, the toxin of the flatfish *Pardachirus marmoratus*. *Biochimica et Biophysica Acta (BBA) - Protein Structure* 626 (2):299–306. doi: 10.1016/0005-2795(80)90124-5.

Qian, Z.-J., J.-Y. Je, and S.-K. Kim. 2007. Antihypertensive effect of angiotensin I converting enzyme-inhibitory peptide from hydrolysates of bigeye tuna dark muscle, *Thunnus obesus*. *Journal of Agricultural and Food Chemistry* 55 (21):8398–8403. doi: 10.1021/jf0710635.

Raihan, T., M. F. Rabbee, P. Roy, S. Choudhury, K.-H. Baek, and A. K. Azad. 2021. Microbial metabolites: The emerging hotspot of antiviral compounds as potential candidates to avert viral pandemic alike COVID-19. *Frontiers in Molecular Biosciences* 8:732256. doi: 10.3389/fmolb.2021.732256.

Rashid, M. A., K. R. Gustafson, L. K. Cartner, N. Shigematsu, L. K. Pannell, and M. R. Boyd. 2000. Microspinosamide, a new HIV-inhibitory cyclic depsipeptide from the marine sponge *Sidonops microspinosa*. *Journal of Natural Products* 64 (1):117–121. doi: 10.1021/np0002379.

Ratnayake, A. S., T. S. Bugni, X. Feng, M. K. Harper, J. J. Skalicky, K. A. Mohammed, C. D. Andjelic, L. R. Barrows, and C. M. Ireland. 2006. Theopapuamide, a cyclic depsipeptide from a Papua New Guinea lithistid sponge *Theonella swinhoei*. *Journal of Natural Products* 69 (11):1582–1586. doi: 10.1021/np060229d.

Ravichandr, S., K. Kumaravel, G. Rameshkuma, and T. T. Ajith Kumar. 2010. Antimicrobial peptides from the Marine fishes. *Research Journal of Immunology* 3 (2):146–156. doi: 10.3923/rji.2010.146.156.

Riccio, G., N. Ruocco, M. Mutalipassi, M. Costantini, V. Zupo, D. Coppola, D. de Pascale, and C. Lauritano. 2020. Ten-year research update review: Antiviral activities from marine organisms. *Biomolecules* 10 (7):1007. doi: 10.3390/biom10071007.

Rinehart, K. L., J. B. Gloer, R. G. Hughes, H. E. Renis, J. P. McGovren, E. B. Swynenberg, D. A. Stringfellow, S. L. Kuentzel, and L. H. Li. 1981. Didemnins: Antiviral and antitumor depsipeptides from a Caribbean tunicate. *Science* 212 (4497): 933–935. doi: 10.1126/science.7233187.

Rombout, J. H. W. M., G. Yang, and V. Kiron. 2014. Adaptive immune responses at mucosal surfaces of teleost fish. *Fish & Shellfish Immunology* 40 (2):634–643. doi: 10.1016/j.fsi.2014.08.020.

Rowley, D. 2003. Halovirs A–E, new antiviral agents from a marine-derived fungus of the genus *Scytalidium*. *Bioorganic & Medicinal Chemistry* 11 (19):4263–4274. doi: 10.1016/s0968-0896(03)00395-x.

Shin, H. J., M. A. Rashid, L. K. Cartner, H. R. Bokesch, J. A. Wilson, J. B. McMahon, and K. R. Gustafson. 2015. Stellettapeptins A and B, HIV-inhibitory cyclic depsipeptides from the marine sponge *Stelletta sp*. *Tetrahedron Letters* 56 (28):4215–4219. doi: 10.1016/j.tetlet.2015.05.058.

Silphaduang, U., and E. J. Noga. 2001. Peptide antibiotics in mast cells of fish. *Nature* 414 (6861):268–269. doi: 10.1038/35104690.

Singh, I. P., and H. S. Bodiwala. 2010. Recent advances in anti-HIV natural products. *Natural Product Reports* 27 (12):1781. doi: 10.1039/c0np00025f.

Smith, V. J., A. P. Desbois, and E. A. Dyrynda. 2010. Conventional and unconventional antimicrobials from fish, marine invertebrates and micro-algae. *Marine Drugs* 8 (4):1213–1262. doi: 10.3390/md8041213.

Sukmarini, L. 2022. Antiviral peptides (AVPs) of marine origin as propitious therapeutic drug candidates for the treatment of human viruses. *Molecules* 27 (9):2619. doi: 10.3390/molecules27092619.

Sumon, T. A., M. A. Hussain, M. Hasan, A. Rashid, M. H. Abualreesh, W. J. Jang, S. M. Sharifuzzaman, C. L. Brown, E.-W. Lee, and M. T. Hasan. 2021. Antiviral peptides from aquatic organisms: Functionality and potential inhibitory effect on SARS-CoV-2. *Aquaculture* 541:736783. doi: 10.1016/j.aquaculture.2021.736783.

Sureram, S., I. Arduino, R. Ueoka, M. Rittà, R. Francese, R. Srivibool, D. Darshana, J. Piel, S. Ruchirawat, L. Muratori, D. Lembo, P. Kittakoop, and M. Donalisio. 2022. The peptide A-3302-B isolated from a marine bacterium *Micromonospora sp*. inhibits HSV-2 infection by preventing the viral egress from host cells. *International Journal of Molecular Sciences* 23 (2):947. doi: 10.3390/ijms23020947.

Tamamura, H., A. Otaka, T. Murakami, T. Ishihara, T. Ibuka, M. Waki, A. Matsumoto, N. Yamamoto, and N. Fujii. 1996. Interaction of an anti-HIV peptide, T22, with GP120 and CD4. *Biochemical and Biophysical Research Communications* 219 (2):555–559. doi: 10.1006/bbrc.1996.0272.

Tonk, M., D. Růžek, and A. Vilcinskas. 2021. Compelling evidence for the activity of antiviral peptides against SARS-CoV-2. *Viruses* 13 (5):912. doi: 10.3390/v13050912.

Uzair, B., Z. Mahmood, and S. Tabassum. 2011. Antiviral activity of natural products extracted from marine organisms. *Bioimpacts* 1 (4):203–211.

Varona, J. F., P. Landete, J. A. Lopez-Martin, V. Estrada, R. Paredes, P. Guisado-Vasco, L. F. de Orueta, M. Torralba, J. Fortun, R. Vates, J. Barberan, B. Clotet, J. Ancochea, D. Carnevali, N. Cabello, L. Porras, P. Gijon, A. Monereo, D. Abad, S. Zuñiga, I. Sola, J. Rodon, J. Vergara-Alert, N. Izquierdo-Useros, S. Fudio, M. J. Pontes, B. de Rivas, P. G. de Velasco, A. Nieto, J. Gomez, P. Aviles, R. Lubomirov, A. Belgrano, B. Sopesen, K. M. White, R. Rosales, S. Yildiz, A.-K. Reuschl, L. G. Thorne, C. Jolly, G. J. Towers, L. Zuliani-Alvarez, M. Bouhaddou, K. Obernier, B. L. McGovern, M. L. Rodriguez, L. Enjuanes, J. M. Fernandez-Sousa, N. J. Krogan, J. M. Jimeno, and A. Garcia-Sastre. 2022. Preclinical and randomized phase I studies of plitidepsin in adults hospitalized with COVID-19. *Life Science Alliance* 5 (4):e202101200. doi: 10.26508/lsa.202101200.

Vilas, B., L. C. Pereira, M. L. Campos, R. L. A. Berlanda, N. de Carvalho Neves, and O. L. Franco. 2019. Antiviral peptides as promising therapeutic drugs. *Cellular and Molecular Life Sciences: CMLS* 76 (18):3525–3542. doi: 10.1007/s00018-019-03138-w.

Vilas, B., L. C. Pereira, L. M. P. de Lima, L. Migliolo, G. dos Santos Mendes, M. G. de Jesus, O. L. Franco, and P. A. Silva. 2017. Linear antimicrobial peptides with activity against herpes simplex virus 1 and Aichi virus. *Peptide Science* 108 (2):22871. doi: 10.1002/bip.22871.

Vishvakarma, V. K., M. B. Singh, P. Jain, K. Kumari, and P. Singh. 2022. Hunting the main protease of SARS-CoV-2 by plitidepsin: Molecular docking and temperature-dependent molecular dynamics simulations. *Amino Acids* 54 (2):205–213. doi: 10.1007/s00726-021-03098-1.

Wagner, E. K., M. J. Hewlett, D. C. Bloom, and D. Camerini. 1999. Basic virology. In M. A. Boston (Eds) *Library of Congress Cataloging-in-Publication Data*, p. 512. Wiley Blackwell, Hoboken, NJ.

Wang, C.-K., L.-Y. Shih, and K. Y. Chang. 2017. Large-scale analysis of antimicrobial activities in relation to amphipathicity and charge reveals novel characterization of antimicrobial peptides. *Molecules* 22 (11):2037. doi: 10.3390/molecules22112037.

Wang, G., X. Li, and Z. Wang. 2016. APD3: The antimicrobial peptide database as a tool for research and education. *Nucleic Acids Research* 44 (D1):D1087–D1093. doi: 10.1093/nar/gkv1278.

Wang, Y.-D., C.-W. Kung, and J.-Y. Chen. 2010. Antiviral activity by fish antimicrobial peptides of epinecidin-1 and hepcidin 1–5 against nervous necrosis virus in medaka. *Peptides* 31 (6):1026–1033. doi: 10.1016/j.peptides.2010.02.025.

Wang, Z., J. Jia, L. Wang, F. Li, Y. Wang, Y. Jiang, X. Song, S. Qin, K. Zheng, J. Ye, Z. Ren, Y. Wang, and S. Qi. 2020. Anti-HSV-1 activity of Aspergillipeptide D, a cyclic pentapepetide isolated from fungus *Aspergillus sp.* SCSIO 41501. *Virology Journal* 17 (1):41. doi: 10.1186/s12985-020-01315-z.

White, K. M., R. Rosales, S. Yildiz, T. Kehrer, L. Miorin, E. Moreno, S. Jangra, M. B. Uccellini, R. Rathnasinghe, L. Coughlan, C. Martinez-Romero, J. Batra, A. Rojc, M. Bouhaddou, J. M. Fabius, K. Obernier, M. Dejosez, M. J. Guillén, A. Losada, P. Avilés, M. Schotsaert, T. Zwaka, M. Vignuzzi, K. M. Shokat, N. J. Krogan, and A. García-Sastre. 2021. Plitidepsin has potent preclinical efficacy against SARS-CoV-2 by targeting the host protein eEF1A. *Science* 371 (6532):926–931. doi: 10.1126/science.abf4058.

Wibowo, J. T., M. Y. Kellermann, M. Köck, M. Y. Putra, T. Murniasih, K. I. Mohr, J. Wink, D. F. Praditya, E. Steinmann, and P. J. Schupp. 2021. Anti-infective and antiviral activity of valinomycin and its analogues from a sea cucumber-associated bacterium, streptomyces sp. SV 21. *Marine Drugs* 19 (2):81. doi: 10.3390/md19020081.

Xie, H., J. Wei, and Q. Qin. 2016. Antiviral function of Tachyplesin I against iridovirus and nodavirus. *Fish & Shellfish Immunology* 58:96–102. doi: 10.1016/j.fsi.2016.09.015.

Xu, Y., R. D. Kersten, S.-J. Nam, L. Lu, A. M. Al-Suwailem, H. Zheng, W. Fenical, P. C. Dorrestein, B. S. Moore, and P.-Y. Qian. 2012. Bacterial biosynthesis and maturation of the didemnin anti-cancer agents. *Journal of the American Chemical Society* 134 (20):8625–8632. doi: 10.1021/ja301735a.

Yasin, B., M. Pang, J. S. Turner, Y. Cho, N. N. Dinh, A. J. Waring, R. I. Lehrer, and E. A. Wagar. 2000. Evaluation of the inactivation of infectious herpes simplex virus by host-defense peptides. *European Journal of Clinical Microbiology & Infectious Diseases* 19 (3):187–194. doi: 10.1007/s100960050457.

Youssef, F. S., M. L. Ashour, A. N. B. Singab, and M. Wink. 2019. A comprehensive review of bioactive peptides from marine fungi and their biological significance. *Marine Drugs* 17 (10):559. doi: 10.3390/md17100559.

Zampella, A., M. V. D'Auria, L. G. Paloma, A. Casapullo, L. Minale, C. Debitus, and Y. Henin. 1996. Callipeltin A, an anti-HIV cyclic depsipeptide from the new *Caledonian lithistida* sponge *Callipelta sp. Journal of the American Chemical Society* 118 (26):6202–6209. doi: 10.1021/ja954287p.

Zampella, A., V. Sepe, P. Luciano, F. Bellotta, M. C. Monti, M. V. D'Auria, T. Jepsen, S. Petek, M.-T. Adeline, O. Laprévôte, A.-M. Aubertin, C. Debitus, C. Poupat, and A. Ahond. 2008. Homophymine A, an anti-HIV Cyclodepsipeptide from the sponge *Homophymia sp. The Journal of Organic Chemistry* 73 (14):5319–5327. doi: 10.1021/jo800583b.

Zhan, S., J. J. Aweya, F. Wang, D. Yao, M. Zhong, J. Chen, S. Li, and Y. Zhang. 2019. Litopenaeus vannamei attenuates white spot syndrome virus replication by specific antiviral peptides generated from hemocyanin. *Developmental & Comparative Immunology* 91:50–61. doi: 10.1016/j.dci.2018.10.005.

Zhang, J., L.-P. Yu, M.-F. Li, and L. Sun. 2014. Turbot (Scophthalmus maximus) hepcidin-1 and hepcidin-2 possess antimicrobial activity and promote resistance against bacterial and viral infection. *Fish Shellfish Immunology* 38 (1):127–134. doi: 10.1016/j.fsi.2014.03.011.

Zhao, Y., B. Li, S. Dong, Z. Liu, X. Zhao, J. Wang, and M. Zeng. 2009. A novel ACE inhibitory peptide isolated from Acaudina molpadioidea hydrolysate. *Peptides* 30 (6):1028–1033. doi: 10.1016/j.peptides.2009.03.002.

Zheng, S.-C., J.-Y. Xu, and H.-P. Liu. 2019. Cellular entry of white spot syndrome virus and antiviral immunity mediated by cellular receptors in crustaceans. *Fish Shellfish Immunology* 93:580–588. doi: 10.1016/j.fsi.2019.08.011.

Zhou, X., W. Fang, S. Tan, X. Lin, T. Xun, B. Yang, S. Liu, and Y. Liu. 2016. Aspernigrins with anti-HIV-1 activities from the marine-derived fungus *Aspergillus niger* SCSIO Jcsw6F30. *Bioorganic & Medicinal Chemistry Letters* 26 (2):361–365. doi: 10.1016/j.bmcl.2015.12.005.

Zorzi, A., K. Deyle, and C. Heinis. 2017. Cyclic peptide therapeutics: Past, present and future. *Current Opinion in Chemical Biology* 38:24–29. doi: 10.1016/j.cbpa.2017.02.006.

6 Mechanism, Sources, and Treatment Methods of Antimicrobial Resistance Present in Wastewater Treatment

Rajesh Nithyanandam, Krishnan Ganesh Prasath, Maadhangi Muthukannan, Divya Sri Muthu, and Keshika Dhanesh

6.1 INTRODUCTION

Resistance or antimicrobial resistance (AMR) is described as the capability of microorganisms to resist and survive in an environment with the presence of any antimicrobial agent. Microorganisms can be fought by using antimicrobial compounds, such as food preservatives and disinfectants or antibiotics, to inhibit growth, prevent replication, or kill them. Due to overuse and abuse in so many industries, resistant microbes have evolved and have become a serious global health concern. It is also a serious threat to the health of humankind at large. Antimicrobial resistance (AMR) refers to the resistance developed by bacteria, viruses, fungi, and parasites to the effectiveness of medicines previously administered to them. If these organisms refuse to accept the antimicrobial medicines, then this leads to AMR, increasing the risk of infection and making conventional treatments ineffective. AMR undermines the effectiveness of antibiotics, antivirals, and other medications, leading to increased illness, death, and healthcare costs.

6.2 MECHANISM OF AMR AND THE MICROBES INVOLVED

Bacteria have developed an AMR mechanism over billions of times, allowing them to withstand the effects of antibiotics, despite the continued use of antibiotics. The classical AMR mechanisms generally include chromosomal resistance, alterations in cell membrane permeability, enzyme degradation, enzymatic modification or degradation of antibiotic, restriction of antibiotic entry to prevent their accumulation in the cells, target modification or mutation alteration of metabolic pathways, active efflux pump system modification to remove antibiotics from cells before they reach the adequate levels (Chiş et al. 2022), and biofilm production that grants additional protection to the bacteria (Uruén et al. 2020). These resistant mechanisms have been summarized in Figure 6.1. Furthermore, bacteria can acquire resistance genes by horizontal gene transfer using plasmids and other mobile genetic components from adjacent cells or even different species. These acquired genes usually include several intricate resistance mechanisms inside a single unit, allowing the fast spread of multidrug-resistant (MDR) bacteria throughout microbial populations. Microorganisms possess a vast array of resistance methods due to their efficient horizontal genetic information transfer capability.

DOI: 10.1201/9781003269823-6

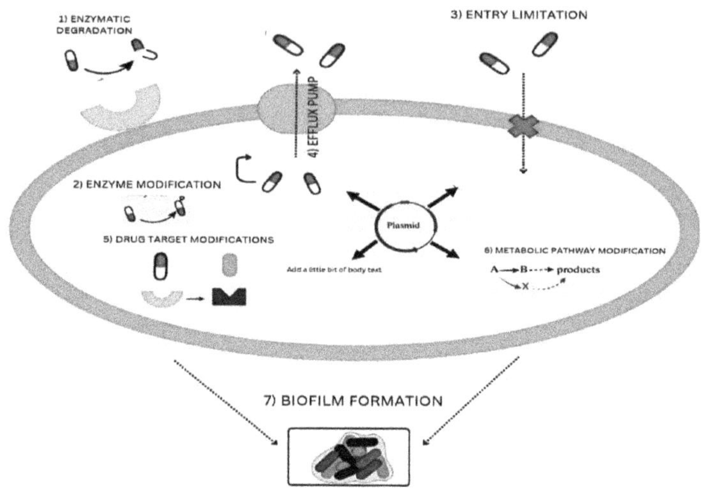

FIGURE 6.1 Mechanisms of antimicrobial resistance in bacteria.

Figure 6.1 shows the mechanism of AMR that contains chromosomal resistance, alterations in cell membrane permeability, enzyme degradation, enzymatic modification or degradation of antibiotic, restriction of antibiotic entry to prevent their accumulation in the cells, target modification or mutation alteration of metabolic pathways, active efflux pump system modification to remove antibiotics from cells before they reach the adequate levels and biofilm production that grants additional protection to the bacteria.

6.2.1 Categories of AMR Mechanism

Intrinsic Resistance: Certain structural and functional traits of certain bacterial genera (or species) confer resistance against some drugs. These bacterial groupings are typically unsuccessful because they lack an antibiotic's target site.

For example: The absence of a cell wall in Mycoplasma species makes them resistant to β-lactam antibiotics and glycopeptides. Additionally, the presence of an outer membrane prevents an antibiotic from penetrating bacterial cells. It could also be due to the existence of an export system (like the AcrAB-TolC system) or the capacity of some bacterial species to generate antibiotic-inactivating enzymes (like *E. coli*'s AmpC β-lactamase) (Schwarz, Cloeckaert, and Roberts 2019).

Acquired Resistance: When genetic codes from other bacterial strains are transferred to naturally susceptible bacteria, the bacteria become resistant to specific antibiotics. The three main processes of acquired resistance are as follows (Schwarz, Cloeckaert, and Roberts 2019).

6.2.1.1 Enzyme Modification or Antimicrobial Agent Inactivation

Enzyme modification and antibiotic inactivation are observed in both Gram-positive and Gram-negative bacteria. Enzymatic modifications involve the addition of acetyl, adenyl, or phosphate groups from bacterial enzymes to a particular antibiotic site. This process changes the antibiotic chemically and renders the antimicrobial agents inactive, preventing them from binding to the target site. For instance, phosphorylation happens in macrolides, while aminoglycosides experience acetylation, adenylation, or phosphorylation (Ramirez and Tolmasky 2010). Enzymatic inactivation occurs when bacterial enzymes directly attach to antibiotics, which is mostly caused by the antibiotics' hydrolytic cleavage activity (e.g., β-lactamases against penicillins and cephalosporins) (Abushaheen et al. 2020).

6.2.1.2 Reduced Intracellular Accumulation of Antimicrobial Agents

Bacteria employ two strategies to lessen the buildup of antibacterial agents within their cells: reduced influx and enhanced efflux. Porins, an outer membrane protein, are thought to be the point of entry for antibiotics, including carbapenems that enter *Pseudomonas aeruginosa* through the OmpD and tetracyclines and β-lactam that enter *E. coli* through the OmpF.

Influx reduction, often referred to as the permeability barrier against antibiotics in Gram-negative bacteria, has been observed as a consequence of downregulation, structural change, or even functional deletion of the porin genes.

However, enhanced efflux mechanisms, such as multidrug transporters, that are typically employed by nearly all bacterial cells to move toxic substances from their metabolism are actively expelling antibiotics from the cells.

Examples of these include members of the resistance nodulation cell division family, which is primarily associated with Gram-negative bacteria and includes the AcrAB-TolC transporter in *E. coli* (Schwarz, Cloeckaert, and Roberts 2019; Poole 2005). Certain transporters, such as the multidrug and toxic-compound extrusion family, ATP-binding cassette family, and major facilitator superfamily, are implicated in bacterial resistance (Poole 2005; Abushaheen et al. 2020).

6.2.1.3 Alterations at the Target Sites of the Antimicrobial Agents

At the target sites, modifications consist of Fluoroquinolone resistance caused by mutations of the quinolone-resistance-determining region in the DNA gyrase (topoisomerase II and topoisomerase IV) in both Gram-positive and Gram-negative bacteria (Abushaheen et al. 2020; Redgrave et al. 2014). In both Gram-positive and Gram-negative bacteria, chemical modification by methylation is thought to be highly effective in causing resistance.

Examples of this include Erm methylases against macrolides, lincosamides, and streptogramin B antibiotics. Sulfonamide and trimethoprim resistance are frequent examples of drug-resistant targets replacing origin targets that are susceptible to certain antibiotics. The genes sul1, sul2, and sul3 in Gram-negative bacteria encode dihydropteroate synthases, which are resistant to sulfonamides (Schwarz, Cloeckaert, and Roberts 2019; Abushaheen et al. 2020).

Furthermore, *Staphylococcus* species contain the genes mecA and mecC. It encoded a different penicillin-binding protein with a markedly decreased affinity for all β-lactam drugs (Schwarz, Cloeckaert, and Roberts 2019; Becker et al. 2014).

6.2.1.4 Biofilm Formation and Antimicrobial Resistance

Biofilms are safe, frequently diverse populations of bacteria that settle and proliferate on the surfaces of plumbing fixtures, medical devices, and human tissue (Sharma, Misba, and Khan 2019). They consist of dense populations of microorganisms immobilized by an extracellular matrix made of proteins, extracellular DNA (eDNA), and exopolysaccharides (EPS), which are released by bacteria (Sharma, Misba, and Khan 2019). Biofilms offer a source of nutrients and defense against disinfection procedures in hospital plumbing fittings and water distribution systems. In regions with low flow rates, biofilm formation is encouraged and standstill that permits microorganisms to adhere to the surface of the infrastructure. In contrast to planktonic bacteria, bacterial biofilm communities have distinct metabolic activities, including higher rates of EPS synthesis, activation or inhibition of genes linked to biofilm formation, and slower rates of growth (Sharma, Misba, and Khan 2019). By inhibiting the diffusion reaction, EPS plays a part in providing aminoglycosides with tolerance by squelching their activity (Abe, Nomura, and Suzuki 2021). The formation of biofilms shields the communities of microorganisms from damaging pH, osmolarity, nutritional deprivation, and shear pressures (Sharma, Misba, and Khan 2019). When compared to planktonic bacteria, microorganisms in biofilms exhibit a 1,000-fold increase in resistance to antimicrobial exposure due to their ability to obstruct antibiotic entry (Soto-Giron et al. 2016). Species like *L. pneumophila* may enter a viable nonculturable (VBNC) stationary phase once a

TABLE 6.1
List of Bacteria Involved in the Development of AMR (Ahmed et al. 2024)

Bacteria	Development of AMR
1. *Staphylococcus aureus*	Resistant to numerous antibiotics, including methicillin, through mutations in mecA and mecC genes and horizontal gene transfer.
2. *Klebsiella pneumoniae* and *Escherichia coli*	By obtaining carbapenemase genes, it becomes resistant to carbapenem medications.
3. *E. coli*	Resistant to a wide range of medicines, such as cephalosporins and penicillins, by acquiring extended spectrum β-lactamase (ESBL) genes, usually via plasmids.
4. *Mycobacterium tuberculosis*	Resistant to various anti-tuberculosis drugs.
5. *Acinetobacter baumannii*	Resistant to several antibiotics due to a combination of mutations and the acquisition of resistance genes.
6. *Neisseria gonorrhoeae*	Multidrug resistance causes treating sexually transmitted illness of gonorrhea.

biofilm community reaches maturity to withstand antibiotic stress (Whiley 2017; Sharma, Misba, and Khan 2019). Pathogens like *P. aeruginosa*, *Mycobacterium* spp., and *S. maltophilia* have the ability to create biofilms, which can facilitate the attachment of other pathogens like *Salmonella* spp., *Campylobacter* spp., and *S. aureus* that are commonly prevalent in the broader hospital environment (Soto-Giron et al. 2016).

These techniques can be modified as needed to withstand continuous improvements in medical science and the use of antibiotics. Over the past few decades, a number of bacteria have developed AMR through various ways, which can be observed from Table 6.1.

The concerning increase of MDR microbial strains highlights the harmful bacteria, viruses, fungi, and protozoa's capacity to quickly evade chemical compounds intended to kill them.

6.3 SOURCES OF ANTIMICROBIAL RESISTANCE (AMR)

AMR is driven by what is called as resistance network, which are complex interacting factors (Holmes et al. 2016). These networks form a clinical link between various factors like veterinary medicine, animal husbandry, overexploitation of antibiotics, etc. Antibiotics can take several routes into the environment through these activities.

6.3.1 Agriculture and Livestock as a Source

In both developed and developing countries worldwide, each serves the dual purpose of growth supplement and growth enhancer to their animal population. The same fact that applies to humans holds good for animals as well: antibiotic-resistant bacteria (ARB) emerge in livestock treated with antibiotics. Through animal feces, ARB circulate throughout the ecosystem, which is harmful to people and easily transmitted through food chains. This might cause difficult-to-cure and long-lasting infections among the humans (Manyi-Loh et al. 2018; Petrovich et al. 2020). These contaminants get into the environment via aquatic bodies, such as lakes, ponds, rivers, and sewage systems. They ultimately get into the ocean and end the lives of the various aquatic as well as other organisms. In Morocco, Israel, the USA, Australia, China, and Mexico, wastewater is reused for irrigation purposes. The agriculture sector has been taken as one of the important vectors in the dissemination of AMR. There is a chance that soil may be contaminated with ARB through irrigation with treated wastewater even though in trace quantities. Also, the possible ARB can accumulate in soil and proliferate under high organic matter concentration conditions (Serna-Galvis et al. 2017).

6.3.2 Feces as a Source

Generally, after a brief residency in human and animal species, the ingested antibiotics are continuously released into the natural ecosystems by excretion through urine and feces. These kinds of drugs in the wastes reach the aquatic environment through landfill leachates, runoff, and leaching from fertilized farmland manure, sewage disposed in agricultural areas, and direct discharge of wastewater treatment plant (WWTP) effluent to surface water or groundwater. As the conventional wastewater treatment methods are inefficient for the removal of human medicine-used antibiotics, the treated wastewater from WWTPs is considered the key contributor to the occurrence of the latter in the environment (Watkinson et al. 2009; Homem and Santos 2011). Before excretion from human and animal bodies, antibiotics undergo a structural transformation into metabolites. Later, during excretion, further biotic and abiotic processes occur at technical plants that change the structural nature of both parent compounds and metabolites. This makes them differ in their structure and physiological morphologies leading to changes from their parent compounds.

6.3.3 Misuse of Drugs

Alexander Fleming, the man who discovered antibiotics, warned the public in 1945 not to use antibiotics too much after finding out more about the risks involved in the misuse of the drugs. In the worldwide perspective, antibiotics are either overused or abused even with repeated cautions made by various health organizations. It is evident from this point that the ability to resist the use of antibiotics is attributed to the misuse of such drugs. A definite association between the usage rate and the emergence of microbial resistance was determined by the epidemiological investigations. The first manner of resistance development to drugs was that the naturally existing drugs triggered mutations or horizontal gene transfer in the microbes. Yet, contemporary medicine further hastened this conversion by eliminating competitors sensitive to drugs, allowing resistant microbes to grow their population and spread through natural selection. Antibiotics taken too frequently or for improper purposes can indeed alter bacteria in a manner that renders them resistant to those drugs (Sulis et al. 2020; Serna-Galvis et al. 2017).

6.3.4 Inappropriate Diagnosis and Prescribing Patterns

Healthcare practitioners sometimes diagnose an infection or prescribe an antibiotic "just in case" or a wide-spectrum antibiotic when a specific narrow-spectrum antibiotic could be more suited based on incomplete or erroneous information. Selection pressure brought on by these conditions may lead to AMR (Denning et al. 2017). AMR may also arise because of improper medication prescribing practices. In case the physician does not know whether a virus or bacteria is aggravating an infection, he may prescribe antibiotics. However, the antibiotics will be useless against viruses, and their resistance may appear.

6.3.5 Antibiotics in WWTPs from Hospital Effluents

Due to their widespread clinical usage, hospital effluents are among the most significant sources of antibiotics (Petrovich et al. 2020). The hospital effluents negatively impact the antibiotic concentrations from these establishments. As a result, they have emerged as key locations for the spread of antibiotic-resistant genes (ARGs), and it is anticipated that their concentrations of ARB and antibiotic residues will be higher than those of other urban wastewater sources (Petrovich et al. 2020; S. Harris et al. 2013). Therefore, guidelines for the treatment of hospital effluents before being released into municipal WWTPs must be established and followed, as well as recommendations for ARG concentration criteria.

6.3.6 Antibiotics in WWTPs from Pharmaceutical Manufacturing Effluents in WWTPs

The global antibiotic load in WWTPs has been found to be primarily contributed by the pharmaceutical manufacturing business. High quantities of oxytetracycline, lincomycin, or fluoroquinolone antibiotics have been found in treated pharmaceutical effluent from Asian nations like China, India, and South Korea. In addition to these, unintentional releases of active pharmaceutical compounds from production facilities have also been observed to foster the development of drug-resistant bacteria. Serious worldwide repercussions may result from this (Larsson 2014).

6.3.7 Heavy Metals in WWTPs

The primary source of AMR in treated effluent is activated sludge, which is subjected to a number of stressors that increase AMR, such as heavy metals, disinfectants, reactive oxygen species, and sublethal antimicrobial concentrations (Petrovich et al. 2020; Narciso-da-Rocha et al. 2018; Marano et al. 2021). Therefore, the effects on the water environment of discharging AMR through wastewater or treated effluent varied significantly. Environmental contaminants that are widely known include heavy metal ions. Heavy metal ions have been found in wastewater, according to a number of reports. Through a number of mechanisms, including co- and cross-resistance, heavy metal ion pollution co-selects antibiotics and metal resistance (Dickinson et al. 2019). Antibiotic molecules form complexes with metal ions, including Zn^{2+}, Co^{2+}, and Cu^{2+} as a result of their reactions (Lambs et al. 1988; Khurana, Pulicharla, and Kaur Brar 2021). These complexes are nonbiodegradable and permanent.

6.3.8 Impact of Microplastics in WWTPs

Microplastics are released into WWTPs from production sites, runoff, textiles, pharmaceuticals, and personal hygiene items (Pham, Clark, and Li 2021). Because they break down gradually in the environment, they persist and interact with many chemicals and microbes, posing a threat to the health of humans (Arias-Andres et al. 2018; Pham, Clark, and Li 2021). They immediately endanger both public health and aquatic habitats. Tetracycline and other antibiotics have recently been found to be transported by these microplastics (Syranidou and Kalogerakis 2022; Naraginti, Li, and Puma 2018). Biofilm develops on the surfaces of WWTPs under favorable conditions, creating an environment that is ideal for bacterial growth and spread. Furthermore, examination of the biofilm has demonstrated a wide range of ARGs, including aminoglycoside (Naraginti, Li, and Puma 2018). There have been reports of genetic material transfer resulting in superpathogens. This is because the biofilm matrix provides the bacteria with higher mechanical force resistance and changes in pH and shortage of food, giving them a platform for gene transfer and development. The plasmids occurring in the biofilm on plastics were found to possess three times as much antibiotic resistance as those isolated from planktonic populations.

6.3.9 Antimicrobial Farm Effluents

Each method of antibiotic administration, with the exception of injection, has a direct impact on the aquatic species and the aquatic ecosystem itself. Results have shown that 75% of the antibiotics given to the fish are excreted back into the pond water. Antibiotic residues are accumulated in the pond water, pond sediment, animal tissue, and cultured aquatic products, hence negative effects on the environment and human health (Lulijwa, Rupia and Alfaro 2020; B. Chen et al. 2018). This also leads to the accumulation of antibiotics in the aquaculture environment. Residues present in the aquaculture products have a high potential of being extremely hazardous for the consumers and can alter the very diverse microbiota that dwells inside the gastrointestinal tract of the human

FIGURE 6.2 Sources of AMR: feces; improper drug usage; heavy metals; microplastics; improper diagnosis and prescription; antimicrobial farm effluents.

body since this may lead to unfavorable outcomes (Monteiro et al. 2018). Lower concentration chloramphenicol residues have the associated risk of cancer and aplastic anemia. Other toxic effects of residual antibiotics include gentamicin mutagenicity and nephropathy, immunopathology, and carcinogenic effects of sulfamethazine, oxytetracycline, furazolidone, and penicillin hypersensitivity. These residuals are excreted in the environment as manure, wherein they directly enter the food chain and produce lethal consequences. Other residual antibiotics, such as tetracyclines, quinolones, and sulfonamides, disruptively affect the early developmental stages of zooplankton and interfere with chlorophyll production in phytoplankton. The effects are thus observed in phytoplankton and zooplankton diversity. If this disruption leads to alterations in the food chain, then it will impact all levels of the ecosystem. In the case of an aquaculture setting and aquatic animals, as well as their products, the residual level of quite a few antibiotics is determined by a number of factors such as the concentration of exposure and physicochemical features. It is acknowledged that the antibiotics with lower water solubility—for instance, quinolones, particularly, norfloxacin, ofloxacin, and enrofloxacin buildup more in aquatic environments (Figures 6.2 and 6.3) (Table 6.2).

6.4 ERADICATION OF AMR

Antimicrobial medicines are classed as a newly developed concern of micropollutants, and their long-term exposure in the environment shows severe ecotoxic effects even at low concentrations. Antimicrobials exert selective pressure on bacteria, and this process has adverse health implications due to the generation of resistance within them. In a scientific study, antibiotic use is gradually increasing globally (30%), due to rising demand in low- and middle-income nations.

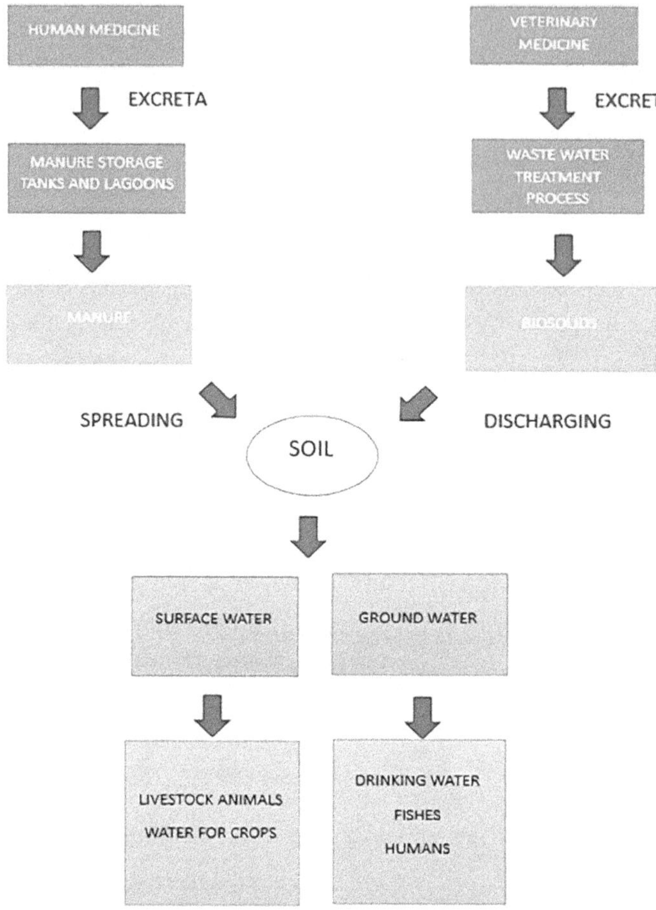

FIGURE 6.3 Routes of AMR through human and veterinary medicine in the environment.

The antimicrobials are not fully metabolized in the course of therapeutic use and an estimated 30%–90% end up being excreted as active substances into sewage water, follows in the presence of different classes of antibiotics which is being widely detected in various urban WWTPs and the receiving environment around the world (M. Chen et al. 2006; W. C. Li 2014). Ecosystems that contain human excreta, like compost toilets and WWTPs, in addition to animal farms and aquaculture, are believed to serve as breeding grounds for resistant bacteria (Zhou et al. 2018; Karkman, Pärnänen, and Larsson 2019).

6.4.1 Ecosystem and Public Health Concerns

It is possible for resistant pathogens to infiltrate human bodies through direct or indirect paths. The primary mechanism through which their resistance genes spread throughout the environment is through horizontal gene transfer processes (Kumar and Pal 2018). The AMR ratio in the output water (effluents) may rise as a result of the WWT plants' reduction of the overall load of bacterial strains. This must be taken into account to strengthen the current WWT plants and spread the elimination of antibiotic resistance. Thus, the effects of AMR are linked to the possibility of developing an AMR and result in therapeutic failures (S. J. Harris, Cormican, and Cummins 2012). Because the selected approach of these medicines is ineffective, the infecting microorganisms are resistant and the majority of AMR bacteria generate an "inferior response" to diagnosis. This leads to the

TABLE 6.2
Overview of the Many Samples that are Used for the Collection of the Studies for this Systematic Review

S. No	Sources	Microorganism	Tested Samples	Observations	References
1.	Sewage and environmental water	*E. coli* and *Salmonella* sp.	Wastewater, water from river, pond, and swimming pool	In all environmental samples, the *E. coli* and *Salmonella* spp. which are examples of AMR bacteria were detected.	Hossain et al. (2020)
2.	Sewage and environmental waters	*E. coli* strains producing β-lactamase	Sewage water, wastewater treatment plant (WWTP) (influent and effluent) and surface water (ambient water)	It was observed that compared to effluent wastewater, influent wastewater had higher relative abundance of ARB and MDR microbes. Environmental surface water has been detected to carry strains of *Escherichia coli* producing extended-spectrum β-lactamase (ESBL).	Haberecht et al. (2019)
3.	Water systems	*Legionella* sp.	Hospitals	MICs: greatest MIC to Clarithromycin	Erdogan et al. (2010)
4.	Tap water	*Pseudomonas aeruginosa*	Hospital	Opr-mediated resistance to IPM (imipenem) is present in seven isolates.	Tran-Dinh et al. (2018)
5.	Wastewater, environmental water and human feces	*Bacteriodes fragilis*	Wastewater from hospital, influent and effluent in wastewater treatment process, and human feces.	Among all the studied isolates, influent and effluent WWTP water isolates generated the highest frequency of drug resistance. The *Bacteroides fragilis* group (BFG) bacteria isolated from WWTPs had more resilient profiles compared to the rat or human excreta isolates.	Niestępski et al. (2019)
6.	Environmental and wastewater	*E. coli*	Hospital wastewater, WWTP samples, and downstream water	The β-lactamase genes such as blaIMP–1, blaIMP–2 and Oxacillinases (OXA) blaOXA–23 were only identified in hospital sewage, whereas blaOXA–48, blaCTX–M–8 and blaSFC–1, Verona imipenemase (VIM) blaVIM–1 and blaVIM–13 were only detected in downstream river water but not in WWTP.	Khan, Söderquist, and Jass (2019)
7.	Taps Shower Drinking fountain	*Pseudomonas aeruginosa*	Hospital Residential care home	7.72% imipenem-resistant. 13.2% of people are resistant to more than one antibiotic.	Schiavano et al. (2017)

(*Continued*)

TABLE 6.2 (*Continued*)
Overview of the Many Samples that are Used for the Collection of the Studies for this Systematic Review

S. No	Sources	Microorganism	Tested Samples	Observations	References
8.	Sewage	*Acinetobacter cumulans*	Hospital sewage		Qin et al. (2019)
9.	Wastewater	*Pseudomonas aeruginosa*	Hospital	Isolates were copper-tolerant	Jeanvoine et al. (2019)
10.	Sewage and environmental wastewater	*Carbapenemase producing Entrobacteriaceae*	Hospital and municipal wastewater (pre- and post-hospital)	Hospital effluent has been detected to have carbapenemase-producing Enterobacteriaceae (CPE) at high rates.	Cahill et al. (2019)
11.	Sewage	*Enterococcus* spp.	Municipal and hospital wastewater	Vancomycin and erythromycin were ineffective against 91% and 100% of the isolated *Enterococcus* spp. (*E. faecalis* and *E. durans*) from the hospital wastewater and final effluent wastewater, respectively.	Iweriebor et al. (2015)
12.	Sewage and environmental waters	*K. pneumonia*	Hospital effluent, sanitary effluent, different sites within WWTP and upstream and downstream river water.	Hospital effluent, WWTP, and river samples had higher isolates of *K. pneumonia* and *E. coli* that produce ESBLs, respectively. Surface water samples, sanitary effluent, hospital effluent, and outflow sewage all contained isolates resistant to quinidine. MDR bacteria were found in river streams and hospital effluent.	Conte et al. (2017)

Resistance genes, named antimicrobial resistance genes, are ARG; antimicrobial resistant bacteria are ARB, and antimicrobial resistant genes are known as AMR.

following conditions: (1) results in increased virulence, disease outbreak and transmission, and pathogenicity; (2) prolongs morbidity and lengthens hospital stays; (3) raises the expense of health care for medication and increases the risk of secondary bacteremia; and (4) increases the number of surgeries and mortality rates due to the spreading of infections (S. J. Harris, Cormican, and Cummins 2012; Cosgrove n.d.).

Pseudomonas aeruginosa and *Acinetobacter baumannii* are two examples of ARB that contribute significantly to nosocomial infections and increase hospital patient traffic. Among the common gut flora, *Enterococcus* spp. have reduced colonization and virulence, and a nosocomial infection may arise in "immune-compromised patients" and individuals with severe sickness (Mesaros et al. 2007).

6.4.2 Effective Post-Treatment Methodologies

The efficiency of different post-treatment methods is categorized as tertiary hospital wastewater management procedures. And, the final stage can include the tertiary WWT systems, which are designed to eliminate pathogens, leftover nutrients, and organic debris. To achieve this goal, various disinfection technologies (post-treatment) can be used to eliminate genes and bacteria resistant to antibiotics from hospital wastewater before it is released into the environment (Kumar and Pal 2018; Gruchlik, Linge, and Joll 2018). Consequently, the post-treatment strategy involves applying modern treatment technologies to remove antibiotics, bacteria resistant to antibiotics, and genes resistant to antibiotics from hospital wastewater. Available technologies for the removal of these various types of micropollutants present in hospital wastewater are AOPs, bioremediation via CW, membrane separation, adsorption onto activated carbon, and waste stabilization ponds (WSPs). Examples are antibiotics, antibiotic resistance bacteria, and antibiotic resistance genes.

6.4.2.1 Activated Carbon Adsorption

The best method for treating wastewater is adsorption, which has the advantages of being inexpensive, having a large surface area, having the capacity to regenerate for future use, and having a higher capability for eliminating micropollutants (Werkneh, Gebretsadik, and Gebru 2023). Many studies have been conducted using granular activated carbon (GAC) and powdered activated carbon (PAC) to examine the adsorption of ARGs and ARB in activated carbon (Luo et al. 2014). In a series of treatment tests, Li investigated the adsorption of sulfamethoxazole (SMX) utilizing PAC in the membrane bioreactor (MBR) (X. Li, Hai, and Nghiem 2011). The results of this investigation showed that there was very little adsorption to the treatment sludge. SMX's hydrophobicity (which causes a stronger affinity for PAC adsorption), loading rate, and PAC dosage all affect how well it removes itself from a substance. With a PAC dosage of 1 g/L, it obtained clearance rates of about $82 \pm 11\%$. Adsorption onto GAC performed better in terms of removal rate than ozonation, according to a pilot-scale investigation conducted on 80% of the chemicals (Östman et al. 2019).

6.4.2.2 Approaches for Phytoremediation

Using green plants and the microorganisms they are connected with, phytoremediation is a new plant-assisted bioremediation technique that removes, stabilizes, and transforms contaminants from contaminated environments (Monsalves et al. 2022). The chosen plants have the ability to either cleanse or accumulate the contaminants and viable alternatives for treating hospital wastewater are constructed wetlands (CWs) and WSPs (Redae, 2011).

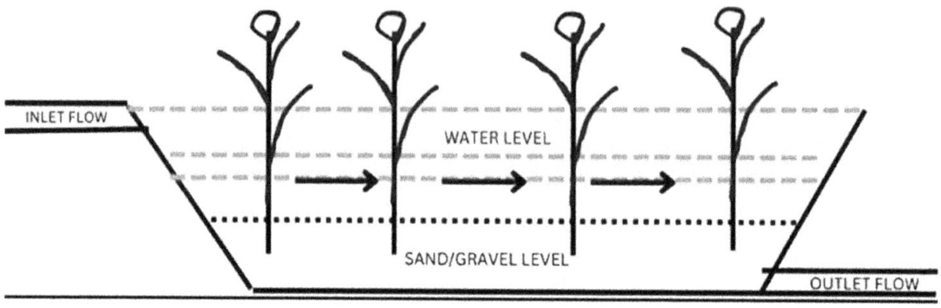

FIGURE 6.4 Cross-sectional view of constructed wetlands.

6.4.3 Physical Treatment

6.4.3.1 Construction of Wetlands

To manage wastewater effluents, CWs are complex, integrated systems including water, vegetation, animals, microbes, gravel, and the surrounding environment. The method is advantageous owing to the cost-effective ways to clean up industrial effluents, petrochemicals, drainage from acid mines, agricultural effluents, landfill leachates, and municipal wastewaters (Aslam et al. 2021). CWs removal efficiency depends on the bacteria and organic matter (assessed through chemical oxygen demand and biochemical oxygen demand) from urban wastewaters (Figure 6.4).

In a pilot-scale hot compressed water with subsurface flow, the hospital wastewater treatment for the elimination of microorganisms resistant to antibiotics (Monsalves et al. 2022). Several studies in this study, including susceptibility testing, bacteriological identifications, and indicator organism enumeration, were completed in accordance with conventional protocols. Additionally, 81.1% of the antibacterial-resistant *E. coli* strains tested positive for ampicillin, and 72.7% for amoxicillin-clavulanic acid and cotrimoxazole. The results of the susceptibility testing of hospital wastewater indicate that the isolated strains of *Salmonella* were found to be 100% resistant to ampicillin and 75% resistant to doxycycline, ceftazidime, erythromycin, cefoxitin, and chloramphenicol. Higher rate of elimination (80%–93.2%) of microorganisms resistant to antibiotics was attained in this investigation (Gruchlik, Linge, and Joll 2018).

6.4.3.2 Waste Stabilization Ponds

Large shallow ponds called WSPs use biological and physical processes to eliminate pathogens, contaminants, and organic compounds found in raw wastewater. It operates by reaching a retention time after a few hours or days, which improves treatment performance and produces well-treated effluent that can be released safely.

By using the WSPs method, many studies were able to remove some antibiotics from wastewater with relatively greater removal efficiencies, such as trimethoprim and SMX. A well-constructed and run WSP can remove fecal coliforms with 99.9% effectiveness and helminths with 100% rate at pH levels higher than 9 [61,121].

6.4.4 Chemical Treatment

6.4.4.1 UV Irradiation

The most popular method for treating hospital wastewater on a large scale is the UV/H_2O-based method. When it comes to wastewater treatment, UV light has been accepted by several utilities as the most suitable option, as opposed to chemical disinfection. Studies that have been conducted employing UV disinfection for hospital wastewater treatment have produced varying results (Werkneh and Islam 2023).

The removal of several medications from hospital wastewater had previously been treated with basic clarifiers and MBR systems through the use of post-treatment technologies. According to this study, compared to ozone and PAC-based treatments, the UV therapy with 2,400 J/m^2 at 254 nm had a lower elimination efficiency (i.e., 33%) (Kovalova et al. 2013).

6.4.4.2 Chlorination Techniques

When compared to other wastewater disinfectants that are currently on the market, chlorine and/or chlorine-based compounds have achieved universal popularity due to their affordability, ease of use, and the potential problems generated by the toxic by-products they produce. The intestinal bacteria can be effectively eliminated by the disinfectants. It must be accomplished against viruses, bacteria, spores, and protozoan cysts at a reduced clearance rate. The amount of ampicillin-resistant *E. coli* is declining. *E. coli* is caused by bacteria that are less resistant to chlorine disinfection than bacteria that are sensitive to antibiotics (Werkneh and Islam 2023; Templeton et al. 2013).

6.4.4.3 Ozonation Treatment

To improve the quality of treated sewage and reduce the possible risk to human health from receiving surface waters, the removal of bacteria by ozonation in combination with charcoal or slow sand filtration for modern sewage treatment was studied on a pilot scale at the sewage treatment plant Eriskirch, Baden Wuerttemberg, Germany (Lüddeke et al. 2015) (Figure 6.5).

Ozonation is commonly known that the various barriers used in WWTPs, particularly toward the end of disinfection, cause 90%–99% of the microorganisms to die. This can lead to the release of

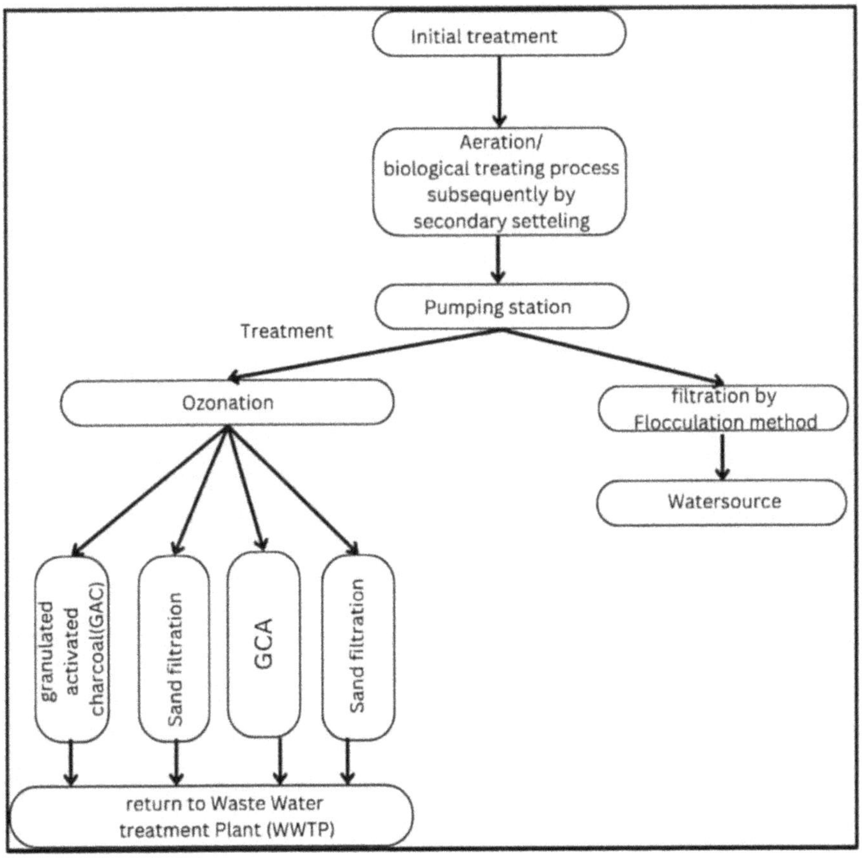

FIGURE 6.5 Flowchart on wastewater treatment by ozonation and filtration.

a significant amount of dead and broken MOs as well as eDNA into the environment that receives the discharged effluent (Sivalingam, Poté, and Prabakar 2020). Regretfully, cell-free ARGs are recognized as a major type of ARG pollution in the effluents of WWTPs (Hu et al. 2019). As a result, eliminating these orphaned genes is crucial and an important step in disinfection. It is commonly known that an ozone concentration of 3 mg/L will remove 90% of ARBs and ARGs (Hu et al. 2019). This is explained by ozone's potent oxidizing power, which reacts with a range of cell materials and lessens the chance of targeting DNA or ARGs.

6.4.5 Biological Removal Methods

In WWTPs, the majority of antibiotics may be partially removed, releasing nondegraded antibiotics into the environment. This section provides a thorough summary of how antibiotics degrade in various biological treatment procedures when paired with modern oxidation techniques.

6.4.5.1 Conventional Activated Sludge (CAS)

CAS is the most commonly used technology for biological wastewater treatment. Biodegradation and adsorption are the most common methods of antibiotic removal in CAS. The microbial population secretes comparable enzymes that can destroy the former antibiotics, whereas the latter bacteria can use antibiotics as carbon sources and energy substrates to develop. As a result, microbes are capable of breaking down and changing antibiotics on their own (Tiwari et al. 2017). The metabolic pathway and breakdown of SMX by *Pseudomonas* psychrophilic HA-4, and it was found that *Pseudomonas* can use the antibiotic as its only source of energy and carbon to break down SMX (Jiang et al. 2014).

According to research by Torresi, MLs' adsorption coefficients in biofilms were higher than sludge in CAS (Torresi et al. 2016).

6.4.5.2 Membrane Bioreactor Processes

Membrane bioreactor (MBR) has received a lot of interest lately in wastewater treatment methods. It integrates membrane separation, biodegradation, and adsorption processes. The main method of reducing antibiotics from MBR, including SMX and TMP (trimethoprim), is biodegradation. With biodegradation, TMP and SMX were eliminated in an anaerobic membrane bioreactor on a lab scale with removal efficiencies of 94.2±5.5% and 67.8±13.9%, respectively (Xiao et al. 2017).

6.4.5.3 Bioelectrochemical System

Bioelectrochemical system (BES) is a process through which oxidation and reduction occur that can transform the chemical energy of organic matter into electrical energy, hydrogen, or other products with added value by employing electrochemically active microorganisms as a catalyst. BES is a developing technology used for energy recovery and the elimination of organic pollutants. It includes microbial electrolysis cells and microbial fuel cells (MFCs) (Xue et al. 2019). BES prevents the biocathodes from producing extremely harmful by-products. Using electrode cathode as an extracellular electron source in MFCs, for instance, nitroaromatic antibiotics (like CAP) are lowered by speeding up the chlorination reaction into nitro group and amino (Kong et al. 2015). According to, 3-amino-5-methylisoxazole, a very hazardous intermediary product of SMX elimination, was efficiently broken down in BESs (Wang et al. 2016).

6.5 CONCLUSION

In conclusion, AMR poses a significant global health threat due to abuse and overuse of antibiotics in human and veterinary medicine, coupled with bad environmental conditions resulting from the residues of these drugs in wastewater. Some of the mechanisms of AMR, such as genetic mutation and horizontal gene transfer, are important facilitators of the spread of resistance within microbial populations. The main cause of AMR is the WWTPs, which serve as breeding sites for resistant infections. Efficient tactics ought to be utilized in the effort to overcome the challenge through optimization of the physical, chemical, and biological treatments utilized in the treatment of wastewater for minimization of the resistant bacteria and the antibiotic resistance genes. Its minimization of effects and protection of public health will also be enhanced by the appropriate use of antibiotics and improved surveillance of AMR in different environmental settings.

REFERENCES

Abe, Kimihiro, Nobuhiko Nomura, and Satoru Suzuki. 2021. Biofilms: Hot Spots of Horizontal Gene Transfer (HGT) in Aquatic Environments, with a Focus on a New HGT Mechanism. *FEMS Microbiology Ecology*, 96 (5): fiaa031. doi:10.1093/FEMSEC/FIAA031.

Abushaheen, Manar Ali, Muzaheed, Amal Jamil Fatani, Mohammed Alosaimi, Wael Mansy, Merin George, Sadananda Acharya, et al. 2020. Antimicrobial Resistance, Mechanisms and Its Clinical Significance. *Disease-a-Month*, 66 (6): 100971. doi:10.1016/j.disamonth.2020.100971.

Ahmed, Sirwan Khalid, Safin Hussein, Karzan Qurbani, Radhwan Hussein Ibrahim, Abdulmalik Fareeq, Kochr Ali Mahmood, and Mona Gamal Mohamed. 2024. Antimicrobial Resistance: Impacts, Challenges, and Future Prospects. *Journal of Medicine, Surgery, and Public Health*, 2 (April): 100081. doi:10.1016/j.glmedi.2024.100081.

Arias-Andres, Maria, Uli Klümper, Keilor Rojas-Jimenez, and Hans Peter Grossart. 2018. Microplastic Pollution Increases Gene Exchange in Aquatic Ecosystems. *Environmental Pollution*, 237 (June): 253–261. doi:10.1016/J.ENVPOL.2018.02.058.

Aslam, Bilal, Mohsin Khurshid, Muhammad Imran Arshad, Saima Muzammil, Maria Rasool, Nafeesa Yasmeen, Taif Shah, et al. 2021. Antibiotic Resistance: One Health One World Outlook. *Frontiers in Cellular and Infection Microbiology*, 11: 771510. doi:10.3389/fcimb.2021.771510.

Becker, Karsten, Britta Ballhausen, Robin Köck, and André Kriegeskorte. 2014. Methicillin Resistance in Staphylococcus Isolates: The 'Mec Alphabet' with Specific Consideration of MecC, a Mec Homolog Associated with Zoonotic S. Aureus Lineages. *International Journal of Medical Microbiology*, 304 (7): 794–804. doi:10.1016/J.IJMM.2014.06.007.

Beyene, Hunachew, and Getachew Redaie. 2011. Assessment of Waste Stabilization Ponds for the Treatment of Hospital Wastewater: The Case of Hawassa University Referral Hospital. *World Applied Sciences Journal*, 15 (1): 142–150.

Burridge, Les, Judith S. Weis, Felipe Cabello, Jaime Pizarro, and Katherine Bostick. 2010. Chemical Use in Salmon Aquaculture: A Review of Current Practices and Possible Environmental Effects. *Aquaculture*, 306 (1–4): 7–23. doi:10.1016/j.aquaculture.2010.05.020.

Cahill, Niamh, Louise O'Connor, Bláthnaid Mahon, Áine Varley, Elaine McGrath, Phelim Ryan, Martin Cormican Carina Brehony, Keith A. Jolley, Martin C. Maiden, Sylvain Brisse, Dearbháile Morris. 2019. Hospital effluent: A reservoir for carbapenemase-producing Enterobacterales?. *Science of the Total Environment* 672: 618–624. doi:10.1016/j.scitotenv.2019.03.428.

Chen, Baowei, Lan Lin, Ling Fang, Ying Yang, Enzhong Chen, Ke Yuan, Shichun Zou, Xiaowei Wang, and Tiangang Luan. 2018. Complex Pollution of Antibiotic Resistance Genes Due to Beta-Lactam and Aminoglycoside Use in Aquaculture Farming. *Water Research*, 134 (May): 200–208. doi:10.1016/j.watres.2018.02.003.

Chen, Mei, Klas Ohman, Chris Metcalfe, Michael G. Ikonomou, Prasanna L. Amatya, and Jeffrey Wilson. 2006. Pharmaceuticals and Endocrine Disruptors in Wastewater Treatment Effluents and in the Water Supply System of Calgary, Alberta, Canada. *Water Quality Research Journal of Canada*, 41 (4): 351–364. doi:10.2166/wqrj.2006.039.

Chiş, Adriana Aurelia, Luca Liviu Rus, Claudiu Morgovan, Anca Maria Arseniu, Adina Frum, Andreea Loredana Vonica-Ţincu, Felicia Gabriela Gligor, Maria Lucia Mureşan, and Carmen Maximiliana Dobrea. 2022. Microbial Resistance to Antibiotics and Effective Antibiotherapy. *Biomedicines*, 10 (5): 1121. doi:10.3390/biomedicines10051121.

Conte, Danieli, Jussara Kasuko Palmeiro, Keite da Silva Nogueira, Thiago Marenda Rosa de Lima, Marco André Cardoso, Roberto Pontarolo, Flávia Lada Degaut Pontes, and Libera Maria Dalla-Costa. 2017. Characterization of CTX-M Enzymes, Quinolone Resistance Determinants, and Antimicrobial Residues from Hospital Sewage, Wastewater Treatment Plant, and River Water. *Ecotoxicology and Environmental Safety*, 136 (February): 62–69. doi:10.1016/j.ecoenv.2016.10.031.

Cosgrove, Sara E. 2006. The Relationship between Antimicrobial Resistance and Patient Outcomes: Mortality, Length of Hospital Stay, and Health Care Costs. *Clinical Infectious Diseases*, 42 (Suppl 2): S82–S89.

Denning, David W., David S. Perlin, Eavan G. Muldoon, Arnaldo Lopes Colombo, Arunaloke Chakrabarti, Malcolm D. Richardson, and Tania C. Sorrell. 2017. Delivering on Antimicrobial Resistance Agenda Not Possible without Improving Fungal Diagnostic Capabilities. *Emerging Infectious Diseases*, 23 (2): 177–283. doi:10.3201/eid2302.152042.

Dickinson, A. W., A. Power, M. G. Hansen, K. K. Brandt, G. Piliposian, P. Appleby, P. A. O'Neill, et al. 2019. Heavy Metal Pollution and Co-Selection for Antibiotic Resistance: A Microbial Palaeontology Approach. *Environment International*, 132 (November): 105117. doi:10.1016/j.envint.2019.105117.

Dires, Simachew, Tarekegn Birhanu, Argaw Ambelu, and Geremew Sahilu. 2018. Antibiotic Resistant Bacteria Removal of Subsurface Flow Constructed Wetlands from Hospital Wastewater. *Journal of Environmental Chemical Engineering*, 6 (4): 4265–4272. doi:10.1016/j.jece.2018.06.034.

Erdogan, Haluk, Fusun Can, Muge Demirbilek, Funda Timurkaynak, and Hande Arslan. 2010. In Vitro Activity of Antimicrobial Agents against Legionella Isolated from Environmental Water Systems: First Results from Turkey. *Environmental Monitoring and Assessment*, 171 (1–4): 487–491. doi:10.1007/s10661-009-1293-y.

Gruchlik, Yolanta, Kathryn Linge, and Cynthia Joll. 2018. Removal of Organic Micropollutants in Waste Stabilisation Ponds: A Review. *Journal of Environmental Management*, 206: 202–214. doi:10.1016/j.jenvman.2017.10.020.

Haberecht, Hannah B., Nora Jean Nealon, Jake R. Gilliland, Amethyst V. Holder, Connor Runyan, Renee C. Oppel, Hend M. Ibrahim, et al. 2019. Antimicrobial-Resistant Escherichia Coli from Environmental Waters in Northern Colorado. *Journal of Environmental and Public Health*, 18: 3862949. doi:10.1155/2019/3862949.

Harris, Suvi, C. Morris, Dearbháile Morris, Martin Cormican, and Enda Cummins. 2013. The Effect of Hospital Effluent on Antimicrobial Resistant *E. Coli* within a Municipal Wastewater System. *Environmental Sciences: Processes and Impacts*, 15 (3): 617–622. doi:10.1039/c2em30934c.

Harris, Suvi J., Martin Cormican, and Enda Cummins. 2012. Antimicrobial Residues and Antimicrobial-Resistant Bacteria: Impact on the Microbial Environment and Risk to Human Health-A Review. *Human and Ecological Risk Assessment*, 18 (4): 767–809. doi:10.1080/10807039.2012.688702.

Holmes, Alison H., Luke S.P. Moore, Arnfinn Sundsfjord, Martin Steinbakk, Sadie Regmi, Abhilasha Karkey, Philippe J. Guerin, and Laura J.V. Piddock. 2016. Understanding the Mechanisms and Drivers of Antimicrobial Resistance. *The Lancet*, 387 (10014): 176–187. doi:10.1016/S0140-6736(15)00473-0.

Homem, Vera, and Lúcia Santos. 2011. Degradation and Removal Methods of Antibiotics from Aqueous Matrices – A Review. *Journal of Environmental Management*, 92 (10): 2304–2347. doi:10.1016/J.JENVMAN.2011.05.023.

Hossain, Md, Md Islam, Md Sobur, Sumaiya Zaman, Ashrafun Nahar, Marzia Rahman, and Md Rahman. 2020. Exploring Poultry Farm Environment for Antibiotic Resistance *Escherichia Coli*, *Salmonella* Spp., and *Staphylococcus* Spp. Having Public Health Significance. *Journal of Bangladesh Agricultural University*, 18 (3): 615–622. doi:10.5455/jbau.98074.

Hu, Yaru, Tianyang Zhang, Lei Jiang, Yi Luo, Shijie Yao, Dong Zhang, Kuangfei Lin, and Changzheng Cui. 2019. Occurrence and Reduction of Antibiotic Resistance Genes in Conventional and Advanced Drinking Water Treatment Processes. *Science of the Total Environment*, 669 (June): 777–784. doi:10.1016/j.scitotenv.2019.03.143.

Iweriebor, Benson Chuks, Sisipho Gaqavu, Larry Chikwelu Obi, Uchechukwu U. Nwodo, and Anthony I. Okoh. 2015. Antibiotic Susceptibilities of Enterococcus Species Isolated from Hospital and Domestic Wastewater Effluents in Alice, Eastern Cape Province of South Africa. *International Journal of Environmental Research and Public Health*, 12 (4): 4231–4246. doi:10.3390/ijerph120404231.

Jeanvoine, Audrey, Alexandre Meunier, Hélène Puja, Xavier Bertrand, Benoît Valot, and Didier Hocquet. 2019. Contamination of a Hospital Plumbing System by Persister Cells of a Copper-Tolerant High-Risk Clone of Pseudomonas Aeruginosa. *Water Research*, 157 (June): 579–586. doi:10.1016/j.watres.2019.04.011.

Jiang, Benchao, Ang Li, Di Cui, Rui Cai, Fang Ma, and Yingning Wang. 2014. Biodegradation and Metabolic Pathway of Sulfamethoxazole by Pseudomonas Psychrophila HA-4, a Newly Isolated Cold-Adapted Sulfamethoxazole-Degrading Bacterium. *Applied Microbiology and Biotechnology*, 98 (10): 4671–4681. doi:10.1007/s00253-013-5488-3.

Karkman, Antti, Katariina Pärnänen, and D. G. Joakim Larsson. 2019. Fecal Pollution Can Explain Antibiotic Resistance Gene Abundances in Anthropogenically Impacted Environments. *Nature Communications*, 10 (1): 80. doi:10.1038/s41467-018-07992-3.

Khan, Faisal Ahmad, Bo Söderquist, and Jana Jass. 2019. Prevalence and Diversity of Antibiotic Resistance Genes in Swedish Aquatic Environments Impacted by Household and Hospital Wastewater. *Frontiers in Microbiology*, 10 (April): 688. doi:10.3389/fmicb.2019.00688.

Khurana, Pratishtha, Rama Pulicharla, and Satinder Kaur Brar. 2021. Antibiotic-Metal Complexes in Wastewaters: Fate and Treatment Trajectory. *Environment International*, 157 (December): 106863. doi:10.1016/j.envint.2021.106863.

Kong, Deyong, Bin Liang, Hui Yun, Haoyi Cheng, Jincai Ma, Minhua Cui, Aijie Wang, and Nanqi Ren. 2015. Cathodic Degradation of Antibiotics: Characterization and Pathway Analysis. *Water Research*, 72 (April): 281–292. doi:10.1016/j.watres.2015.01.025.

Kovalova, Lubomira, Hansruedi Siegrist, Urs Von Gunten, Jakob Eugster, Martina Hagenbuch, Anita Wittmer, Ruedi Moser, and Christa S. McArdell. 2013. Elimination of Micropollutants during Post-Treatment of Hospital Wastewater with Powdered Activated Carbon, Ozone, and UV. *Environmental Science and Technology*, 47 (14): 7899–7908. doi:10.1021/es400708w.

Kumar, Awanish, and Dharm Pal. 2018. Antibiotic Resistance and Wastewater: Correlation, Impact and Critical Human Health Challenges. *Journal of Environmental Chemical Engineering*, 6 (1): 52–58. doi:10.1016/j.jece.2017.11.059.

Lambs, Luc, Brigitte Decock-Le Rbvirend, Henryk Kozlowski, and Guy Berthon. 1988. Metal Ion-Tetracycline Interactions in Biological Fluids. 9. Circular Dichroism Spectra of Calcium and Magnesium Complexes with Tetracycline, Oxytetracycline, Doxycycline, and Chlortetracycline and Discussion of Their Binding Modes. *Inorganic Chemistry*, 27 (17): 3001–3012.

Larsson, D. G. Joakim. 2014. Pollution from Drug Manufacturing: Review and Perspectives. *Philosophical Transactions of the Royal Society B: Biological Sciences. Royal Society of London*, 369 (1656): 20130571. doi:10.1098/rstb.2013.0571.

Li, Wai Chin. 2014. Occurrence, Sources, and Fate of Pharmaceuticals in Aquatic Environment and Soil. *Environmental Pollution*, 187: 193–201. doi:10.1016/j.envpol.2014.01.015.

Li, Xueqing, Faisal I. Hai, and Long D. Nghiem. 2011. Simultaneous Activated Carbon Adsorption within a Membrane Bioreactor for an Enhanced Micropollutant Removal. *Bioresource Technology*, 102 (9): 5319–5324. doi:10.1016/j.biortech.2010.11.070.

Lüddeke, Frauke, Stefanie Heß, Claudia Gallert, Josef Winter, Hans Güde, and Herbert Löffler. 2015. Removal of Total and Antibiotic Resistant Bacteria in Advanced Wastewater Treatment by Ozonation in Combination with Different Filtering Techniques. *Water Research*, 69 (February): 243–251. doi:10.1016/j.watres.2014.11.018.

Lugoli, Federica, Massimo I. Leopizzi, Francesco Bagordo, Tiziana Grassi, Marcello Guido, and A. De Donno. 2011. Widespread Microbiological Groundwater Contamination in the South-Eastern Salento (Puglia-Italy). *Journal of Environmental Monitoring*, 13 (1): 192–200. doi:10.1039/c0em00193g.

Lulijwa, Ronald, Emmanuel Joseph Rupia, and Andrea C. Alfaro. 2020. Antibiotic Use in Aquaculture, Policies and Regulation, Health and Environmental Risks: A Review of the Top 15 Major Producers. *Reviews in Aquaculture*, 12 (2): 497–1265. doi:10.1111/raq.12344.

Luo, Yunlong, Wenshan Guo, Huu Hao Ngo, Long Duc Nghiem, Faisal Ibney Hai, Jian Zhang, Shuang Liang, and Xiaochang C. Wang. 2014. A Review on the Occurrence of Micropollutants in the Aquatic Environment and Their Fate and Removal during Wastewater Treatment. *Science of the Total Environment*, 473–474: 619–641. doi:10.1016/j.scitotenv.2013.12.065.

Manyi-Loh, Christy, Sampson Mamphweli, Edson Meyer, and Anthony Okoh. 2018. Antibiotic Use in Agriculture and Its Consequential Resistance in Environmental Sources: Potential Public Health Implications. *Molecules*, 23 (4): 795. doi:10.3390/molecules23040795.

Marano, Roberto B.M., Chhedi Lal Gupta, Tamar Cozer, Edouard Jurkevitch, and Eddie Cytryn. 2021. Hidden Resistome: Enrichment Reveals the Presence of Clinically Relevant Antibiotic Resistance Determinants in Treated Wastewater-Irrigated Soils. *Environmental Science and Technology*, 55 (10): 6814–6827. doi:10.1021/acs.est.1c00612.

Mesaros, Narcisa, Patrice Nordmann, Patrick Plésiat, Micheline Roussel-Delvallez, Johan Van Eldere, Youri Glupczynski, Yves Van Laethem, Frédérique Jacobs, Patrick Lebecque,Anne Malfroot, Paul M. Tulkens, Françoise Van Bambeke. 2007. Pseudomonas Aeruginosa: Resistance and Therapeutic Options at the Turn of the New Millennium. *Clinical Microbiology and Infection*, 13 (6): 560–578. doi:10.1111/j.1469-0691.2007.01681.x.

Mompelat, Sophie, Barbara Le Bot, and Olivier Thomas. 2009. Occurrence and Fate of Pharmaceutical Products and By-Products, from Resource to Drinking Water. *Environment International*, 35 (5): 803–814. doi:10.1016/j.envint.2008.10.008.

Monsalves, Naomi, Ana María Leiva, Gloria Gómez, and Gladys Vidal. 2022. Antibiotic-Resistant Gene Behavior in Constructed Wetlands Treating Sewage: A Critical Review. *Sustainability*, 14 (14): 8524. doi:10.3390/su14148524.

Monteiro, Sérgio Henrique, Graziela Moura Andrade, Fabiana Garcia, and Fabiana Pilarski. 2018. Antibiotic Residues and Resistant Bacteria in Aquaculture. https://www.researchgate.net/publication/329322582.

Naraginti, Saraschandra, Yi Li, and Gianluca Li Puma. 2018. Photocatalytic Mineralization and Degradation Kinetics of Sulphamethoxazole and Reactive Red 194 over Silver-Zirconium Co-Doped Titanium Dioxide: Reaction Mechanisms and Phytotoxicity Assessment. *Ecotoxicology and Environmental Safety*, 159 (September): 301–309. doi:10.1016/J.ECOENV.2018.04.062.

Narciso-da-Rocha, Carlos, Jaqueline Rocha, Ivone Vaz-Moreira, Felipe Lira, Javier Tamames, Isabel Henriques, José Luis Martinez, and Célia M. Manaia. 2018. Bacterial Lineages Putatively Associated with the Dissemination of Antibiotic Resistance Genes in a Full-Scale Urban Wastewater Treatment Plant. *Environment International*, 118 (September): 179–188. doi:10.1016/J.ENVINT.2018.05.040.

Niestępski, Sebastian, Monika Harnisz, Ewa Korzeniewska, Ma Guadalupe Aguilera-Arreola, Araceli Contreras-Rodríguez, Zofia Filipkowska, and Adriana Osińska. 2019. The Emergence of Antimicrobial Resistance in Environmental Strains of the Bacteroides Fragilis Group. *Environment International*, 124 (March): 408–419. doi:10.1016/j.envint.2018.12.056.

Östman, Marcus, Berndt Björlenius, Jerker Fick, and Mats Tysklind. 2019. Effect of Full-Scale Ozonation and Pilot-Scale Granular Activated Carbon on the Removal of Biocides, Antimycotics and Antibiotics in a Sewage Treatment Plant. *Science of the Total Environment*, 649 (February): 1117–1123. doi:10.1016/j.scitotenv.2018.08.382.

Petrovich, Morgan L., Adi Zilberman, Aviv Kaplan, Gefen R. Eliraz, Yubo Wang, Kathryn Langenfeld, Melissa Duhaime, et al. 2020. Microbial and Viral Communities and Their Antibiotic Resistance Genes throughout a Hospital Wastewater Treatment System. *Frontiers in Microbiology*, 11 (February): 153. doi:10.3389/fmicb.2020.00153.

Pham, Dung Ngoc, Lerone Clark, and Mengyan Li. 2021. Microplastics as Hubs Enriching Antibiotic-Resistant Bacteria and Pathogens in Municipal Activated Sludge. *Journal of Hazardous Materials Letters*, 2 (November): 100014. doi:10.1016/j.hazl.2021.100014.

Poole, Keith. 2005. Efflux-Mediated Antimicrobial Resistance. *Journal of Antimicrobial Chemotherapy*, 56 (1): 20–51. doi:10.1093/jac/dki171.

Qin, Jiayuan, Martina Maixnerová, Matěj Nemec, Yu Feng, Xinzhuo Zhang, Alexandr Nemec, and Zhiyong Zong. 2019. Acinetobacter Cumulans Sp. Nov., Isolated from Hospital Sewage and Capable of Acquisition of Multiple Antibiotic Resistance Genes. *Systematic and Applied Microbiology*, 42 (3): 319–325. doi:10.1016/j.syapm.2019.02.001.

Ramirez, Maria S., and Marcelo E. Tolmasky. 2010. Aminoglycoside Modifying Enzymes. *Drug Resistance Updates*, 13 (6): 151–171. doi:10.1016/j.drup.2010.08.003.

Redae, Getachew. 2011. Assessment of Waste Stabilization Ponds for the Treatment of Hospital Wastewater: The Case of Hawassa University Referral Hospital. *World Applied Sciences Journal*, 15 (1): 142–150

Redgrave, Liam S., Sam B. Sutton, Mark A. Webber, and Laura J.V. Piddock. 2014. Fluoroquinolone Resistance: Mechanisms, Impact on Bacteria, and Role in Evolutionary Success. *Trends in Microbiology*, 22 (8): 438–445. doi:10.1016/J.TIM.2014.04.007.

Schiavano, Giuditta Fiorella, Elisa Carloni, Francesca Andreoni, Silvia Magi, Maria Chironna, Giorgio Brandi, and Giulia Amagliani. 2017. Prevalence and Antibiotic Resistance of Pseudomonas Aeruginosa in Water Samples in Central Italy and Molecular Characterization of OprD in Imipenem Resistant Isolates. *PLoS One*, 12 (12): 189172. Public Library of Science. doi:10.1371/journal.pone.0189172.

Schwarz, Stefan, Axel Cloeckaert, and Marilyn C. Roberts. 2019. Mechanisms and Spread of Bacterial Resistance to Antimicrobial Agents. In *Antimicrobial Resistance in Bacteria of Animal Origin*, edited by Frank M. Aarestrup, 73–98. Washington, DC: ASM Press. doi:10.1128/9781555817534.ch6.

Serna-Galvis, Efraím A., Franklin Ferraro, Javier Silva-Agredo, and Ricardo A. Torres-Palma. 2017. Degradation of Highly Consumed Fluoroquinolones, Penicillins and Cephalosporins in Distilled Water and Simulated Hospital Wastewater by UV254 and UV254/Persulfate Processes. *Water Research*, 122 (1): 128–138. doi:10.1016/j.watres.2017.05.065.

Sharma, Divakar, Lama Misba, and Asad U. Khan. 2019. Antibiotics versus Biofilm: An Emerging Battleground in Microbial Communities. *Antimicrobial Resistance and Infection Control*, 8: 76. doi:10.1186/s13756-019-0533-3.

Sivalingam, Periyasamy, John Poté, and Kandasamy Prabakar. 2020. Extracellular DNA (EDNA): Neglected and Potential Sources of Antibiotic Resistant Genes (ARGs) in the Aquatic Environments. *Pathogens*, 9 (11): 874. doi:10.3390/pathogens9110874.

Soto-Giron, Maria J., Luis M. Rodriguez-R, Chengwei Luo, Michael Elk, Hodon Ryu, Jill Hoelle, Jorge W. Santo Domingo, and Konstantinos T. Konstantinidis. 2016. Biofilms on Hospital Shower Hoses: Characterization and Implications for Nosocomial Infections. *Applied and Environmental Microbiology*, 82 (9): 2872–83. doi:10.1128/AEM.03529-15.

Sulis, Giorgia, Benjamin Daniels, Ada Kwan, Sumanth Gandra, Amrita Daftary, Jishnu Das, and Madhukar Pai. 2020. Antibiotic Overuse in the Primary Health Care Setting: A Secondary Data Analysis of Standardised Patient Studies from India, China and Kenya. *BMJ Global Health*, 5 (9): e003393. doi:10.1136/bmjgh-2020-003393.

Syranidou, Evdokia, and Nicolas Kalogerakis. 2022. Interactions of Microplastics, Antibiotics and Antibiotic Resistant Genes within WWTPs. *Science of The Total Environment*, 804 (January): 150141. doi:10.1016/J.SCITOTENV.2021.150141.

Templeton, Michael R., Francine Oddy, Wing-kit Leung, and Michael Rogers. 2013. Chlorine and UV Disinfection of Ampicillin-Resistant and Trimethoprim-Resistant Escherichia Coli. *Journal of Environmental Engineering and Science*, 8 (2) 231–237. doi:10.1680/jees.2013.0025.

Tiwari, Bhagyashree, Balasubramanian Sellamuthu, Yassine Ouarda, Patrick Drogui, Rajeshwar D. Tyagi, and Gerardo Buelna. 2017. Review on Fate and Mechanism of Removal of Pharmaceutical Pollutants from Wastewater Using Biological Approach. *Bioresource Technology*, 224 (January): 1–12. doi:10.1016/j.biortech.2016.11.042.

Torresi, Elena, S. Jane Fowler, Fabio Polesel, Kai Bester, Henrik R. Andersen, Barth F. Smets, Benedek Gy Plósz, and Magnus Christensson. 2016. Biofilm Thickness Influences Biodiversity in Nitrifying MBBRs - Implications on Micropollutant Removal. *Environmental Science and Technology*, 50 (17): 9279–9288. doi:10.1021/acs.est.6b02007.

Tran-Dinh, Alexy, Caroline Neulier, Marlène Amara, Nicaise Nebot, Gilles Troché, Nelly Breton, Benjamin Zuber, et al. 2018. Impact of Intensive Care Unit Relocation and Role of Tap Water on an Outbreak of Pseudomonas Aeruginosa Expressing OprD-Mediated Resistance to Imipenem. *Journal of Hospital Infection*, 100 (3): e105–114. doi:10.1016/j.jhin.2018.05.016.

Uruén, Cristina, Gema Chopo-Escuin, Jan Tommassen, Raúl C Mainar-Jaime, and Jesús Arenas. 2020. Antibiotics Biofilms as Promoters of Bacterial Antibiotic Resistance and Tolerance. *Antibiotics*, 10 (1): 3. doi:10.3390/antibiotics.

Wang, Lu, Yulei Liu, Jun Ma, and Feng Zhao. 2016. Rapid Degradation of Sulphamethoxazole and the Further Transformation of 3-Amino-5-Methylisoxazole in a Microbial Fuel Cell. *Water Research*, 88 (January): 322–328. doi:10.1016/j.watres.2015.10.030.

Watkinson, A. J., E. J. Murby, Dana W. Kolpin, and S. D. Costanzo. 2009. The Occurrence of Antibiotics in an Urban Watershed: From Wastewater to Drinking Water. *Science of the Total Environment*, 407 (8): 2711–2723. doi:10.1016/J.SCITOTENV.2008.11.059.

Werkneh, Adhena Ayaliew, Gebretsadkan Gebremedhin Gebretsadik, and Shifare Berhe Gebru. 2023. Review on Environmental Selenium: Occurrence, Public Health Implications and Biological Treatment Strategies. *Environmental Challenges*, 11 (April): 100698. doi:10.1016/j.envc.2023.100698.

Werkneh, Adhena Ayaliew, and Md Aminul Islam. 2023. Post-Treatment Disinfection Technologies for Sustainable Removal of Antibiotic Residues and Antimicrobial Resistance Bacteria from Hospital Wastewater. *Heliyon*, 9 (4): e15360. doi:10.1016/j.heliyon.2023.e15360.

Whiley, Harriet. 2017. Legionella Risk Management and Control in Potable Water Systems: Argument for the Abolishment of Routine Testing. *International Journal of Environmental Research and Public Health*, 14 (1): 12. doi:10.3390/ijerph14010012.

Xiao, Yeyuan, Hazarki Yaohari, Cecilia De Araujo, Chun Chau Sze, and David C. Stuckey. 2017. Removal of Selected Pharmaceuticals in an Anaerobic Membrane Bioreactor (AnMBR) with/without Powdered Activated Carbon (PAC). *Chemical Engineering Journal*, 321: 335–45. doi:10.1016/j.cej.2017.03.118.

Xue, Gang, Mingji Jiang, Hong Chen, Min Sun, Yunfan Liu, Xiang Li, and Pin Gao. 2019. Critical Review of ARGs Reduction Behavior in Various Sludge and Sewage Treatment Processes in Wastewater Treatment Plants. *Critical Reviews in Environmental Science and Technology*, 49 (18): 1623–1674. doi:10.1080/10643389.2019.1579629.

Zhao, Panpan, Meng Ma, Yaqi Hu, Wenyong Wu, and Juan Xiao. 2022. Comparison of International Standards for Irrigation with Reclaimed Water. *Agricultural Water Management*, 1 (December): 107974. doi:10.1016/j.agwat.2022.107974.

Zhou, Zhen Chao, Wan Qiu Feng, Yue Han, Ji Zheng, Tao Chen, Yuan Yuan Wei, Michael Gillings, Yong Guan Zhu, and Hong Chen. 2018. Prevalence and Transmission of Antibiotic Resistance and Microbiota between Humans and Water Environments. *Environment International*, 121 (December): 1155–1161. doi:10.1016/j.envint.2018.10.032.

7 Characteristic and Impact of Antimicrobial Resistance in Wastewater

Salmiah Jamal Mat Rosid, Sarina Mat Rosid, Wan Nazwanie Wan Abdullah, and Nur Atiqah Nasir

7.1 INTRODUCTION

Antimicrobial wastewater is an emerging environmental concern characterized by the presence of antimicrobial agents designed to kill or inhibit the growth of microorganisms. These agents include a wide range of chemicals, such as antibiotics, antiseptics, disinfectants, and preservatives, which are essential in various sectors including health care, agriculture, and industry (Salam et al., 2023; Tong et al., 2021). While these compounds play a crucial role in controlling infections and maintaining hygiene standards, their release into wastewater systems poses significant challenges to environmental and public health.

Wastewater, which encompasses effluents from households, hospitals, industrial facilities, and agricultural runoffs, serves as a major reservoir for antibiotics, resistant bacteria, and resistance genes. Unlike other environmental reservoirs, wastewater is a confluence of diverse sources of antimicrobial agents and microorganisms, creating an ideal environment for the emergence and proliferation of resistant strains. The sources of antimicrobial wastewater are diverse, stemming from multiple facets of modern life. Healthcare facilities such as hospitals, clinics, and pharmaceutical manufacturers are major contributors, releasing a mixture of antibiotics and disinfectants used in treatment and sterilization processes (Mackulak et al., 2021). The improper disposal of unused medications, the extensive use of antibiotics in medical and veterinary practices, and the discharge of untreated or inadequately treated wastewater into natural water bodies are some of the primary factors contributing to the escalating AMR crisis (Samrot et al., 2023). Hospitals and clinics are primary sources of antibiotics and resistant bacteria. Patients undergoing antibiotic treatments can excrete unmetabolized drugs and resistant bacteria in their waste, which enters the sewage system. The high concentration of antibiotics in hospital effluents promotes the selection and proliferation of resistant strains.

One major factor contributing to the AMR issue is the extensive use of antibiotics in agriculture, especially in the raising of cattle and poultry (Manyi-Loh et al., 2018). Antibiotics are often used not only for therapeutic purposes but also as growth promoters, leading to their presence in animal waste. When this waste is used as a fertilizer or when runoff from farms enters water bodies, it introduces resistant bacteria and resistance genes into the environment. Industrial processes, including those in the cosmetics, cleaning products, and textile industries, also use various antimicrobial agents, if not properly treated, can enter water bodies and contribute to the environmental burden adding to the contamination of wastewater. Additionally, the contribution of households to AMR in wastewater is often underestimated (Ugoeze et al., 2024). Inappropriate disposal of antibiotics, such as flushing unused medications down the toilet, and the excretion of unmetabolized antibiotics by individuals undergoing treatment, introduce these substances into the sewage system. Other than that, households contribute through the use of antimicrobial soaps, detergents, and personal care

DOI: 10.1201/9781003269823-7

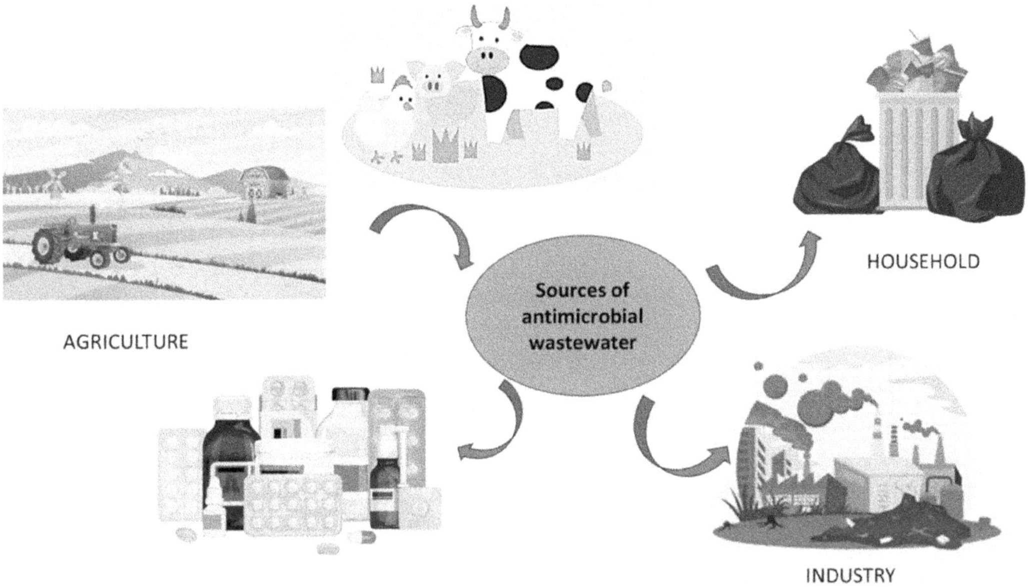

FIGURE 7.1 The diverse sources of antimicrobial wastewater.

products that are commonly washed down the drain, further adding to the complexity of antimicrobial wastewater. Figure 7.1 shows the diverse sources of antimicrobial wastewater.

The formation and dissemination of AMR bacteria might result from persistent exposure of microorganisms to sublethal doses of antimicrobial agents in wastewater. One of the biggest risks to global public health in the 21st century is AMR. The World Health Organization has identified AMR as a critical priority, highlighting the urgent need to address its spread across various environments, including health care, agriculture, and natural ecosystems.

These resistant strains pose a serious risk to public health because they can spread resistance genes to other bacteria, making antimicrobial treatments less effective. Furthermore, antimicrobial agents can disrupt natural microbial communities in water bodies and soils, leading to ecological disruptions that affect nutrient cycling, decomposition, and overall ecosystem health. Some antimicrobial compounds are persistent in the environment, accumulating in aquatic organisms and potentially entering the food chain, which can impact both wildlife and human health.

Next, the mechanism of spread is horizontal gene transfer (HGT) (Michaelis and Grohmann, 2023) and environmental dispersal. Wastewater environments facilitate the transfer of resistance genes between bacteria through HGT mechanisms, such as conjugation, transformation, and transduction. This process enables nonresistant bacteria to acquire resistance genes from resistant strains, thereby amplifying the spread of AMR. Wastewater treatment plants (WWTPs), though essential for reducing pollutant loads, are often not fully effective in eliminating antibiotics and resistant bacteria. Treated effluent, which may still contain these contaminants, is discharged into natural water bodies, spreading AMR beyond the point of origin.

More room and manual effort may be needed for older technologies than for more recent ones. Innovation in the industrial wastewater treatment sector has been fueled by two essential principles: operational efficiency and footprint reduction. To help separate the contaminants from the wastewater, chemicals are used in many industrial wastewater treatment systems. Supply chain problems have made industrial wastewater treatment operations more difficult recently. New technologies are required as a result of product shortages, price increases, and the need for substitutes. Addressing the challenges posed by antimicrobial wastewater requires advanced treatment methods that go beyond conventional wastewater treatment processes. Advanced oxidation processes

(AOPs) such as ozonation, photocatalysis, and Fenton reactions are effective techniques that generate highly reactive radicals capable of degrading antimicrobial compounds (Garrido-Cardenas et al., 2019). Bioremediation, which utilizes specific microorganisms that can metabolize and break down antimicrobial agents, offers a sustainable and ecofriendly treatment option. Membrane filtration technologies, including nanofiltration and reverse osmosis, provide a physical barrier to remove antimicrobial agents from wastewater, though these methods often demand significant energy and maintenance. Constructed wetlands, engineered to harness natural processes involving plants and microbes, also show promise in degrading and removing contaminants, including antimicrobials, from wastewater.

Next, the impact on public health and environment can be divided into three different levels, which are patient perspective, healthcare level, and economic level (Dadgostar, 2019). Two significant effects of AMR on patients are morbidity and mortality. Resistant bacteria will quadruple the risk of mortality and double the risk of getting a significant health problem when compared to nonresistant versions. Naturally, as the intensity of the resistant infections and the host's vulnerability increase, these detrimental effects will become more noticeable. To effectively mitigate the impacts of antimicrobial wastewater, it is crucial to implement robust monitoring and regulatory frameworks. Regular monitoring and comprehensive assessments of antimicrobial concentrations in wastewater and natural water bodies are essential for understanding the extent of contamination and the effectiveness of treatment strategies. Establishing and enforcing regulatory standards to limit the discharge of antimicrobial agents into the environment can help control the spread of AMR and protect public health. Additionally, promoting public awareness and education about the responsible use of antimicrobials in health care, agriculture, and households can play a significant role in reducing the amount of these compounds entering wastewater streams.

In conclusion, antimicrobial wastewater represents a complex environmental challenge that demands an integrated approach combining advanced treatment technologies, regulatory oversight, and public awareness. By addressing both the sources and impacts of antimicrobial agents in wastewater, we can safeguard public health, protect ecosystems, and ensure the sustainable use of these critical compounds. Through coordinated efforts across sectors and disciplines, we can develop effective solutions to manage and mitigate the risks associated with antimicrobial wastewater. Therefore, this chapter will discuss more about the characteristics of antimicrobial development and spread in wastewater in Section 7.2. Subsequently, Section 7.3 outlines the detection and monitoring of AMR. In Section 7.4, the impact of AMR on wastewater treatment processes is further explained. Finally, the challenges and future prospects of AMR were concluded in Section 7.5.

7.2 CHARACTERISTICS OF ANTIMICROBIAL DEVELOPMENT AND SPREAD IN WASTEWATER

By 2050, AMR is expected to be a major global public health hazard of the 21st century, resulting in 10 million deaths annually worldwide. AMR is a complex issue impacting economies, the environment, and the health of humans, animals, and society as a whole. AMR is also sometimes referred to as the "Silent Pandemic," for which prompt intervention is required. AMR and its spread in water environments exhibit several characteristic features that contribute to their significance and potential impact on human health and ecosystems. Figure 7.2 simplifies the factors resulting in AMR in water, which are selection pressure, HGT, persistence, and environmental reservoir.

The factors labeled in Figure 7.2 were explained in detail as follows:

1. **Selection Pressure:** Whether from agricultural runoff, treated wastewater, or pharmaceutical production effluents, the presence of antimicrobial agents in water produces a selective environment that promotes the survival and growth of resistant bacteria. Nonresistant bacteria are killed off, leaving behind resistant strains that can thrive and propagate.

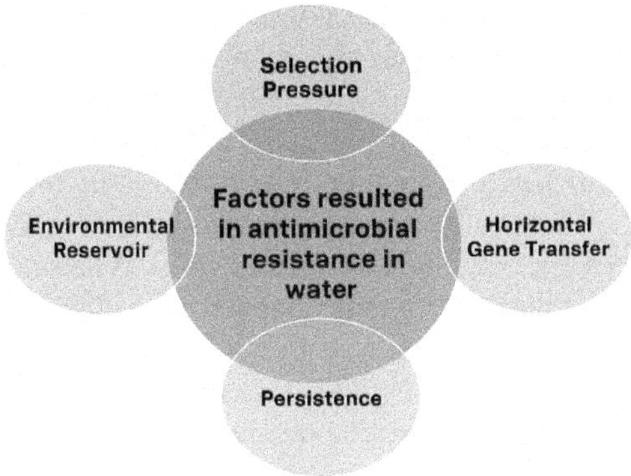

FIGURE 7.2 Factors resulted in antimicrobial resistance in water.

2. **Horizontal Gene Transfer:** Water environments facilitate the exchange of genetic material between different bacteria, a process known as HGT. This includes the transfer of resistance genes between bacteria, even across different species or genera. Water bodies can act as reservoirs where these resistance genes persist and spread.
3. **Persistence:** Antimicrobial residues and resistant bacteria can persist in water environments due to the complex interactions between microorganisms, organic matter, and physical–chemical properties of water. This persistence increases the likelihood of exposure and transmission of resistance traits.
4. **Environmental Reservoir:** Water serves as a crucial reservoir for resistant bacteria and resistance genes. Contaminated water can disseminate resistant bacteria to different geographical locations through water currents, human activities, and wildlife, contributing to the global spread of AMR.

7.2.1 Antimicrobial Resistance in Wastewater

The majority of antibiotics used in food production, agriculture, and medicine, along with resistant bacteria, wind up in wastewater (Begmatov et al., 2024). But there are still many questions about the microbial ecological processes that take place in those wastewater systems and how much they exacerbate or attenuate AMR. The biological therapy procedure continuously mixes bacteria with antibiotics at subinhibitory concentrations, which may foster the development and spread of bacteria resistance. Therefore, it is now widely known that one of the main point sources of antibiotic resistance genes (ARGs) in water resources is wastewater treatment facilities (WWTPs). For example, a number of studies found that the presence of WWTP effluents greatly raised the absolute abundances of ARGs in river water. However, the effects of WWTPs are not just felt in aquatic environments. The influence of WWTPs can occur in various riverine environments, as evidenced by reports indicating that the relative abundances of ARGs downstream of wastewater discharge locations are higher than in the upstream for biofilms and sediments (Lee et al., 2023). The development of ARGs and antibiotic resistance bacteria (ARB) is the primary cause for concern regarding the discharge of antibiotics into the environment. This decreases the therapeutic potential against infections that affect both humans and animals.

In big facilities, wastewater treatment typically occurs in three stages. In the first stage, water is cleaned physically; in the second stage, activated sludge (AS) is used in bioreactors for microbiological treatment; and in the third stage, water is finally treated and disinfected. Organic matter, ammonium, and (in more complicated procedures) phosphorus are eliminated by AS bacteria during the second stage of treatment, which can be carried out using various technologies. As a result of their adsorption on AS particles, which are eliminated along with excess AS, microorganisms present in the wastewater, including ARB, are now eliminated. At this point, however, the large densities of active bacteria and their thorough mixing with selection agents may promote the emergence of novel resistant strains, for example, by integrons capturing additional ARGs in conjunction with HGT. In biological treatment, high concentrations of bacteria and intimate contact with them increase the processes of conjugation, transformation, and transduction, which are all ways that HGT might occur. Wastewater is usually released into receiving ecosystems once it has been treated, such as lakes, rivers, seas, or agricultural soils. When used for irrigation, reclaimed wastewater can introduce AMR into these environments. This allows for the transfer of resistance between bacteria, including pathogenic ones, which poses a significant threat to human health and, in severe cases, can be life-threatening. Thus, ARGs and ARB reach the aquatic environment mostly through the final discharge into the receiving water because WWTPs are not built to remove them.

7.2.2 Antibiotic Resistance Genes and Antibiotic Resistance Bacteria in Wastewater

Large concentrations of antibiotic resistance bacteria (ARB) and antibiotic resistance gene (ARGs) can be observed in WWTP effluents and mostly in aquatic habitats (Begmatov et al., 2024). For instance, Enterobacteriaceae, *Pseudomonas aeruginosa*, *Escherichia coli*, and *Acinetobacter* spp. are among the resistant and multiresistant bacteria that are present in different WWTP processes. In one study, from the AS of two WWTPs in China, 30 ARGs encoding resistance to tetracycline, sulphonamides, quinolones, or macrolides were found, and ten ARGs (sulI, sulII, qnrB, tetB, tetH, tetS, tetG, tetT, tetX, and ermC) were found to be significantly enlarged in comparison to the abundance of 16S rRNA genes (Mao et al., 2015). ARG dynamics are not the same as bacterial fate. After wastewater treatment, the bacterial population dropped by up to 99%, but the relative abundance of ARGs in the effluent increased. Consequently, a significant amount of ARGs seeps into certain groundwater compartments and surface water downstream (Pazda et al., 2019). Figure 7.3 simplifies the major route of antibiotics, ARB, and ARGs from WWTPs and then to the aquatic environment.

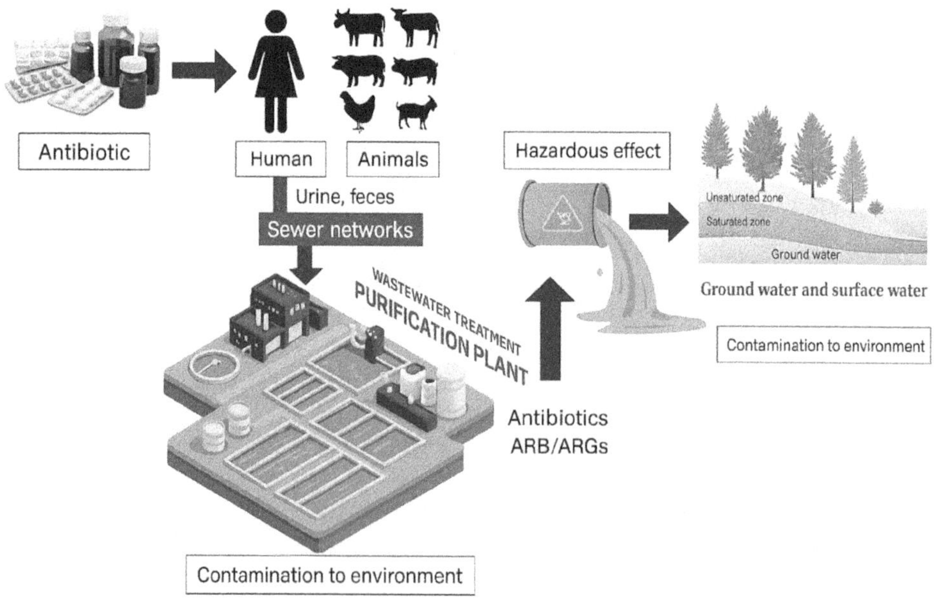

FIGURE 7.3 An outline showing the main path taken by antibiotics, ARB, and ARGs from WWTPs to aquatic ecosystems.

WWTPs receive high amounts of microbial contaminants from industrial, agricultural, and hospital wastewater, which exacerbates the genetic information transfer between commensal and environmental pathogenic bacteria. However, the methods and treatments employed in WWTPs are ineffective in eliminating antibiotic-resistant organisms (AROs). Consequently, it is clear that WWTPs can serve as efficient hotspots for the distribution of ARMs, AROs, and ARGs.

The high levels of antibiotics in wastewater cause selective pressure on the native bacteria found in WWTPs, which leads to the emergence of these ARB and ARGs. Research conducted by Lucas et al. (2016) highlighted that the rise in ARGs and ARB in WWTPs is largely due to antibiotic residues that create selective pressures on microbial communities. ARGs can multiply and spread among the bacteria that are already present in WWTPs during the water treatment process. To combat the proliferation of ARB and ARGs, it is essential to enhance wastewater treatment technologies and prevent the misuse of antibiotics.

Biofilm production accelerates the growth of ARGs and ARB in wastewater treatment facilities (WWTPs). These biofilms foster the growth of AROs by providing environments where high concentrations of bacteria can thrive. The elevated levels of antibiotics in WWTPs create strong selective pressures within these biofilms, leading to the rapid development and spread of ARB and ARGs (Huijbers et al., 2019). The bacterial communities within biofilms in WWTPs harbor dense colonies of bacteria that are resistant to antibiotics. Efforts to curb the spread of AROs in WWTPs should focus on strategies to prevent the formation and growth of these biofilms, which play a key role in the development of ARB and ARGs. Additionally, addressing the overuse of antibiotics, which results in high concentrations in WWTPs, is crucial. The design of WWTPs also significantly influences the proliferation and spread of ARB and ARGs, underscoring the need for plant designs that minimize these risks (Osińska et al., 2020).

There is currently insufficient recognition of the efficacy of eliminating ARB and ARGs in WWTPs, since this is not the real purpose for which these facilities were intended. To effectively control the proliferation of ARB and ARGs, WWTPs need to be redesigned with this specific goal in mind. Despite the introduction of recent technological innovations, some studies have shown that ARGs and ARB can still persist and even increase in WWTPs after treatment. This highlights the need for continued improvement and adaptation of wastewater treatment technologies to address the challenge of AMR effectively (Li et al., 2017). Nonetheless, effort and study should be made to close the huge gap in this subject.

7.2.3 ARB and ARGs' Development and Spreading Mechanism

Numerous methods exist for bacteria to show signs of resistance toward antibacterial medications and drugs. Certain bacterial species are naturally resistant to various antimicrobial drugs. In these conditions, every strain of that particular bacterial species is resistant to every member of those antibiotic classes. Cases of acquired resistance, when populations of bacteria that were previously susceptible to an antibacterial agent developed a resistance to it and multiplied and dispersed due to the agent's selection pressure, are more concerning. Different antibiotic resistance mechanisms can easily spread across different bacterial genera (Tenover, 2006). The formation and dissemination mechanism of ARB and ARGs involves both nonresistant-bacterium and antimicrobial-resistant-bacterium pathways. According to Li et al. (2022), the pathway for ARB and ARGs' development and spreading mechanism may consist of four steps and was illustrated in detail in Figure 7.4.

The persistence and spread of AMR in wastewater systems pose a serious threat to world health since it diminishes the efficiency of antibiotics, making illnesses more difficult to cure and raising the possibility of disease transmission, serious sickness, and death. Efforts to combat AMR in wastewater require a multifaceted approach, including improving wastewater treatment processes, reducing the use of antibiotics, promoting responsible disposal practices, and strengthening

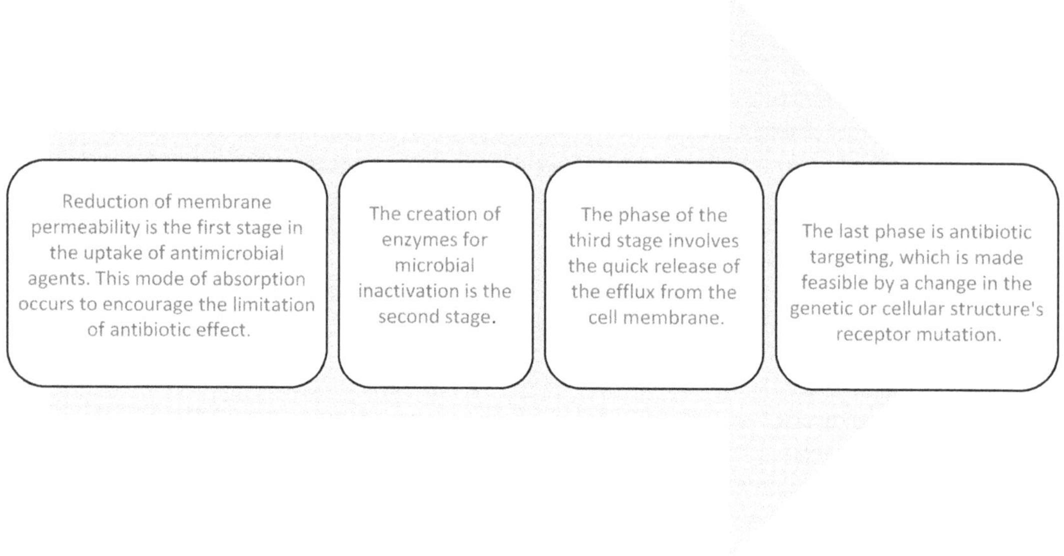

FIGURE 7.4 The pathway for ARB and ARGs' development and spreading mechanism.

monitoring and regulation. Those prompt actions are necessary to restrict the release of antibiotics, ARB, ARGs, and MGEs into the environment from human wastewater streams, hence managing the chronic risk of exposure to AMR for both humans and animals.

7.3 DETECTION AND MONITORING OF ANTIMICROBIAL RESISTANCE

AMR presents a major challenge to health care, and if it is not controlled, it will cause over ten million deaths annually within the next few decades (Murray et al., 2022). AMR is a serious health issue worldwide. Water environments, including natural water bodies, wastewater, and drinking water systems, are major factors in the spread of AMR. Monitoring AMR in water is crucial to understanding its prevalence, spread, and potential impact on human and environmental health.

7.3.1 Reasons for Environmental AMR Monitoring

In veterinary AMR surveillance and clinical settings, the goal is typically to enable targeted interventions for specific pathogens and antimicrobial treatment protocols. Similarly, surveillance data can be utilized to implement measures that alter antimicrobial usage patterns and quantities (Bengtsson-Palme et al., 2023). The necessity to monitor the AMR situation in the general population of urban area or city has frequently served as the driving force behind the sewage and effluent monitoring (Huijbers et al., 2020) in a low-cost manner. As evidenced by the COVID-19 pandemic, this approach has been shown to provide useful and practical details regarding the spread of viruses in humans (Bonanno Ferraro et al., 2022). Sewage monitoring aims to use sewage as a representative sample of people in a restricted region, like a city, rather than directly address environmental issues linked to AMR. It becomes very necessary to analyze the many roles that the environment plays in relation to AMR if one chooses to see monitoring as a tool to evaluate hazards to animal health and human related with environmental development and transmission of resistance. To control the spread of resistant microorganisms, it is important to identify the primary sources that introduce resistant bacteria into the environment. These sources include sewage and sewage treatment plants, industrial sources, and agricultural and aquaculture activities.

It is crucial to identify the circumstances in which people encounter high quantities of bacteria that are resistant in the environment. The research on these exposure processes is limited, and our knowledge of how resistant bacteria spread from environments to humans relies heavily on anecdotal evidence (Stanton et al., 2022). Ways of exposure could include eating fresh fruits and vegetables (Rahman et al., 2022) or participating in recreational swimming (Leonard et al., 2015). It is currently unclear if these pathways are truly the primary ways that diseases spread from the environment to people. Additionally, monitoring for AMR could be conducted to specifically observe any shifts in the prevalence of AMR over time in specific environments. This could enable temporary measures, such as shutting down beaches or advising people to boil their drinking water.

Environmental monitoring can track changes in AMR over time to evaluate the effectiveness of measures aimed at reducing AMR in the environment and potentially guide temporary interventions. Moreover, monitoring might aim to identify environments that could select for resistant bacteria and ARGs. Finally, it could serve as an early warning system for detecting novel resistance mechanisms before they become prevalent in human pathogens.

7.3.2 Culture-Based Methods

Culture-based approaches are useful for AMR surveillance because they allow for the selection of targets such as taxonomic groups that include human pathogens, and the methods for determining clinical resistance levels are relatively well standardized. In selective culturing, water samples are cultured on media containing specific antibiotics to isolate resistant strains of bacteria. Culture-based targets are useful in assessing human health risks. However, culture methods for AMR monitoring are limited because many genera and species in surface water and wastewater could be valuable targets. The increased number of microorganisms in these environments can interfere with the isolation of the desired targets. Furthermore, most environmental microbes are difficult to culture, meaning that culture-based methods will likely miss the broader spectrum of resistance present in a given environmental reservoir. No single target can fully represent the AMR status of a particular environment.

E. coli and *Enterococcus*, which are fecal indicator bacteria and fecal coliforms, are readily chosen due to their established role in regulatory monitoring of water and wastewater. Bacterial targets commonly found in sewage, such as *Aeromonas* and *Pseudomonas aeruginosa,* which have broader environmental niches than traditional fecal indicators, should also be considered. These bacteria can survive and proliferate in the environment, making them valuable indicators of AMR acquired through HGT of ARGs. This process is often influenced by human activities and sustained by pollution in the receiving environments. Culture-based monitoring has been extensively used to study changes in *E. coli* populations throughout wastewater treatment. A thorough study found no evidence of *E. coli* acquiring ARGs during treatment. However, this may be partly due to the tendency of *E. coli* to die off during the treatment process. Consequently, researchers have investigated resistance patterns in organisms like *Aeromonas*, which are capable of growing in WWTPs.

Culture-based methods are commonly used to monitor AMR in wastewater by isolating and identifying bacteria that are capable of growing in the presence of specific antibiotics. The process begins with sample collection, followed by bacterial isolation and incubation for colony identification. Antibiotic testing is then conducted, and the procedure concludes with data analysis.

Culture-based methods are valuable because they allow for the direct observation of live bacteria that are resistant, providing insights into how AMR bacteria survive and propagate within wastewater systems. However, they may not capture all resistant bacteria, especially those that are viable but nonculturable or those with resistance genes not expressed under the test conditions.

7.3.3 Molecular Methods

Molecular methods, including real-time quantitative polymerase chain reaction (qPCR) and random amplified polymorphic DNA, metagenomic and whole-genome sequencing (WGS), are used to identify and detect resistance-associated genes (genotypic resistance). However, the majority of these methods, aside from metagenomics and WGS, rely on target gene amplification. A qPCR offers highly sensitive detection and quantification across a wide range of concentrations, making it a valuable resource for determining and describing the patterns of human-caused changes. This method detects resistance gene and characterizes the AMR profile.

7.3.4 Detection of Antimicrobial Resistance Using Spectrometric Techniques

The best way to manage the patient's treatment outcomes is to diagnose them quickly. If pathogenic diseases cannot be effectively treated, rates of death and morbidity will rise along with the expense of patient care. For more effective treatment and targeted prescriptions against harmful germs, prompt diagnosis is essential. The issue is that traditional approaches take too long to distinguish resistance strains from nonresistance ones. Promising findings have been observed in the rapid and precise detection of AMR using advanced spectroscopic techniques, including mass infrared (IR), nuclear magnetic resonance (NMR), and mass spectrometry (MS).

7.3.4.1 MALDI-TOF Mass Spectrometry

Matrix-assisted laser desorption ionization (MALDI)-TOF/MS allowed for the rapid and accurate identification of therapeutically relevant pathogens, including bacteria and fungi. In clinical microbiology, MALDI is widely used for AMR detection and microbiological identification. As a result, a number of studies examined the viability of identifying antifungal and antibiotic resistance in pathogenic bacteria isolated from bloodstream infections using MALDI-TOF/MS technology (Ramzan et al., 2024).

7.3.4.2 LC-MS

Tandem mass spectrometry coupled with liquid chromatography (LC-MS/MS) offers enhanced sensitivity for identifying small peptides specific to an organism or enzyme compared to MALDI-TOF MS. All known AMR genes may be found at the molecular level using LC-MS/MS, which can also identify different species of bacteria. The majority of proteomics research on AMR focuses on specific proteins that act as AMR drivers (Cecchini et al., 2018).

Using LC-MS/MS analysis, a number of techniques have been developed for the quick identification of AMR in bacteria. One such technique is rapid resistance testing (RRT), which can identify *Enterobacteriaceae* resistance to cephalosporins and carbapenems efficiently. In less than 2h, the protocol can be finished, which entails detecting hydrolysis in a panel of β-lactam antibiotics. Additionally, ten *E. coli* isolates susceptible to imipenem, cefazolin, ampicillin, and cefotaxime were screened (Tadros et al., 2018). LC-MS/MS analysis is proven to be a useful tool in the quick identification of AMR in bacteria, with several techniques focusing on various facets of resistance.

7.3.4.3 IR Spectroscopy

IR spectroscopy can detect AMR by identifying molecular changes linked to resistance. Chemical alterations in *S. aureus* associated with the development of resistance to last-resort antibiotics, such as daptomycin and vancomycin, were detected using an innovative technique that combines chemometric analysis with atomic force microscopy-infrared spectroscopy. Paired clinical isolates were employed to pinpoint molecular changes specific to resistance development. These included resistant clinical and susceptible isolates, where the susceptible strain was collected before the resistant (daughter) strain and antibiotic treatment was obtained from the same patient after treatment failure.

Wijesinghe et al. have investigated the capacity of a less expensive portable benchtop attenuated total reflectance-Fourier transform infrared spectroscopy (ATR)-FTIR spectrometer to detect AMR using a straightforward, clinically oriented sampling procedure. A multiday study of a strain similar to *E. coli* was used as a quality control to ensure that the methodology could be repeated. The information gathered for this work points to the possible implementation of a portable FTIR spectrometer as a quick AST prediction tool to facilitate the effective use of antibiotics (Wijesinghe et al., 2021).

7.3.4.4 Nuclear Magnetic Resonance Spectroscopy

The ability to do atomic-level analysis of molecules and the capability to carry out numerous (NMR) tests on the same sample are the two main advantages of NMR spectroscopy. NMR can therefore be used to measure and identify different chemicals in sample combinations. Microbiological susceptibility and resistance studies have made extensive use of NMR spectroscopy. To ascertain antibiotic minimum inhibitory concentrations, pinpoint resistance routes, and describe microbial populations, NMR-based techniques have been developed recently. The effect of different antibiotic drugs on *E. coli* cultures has been monitored using NMR spectroscopy. Five different classes of antimicrobials were used in the investigation, and intracellular and extracellular fingerprint and footprint analyses were used to examine the effects of the antibiotics. Details about the production of cell walls, DNA, and several crucial proteins were found (Hoerr et al., 2016).

7.4 IMPACT OF ANTIMICROBIAL RESISTANCE ON WASTEWATER TREATMENT PROCESSES

AMR poses a significant challenge to wastewater treatment processes, complicating efforts to manage and mitigate the spread of resistant microorganisms. Antibiotics and other antimicrobial compounds, which are frequently utilized in health care, agriculture, human waste, and industrial effluents, tend to enter wastewater systems (Figure 7.5). The existence of these compounds in wastewater can apply selective pressure on microbial populations, promoting the survival and expansion of resistant strains (Sambaza & Naicker, 2023). Wastewater treatment processes do not have the capability to eliminate antibiotic and AMR determinants present in wastewater. This can lead to a bigger occurrence of resistant bacteria in treated effluent, which could have implications for public health and the environment when it comes to the disposal and reuse of wastewater (Kalli et al., 2023).

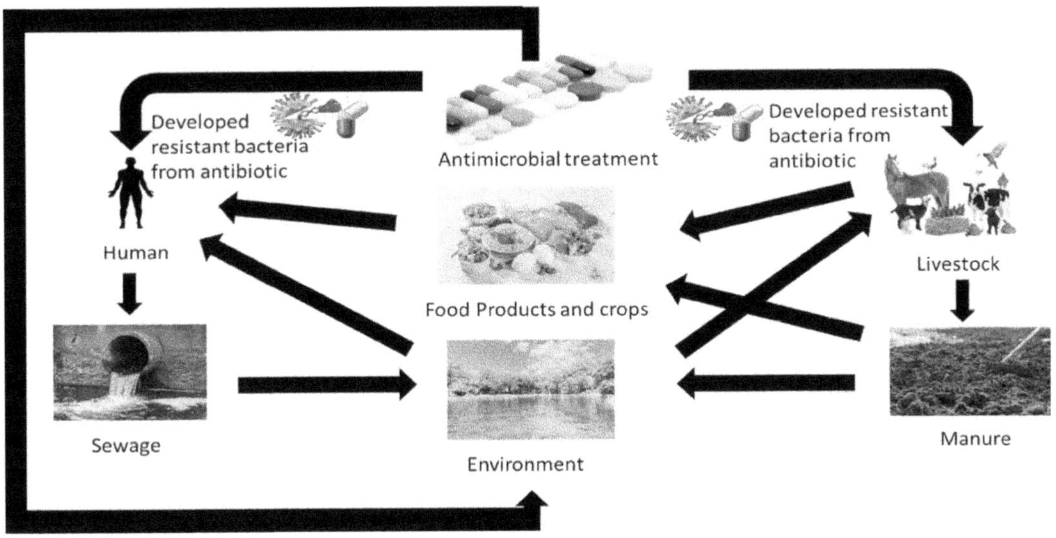

FIGURE 7.5 Spread of antimicrobial resistance.

The existence of antibiotics and antimicrobial compounds in wastewater applies selective pressure on microbial populations, resulting in the emergence and continuation of resistant bacteria. The crowded and diverse microbial environment in wastewater treatment process can also lead to the emergence and spread of new resistant strains through HGT, complicating the microbial dynamics. The main consequence of AMR on wastewater treatment is the diminished effectiveness of conventional treatment methods. This is because the standard wastewater treatment processes may not fully eliminate AMR. A considerable proportion of AMR has the ability to persist during the treatment process and thereafter be released into water bodies, possibly affecting aquatic ecosystems and the quality of drinking water supplies (Honda et al., 2023; Kasuga et al., 2022), hence facilitating the dissemination of antimicrobial-resistant bacteria in the environment.

Furthermore, a portion of AMR bacteria might also become more concentrated in the sludge throughout the process of treating wastewater. Utilizing this sludge as a fertilizer in agriculture may bring bacteria with AMR into the soil, hence increasing the potential for additional environmental pollution. The presence of AMR in the environment also increases the risk of human exposure, either through contact directly with contaminated water or indirectly through the food chain (such as crops irrigated with contaminated water or seafood sourced from contaminated areas). This would increase the difficulty of managing infections in both human and veterinary medicine.

7.4.1 Wastewater Treatment Methods

Typically, conventional wastewater treatment facilities utilize physical, chemical, and biological techniques to get rid of impurities, such as microorganisms, as shown in Figure 7.6. Nevertheless, the presence of antimicrobial-resistant bacteria can significantly affect the treatment process and result in environmental contamination. Antimicrobial-resistant microorganisms impact all stages of wastewater treatment, like in preliminary treatment, primary treatment, secondary treatment, and tertiary treatment.

In preliminary treatment, the influent comprises a diverse range of pollutants, such as antibiotics, and antimicrobial-resistant bacteria (Uluseker et al., 2021). These contaminants originate from hospitals, residences, industrial sites, and agricultural runoff. These pollutants are introduced into the system during the initial phase of treatment. Antimicrobial-resistant bacteria do not have a

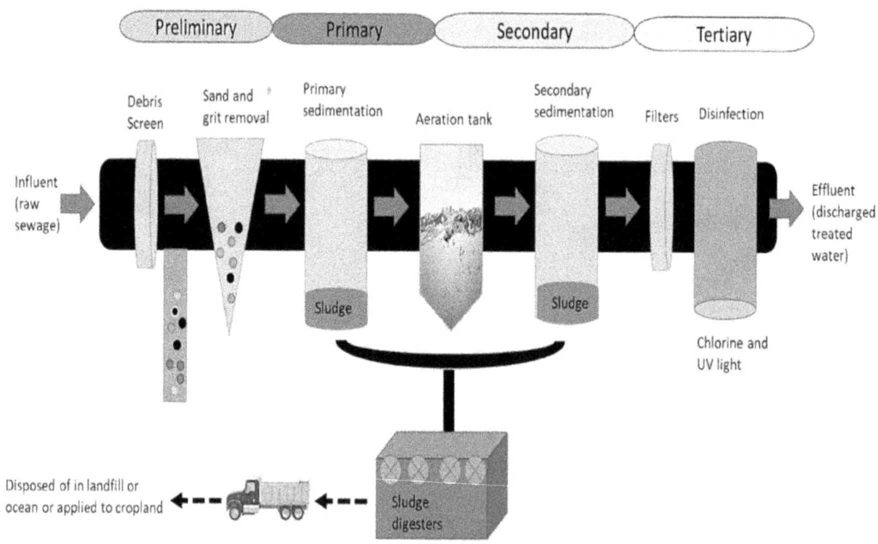

FIGURE 7.6 Wastewater treatment processes.

direct impact on the mechanical processes of screening and grit removal. However, the introduction of antibiotics and resistant bacteria into the system starts to apply selection pressure on microbial populations (Bouki et al., 2013; Zhang et al., 2009).

Primary treatment procedures mostly utilize physical processes such as sedimentation, which may not adequately eliminate antimicrobial compounds or resistant microbes. During the primary treatment process, certain bacteria adhere to the organic substances present in the tanks and are subsequently kept within the primary treatment tanks. The bacteria will engage in direct interaction with the incoming genetic material, creating a chance for additional genetic exchanges among and between the bacteria (Manaia et al., 2018). This, in turn, leads to an increment in the relative abundance of resistant bacteria in the sludge.

The secondary treatment employs biological processes utilizing microbial mechanisms to decompose organic waste, but it may not explicitly address antibiotic-resistant microorganisms. Furthermore, the existence of antibiotics in wastewater might impede or eliminate susceptible bacteria, disturbing the microbial population and diminishing treatment effectiveness. In addition, the biological treatment also creates an environment that promotes the rapid growth of AMR. Throughout the treatment process, the sewage microbiota consists of a combination of bacteria derived from human waste and other bacteria originating from different sources, resulting in a very conducive environment for genetic exchange (Sambaza & Naicker, 2023). The metabolic activities that occur during the biological treatment may develop a completely new population of resistant bacteria.

Finally, the last stage of treatment involves filtration and chemical processes. However, many of these processes are not specifically designed to address the complex challenge posed by antimicrobial-resistant organisms (Sambaza et al., 2019). Consequently, resistant bacteria and other microorganisms can survive the treatment process and be discharged into receiving waters, where they can further propagate and spread resistance genes. This potential for resistant strains to escape treatment systems increases the overall burden of resistance in the environment.

Due to that, advanced treatment technologies, such as ultraviolet (UV) disinfection, membrane bioreactors (MBRs) and ozonation and AOPs, offer improved performance in removing resistant microorganisms compared to conventional methods. Nonetheless, these technologies may still fall short in completely eliminating all resistant strains. For example, MBRs, although capable of producing high-quality effluent, are nonetheless susceptible to the transmission of resistant bacteria, particularly when the wastewater contains elevated levels of resistance genes. UV disinfection can be efficient, but its efficiency is contingent upon variables such as the specific type of UV radiation employed and the existence of organic substances that can provide protection for germs. The integration of ozonation and AOPs can offer improved removal of resistant microorganisms. These methods generate highly reactive species that can degrade antimicrobial agents and disrupt microbial cells, potentially reducing the prevalence of resistance. However, it is important to note that AOPs have limitations in completely eradicating resistance and can be costly and energy-intensive.

7.4.2 Impact on Wastewater Treatment Infrastructure

Infrastructure used for wastewater treatment is also at risk from the presence of AMR in wastewater. In biofilm-based systems, antibiotics can disrupt the stability and composition of the biofilm. On the other hand, resistant bacteria have the ability to create biofilms on surfaces in treatment systems (Mirghani et al., 2022). This allows for the transfer of genes across various bacterial species, which in turn promotes the growth and spread of resistant bacteria. This could result in clogging, reduced efficiency of organic matter degradation and nutrient removal and increased maintenance needs. Solving issues connected to biofilm requires the adoption of extra monitoring and intervention strategies, which could lead to higher operational costs and greater complexity.

Another issue of concern is the possible influence of AMR on the environment. Introducing resistant microbes into aquatic environments can occur when treated wastewater is released into natural water bodies. This can result in a series of consequences on microbial communities, which have the ability to disrupt normal processes and give rise to the development of new mechanisms of resistance (Mutuku et al., 2022). Moreover, bacteria that are resistant to antibiotics have the ability to engage with other microorganisms in the environment, thereby facilitating the transmission of genes that confer resistance through HGT.

The release of wastewater carrying resistant microbes poses a threat to public health. Exposure to polluted water sources, either through recreational activities or drinking contaminated water, can result in diseases that are challenging to treat due to resistance. This is especially alarming in areas where there is limited availability of potent antibiotics, which worsens the impact of AMR.

7.4.3 Summary of Impacts of Wastewater Treatment Processes

To summarize, AMR has a substantial impact on the processes involved in treating wastewater. This presents difficulties for both the efficiency of treatment technologies and the safeguarding of public health and the environment. To tackle this problem, a comprehensive strategy is needed, which involves implementing cutting-edge treatment techniques, conducting thorough surveillance, and fostering cooperation across different sectors. Gaining a deeper comprehension of AMR and its consequences for wastewater treatment will help us develop more long-lasting and efficient ways to counteract this escalating danger.

7.5 CHALLENGES AND FUTURE PROSPECTS

Addressing the attributes and consequences of AMR in wastewater is a complex attempt that requires a well-coordinated and thorough strategy. This strategy should incorporate advanced technologies, extensive monitoring systems, strong regulatory structures, active involvement of the public, and ongoing research and innovation. To reduce the effects of AMR in wastewater, it is essential to focus on the creation and adoption of cutting-edge treatment methods. AOPs, such as ozonation and photocatalysis, have the ability to effectively break down complex antimicrobial compounds and deactivate resistant bacteria that conventional treatment methods may not be able to handle. In addition, bioremediation, which involves the use of microbial communities, and phytoremediation, which utilizes plants, offer environmentally benign and sustainable options for breaking down persistent antimicrobial chemicals.

Implementing worldwide surveillance networks is another crucial element in treating AMR. These networks would allow for extensive surveillance and data gathering in many areas, making it easier to identify and measure AMR in wastewater. Establishing and executing standardized methods for these surveillance activities would guarantee that data is comparable and dependable, facilitating trend analysis and the establishment of early warning systems. An international outlook on AMR will enable prompt responses and the exchange of effective ideas for reducing its impact across different countries and areas.

Considerable focus is needed on the regulatory framework pertaining to AMR in wastewater. Establishing standardized norms and guidelines that are consistent across both national and international boundaries can result in consistent practices and enhanced adherence. Moreover, providing incentives to enterprises and wastewater treatment facilities to enhance and embrace cutting-edge technologies can encourage the wider implementation of optimal methods. This regulatory framework should also tackle the difficulties in enforcing regulations and offer explicit instructions on how to handle the release of antimicrobial substances.

Public awareness and education are crucial elements in the endeavor to address AMR. Involving a wide range of individuals, such as legislators, industry leaders, healthcare professionals, and the general public, in meaningful discussions can promote a shared sense of responsibility. Implementing educational initiatives that focus on encouraging the responsible utilization of antimicrobials and raising awareness about the detrimental environmental consequences of their improper usage can effectively induce behavioral modifications that result in a decrease in the quantity of antimicrobials being discharged into the wastewater system. To effectively manage the dangers associated with AMR in wastewater, a comprehensive and proactive approach is necessary. This approach should address the technological, regulatory, and sociological aspects of AMR. These endeavors will not only safeguard the well-being of the general population but also guarantee the long-term viability of our ecosystems for future cohorts.

REFERENCES

Begmatov, S., Beletsky, A. V., Dorofeev, A. G., Pimenov, N. V., Mardanov, A. V., & Ravin, N. V. (2024). Metagenomic insights into the wastewater resistome before and after purification at large-scale wastewater treatment plants in the Moscow city. *Scientific Reports*, *14*(1), 1–12. https://doi.org/10.1038/s41598-024-56870-0

Bengtsson-Palme, J., Abramova, A., Berendonk, T. U., Coelho, L. P., Forslund, S. K., Gschwind, R., Heikinheimo, A., Jarquín-Díaz, V. H., Khan, A. A., Klümper, U., Löber, U., Nekoro, M., Osińska, A. D., Ugarcina Perovic, S., Pitkänen, T., Rødland, E. K., Ruppé, E., Wasteson, Y., Wester, A. L., & Zahra, R. (2023). Towards monitoring of antimicrobial resistance in the environment: For what reasons, how to implement it, and what are the data needs? *Environment International*, *178*, 108089. https://doi.org/10.1016/j.envint.2023.108089

Bonanno Ferraro, G., Veneri, C., Mancini, P., Iaconelli, M., Suffredini, E., Bonadonna, L., Lucentini, L., Bowo-Ngandji, A., Kengne-Nde, C., Mbaga, D. S., Mahamat, G., Tazokong, H. R., Ebogo-Belobo, J.T., Njouom, R., Kenmoe, S., & La Rosa, G. (2022). A state-of-the-art scoping review on SARS-CoV-2 in sewage focusing on the potential of wastewater surveillance for the monitoring of the COVID-19 pandemic. *Food and Environmental Virology*, *14*(4), 315–354. https://doi.org/10.1007/s12560-021-09498-6

Bouki, C., Venieri, D., & Diamadopoulos, E. (2013). Detection and fate of antibiotic-resistant bacteria in wastewater treatment plants: A review. *Ecotoxicology and Environmental Safety*, *91*, 1–9. https://doi.org/10.1016/j.ecoenv.2013.01.016

Cecchini, T., Yoon, E. J., Charretier, Y., Bardet, C., Beaulieu, C., Lacoux, X., Docquier, J. D., Lemoine, J., Courvalin, P., Grillot-Courvalin, C., & Charrier, J. P. (2018). Deciphering multifactorial resistance phenotypes in *Acinetobacter baumannii* by genomics and targeted label-free proteomics. *Molecular and Cellular Proteomics*, *17*(3), 442–456. https://doi.org/10.1074/mcp.RA117.000107

Dadgostar P. (2019). Antimicrobial resistance: Implications and costs. *Infection and Drug Resistance*, *12*, 3903–3910. https://doi.org/10.2147/IDR.S234610

Garrido-Cardenas, J. A., Esteban-García, B., Agüera, A., Sánchez-Pérez, J. A., & Manzano-Agugliaro, F. (2019). Wastewater treatment by advanced oxidation process and their worldwide research trends. *International Journal of Environmental Research and Public Health*, *17*(1), 170. https://doi.org/10.3390/ijerph17010170

Hoerr, V., Duggan, G. E., Zbytnuik, L., Poon, K. K. H., Große, C., Neugebauer, U., Methling, K., Löffler, B., & Vogel, H. J. (2016). Characterization and prediction of the mechanism of action of antibiotics through NMR metabolomics. *BMC Microbiology*, *16*(1), 82. https://doi.org/10.1186/s12866-016-0696-5

Honda, R., Matsuura, N., Sorn, S., Asakura, S., Morinaga, Y., Huy, T. V., Sabar, M. A., Masakke, Y., Hara-Yamamura, H., & Watanabe, T. (2023). Transition of antimicrobial resistome in wastewater treatment plants: Impact of process configuration, geographical location and season. *NPJ Clean Water*, *6*(46), 1–12. https://doi.org/10.1038/s41545-023-00261-x

Huijbers, P. M. C., Flach, C., & Larsson, D. G. J. (2019). A conceptual framework for the environmental surveillance of antibiotics and antibiotic resistance. *Environment International*, *130*, 104880. https://doi.org/10.1016/j.envint.2019.05.074

Huijbers, P. M. C., Larsson, D. G. J., & Flach, C. F. (2020). Surveillance of antibiotic resistant *Escherichia coli* in human populations through urban wastewater in ten European countries. *Environmental Pollution*, *261*, 114200. https://doi.org/10.1016/j.envpol.2020.114200

Kalli, M., Noutsopoulos, C., & Mamais, D. (2023). The fate and occurrence of antibiotic-resistant bacteria and antibiotic resistance genes during advanced wastewater treatment and disinfection: A review. *Water*, *15*(11), 2084. https://doi.org/10.3390/w15112084

Kasuga, I., Nagasawa, K., Suzuki, M., Kurisu, F., & Furumai, H. (2022). High-throughput screening of antimicrobial resistance genes and their association with class 1 integrons in urban rivers in Japan. *Frontiers in Environmental Science*, *10*, 1–15. https://doi.org/10.3389/fenvs.2022.825372

Lee, J., Ju, F., Beck, K., & Bürgmann, H. (2023). Differential effects of wastewater treatment plant effluents on the antibiotic resistomes of diverse river habitats. *ISME Journal*, *17*(11), 1993–2002. https://doi.org/10.1038/s41396-023-01506-w

Leonard, A. F. C., Zhang, L., Balfour, A. J., Garside, R., & Gaze, W. H. (2015). Human recreational exposure to antibiotic resistant bacteria in coastal bathing waters. *Environment International*, *82*, 92–100. https://doi.org/10.1016/j.envint.2015.02.013

Li, N., Sheng, G., Lu, Y., Zeng, R. J., & Yu, H. (2017). Removal of antibiotic resistance genes from wastewater treatment plant effluent by coagulation. *Water Research*, 111, 204–212. https://doi.org/10.1016/j.watres.2017.01.010

Li, S., Stevy, B., Ho, S., Jiang, J., & Li, F. (2022). Science of the total environment antibiotic resistant bacteria and genes in wastewater treatment plants: From occurrence to treatment strategies. *Science of the Total Environment*, *838*, 156544. https://doi.org/10.1016/j.scitotenv.2022.156544

Lucas, D., Badia-fabregat, M., Vicent, T., Caminal, G., Rodríguez-mozaz, S., & Balc, J. L. (2016). Chemosphere fungal treatment for the removal of antibiotics and antibiotic resistance genes in veterinary hospital wastewater. *Chemosphere*, *152*, 301–308. https://doi.org/10.1016/j.chemosphere.2016.02.113

Mackulak, T., Cverenkárová, K., Vojs Staňová, A., Fehér, M., Tamáš, M., Škulcová, A. B., Gál, M., Naumowicz, M., Špalková, V., & Bírošová, L. (2021). Hospital wastewater-source of specific micropollutants, antibiotic-resistant microorganisms, viruses, and their elimination. *Antibiotics*, *10*(9), 1070. https://doi.org/10.3390/antibiotics10091070

Manaia, C. M., Rocha, J., Scaccia, N., Marano, R., Radu, E., Biancullo, F., Cerqueira, F., Fortunato, G., Iakovides, I. C., Zammit, I., Kampouris, I., Vaz-moreira, I., & Nunes, O. C. (2018). Antibiotic resistance inp wastewater treatment plants: Tackling the black box. *Environmental International*, *115*, 312–324. https://doi.org/10.1016/j.envint.2018.03.044

Manyi-Loh, C., Mamphweli, S., Meyer, E., & Okoh, A. (2018). Antibiotic use in agriculture and its consequential resistance in environmental sources: Potential public health implications. *Molecules*, *23*(4), 795. https://doi.org/10.3390/molecules23040795

Mao, D, Yu, S, Rysz, M, Luo, Y, Yang, F, Li, F, Hou, J, Mu, Q, & Alvarez, P. J. J. (2015). Prevalence and proliferation of antibiotic resistance genes in two municipal wastewater treatment plants. *Water Research*, *85*, 458–466. http://doi.org/10.1016/j.watres.2015.09.010

Michaelis, C., & Grohmann, E. (2023). Horizontal gene transfer of antibiotic resistance genes in biofilms. *Antibiotics*, *12*(2), 328. https://doi.org/10.3390/antibiotics12020328

Mirghani, R., Saba, T., Khaliq, H., Mitchell, J., Do, L., Chambi, L., Diaz, K., Kennedy, T., Alkassab, K., Huynh, T., Elmi, M., Martinez, J., Sawan, S., & Rijal, G. (2022). Biofilms: Formation, drug resistance and alternatives to conventional approaches. *AIMS Microbiology*, *8*(3), 239–277. https://doi.org/10.3934/microbiol.2022019

Murray, C. J., Ikuta, K. S., Sharara, F., Swetschinski, L., Robles Aguilar, G., Gray, A., Han, C., Bisignano, C., Rao, P., Wool, E., Johnson, S. C., Browne, A. J., Chipeta, M. G., Fell, F., Hackett, S., Haines-Woodhouse, G., Kashef Hamadani, B. H., Kumaran, E. A. P., McManigal, B., & Naghavi, M. (2022). Global burden of bacterial antimicrobial resistance in 2019: A systematic analysis. *The Lancet*, *399*(10325), 629–655. https://doi.org/10.1016/S0140-6736(21)02724-0

Mutuku, C., Gazdag, Z., & Melegh, S. (2022). Occurrence of antibiotics and bacterial resistance genes in wastewater: Resistance mechanisms and antimicrobial resistance control approaches. *World Journal of Microbiology and Biotechnology*, *38*(152), 1–27. https://doi.org/10.1007/s11274-022-03334-0

Osińska, A., Korzeniewska, E., Harnisz, M., Felis, E., Bajkacz, S., Jachimowicz, P., Niestępski, S., & Konopka, I. (2020). Small-scale wastewater treatment plants as a source of the dissemination of antibiotic resistance genes in the aquatic environment. *Journal of Hazardous Materials*, *381*, 121221. https://doi.org/10.1016/j.jhazmat.2019.121221

Pazda, M., Kumirska, J., Stepnowski, P., & Mulkiewicz, E. (2019). Science of the total environment antibiotic resistance genes identified in wastewater treatment plant systems – A review. *Science of the Total Environment*, *697*, 134023. https://doi.org/10.1016/j.scitotenv.2019.134023

Rahman, M., Alam, M. U., Luies, S. K., Kamal, A., Ferdous, S., Lin, A., Sharior, F., Khan, R., Rahman, Z., Parvez, S. M., Amin, N., Hasan, R., Tadesse, B. T., Taneja, N., Islam, M. A., & Ercumen, A. (2022). Contamination of fresh produce with antibiotic-resistant bacteria and associated risks to human health: A scoping review. *International Journal of Environmental Research and Public Health*, *19*(1), 390. https://doi.org/10.3390/ijerph19010360

Ramzan, M., Raza, A., Nisa, Z., Abdel-Massih, R. M., Al Bakain, R., Cabrerizo, F. M., Dela Cruz, T. E., Aziz, R. K., & Musharraf, S. G. (2024). Detection of antimicrobial resistance (AMR) and antimicrobial susceptibility testing (AST) using advanced spectroscopic techniques: A review. *TrAC - Trends in Analytical Chemistry*, *172*, 117562. https://doi.org/10.1016/j.trac.2024.117562

Salam, M. A., Al-Amin, M. Y., Salam, M. T., Pawar, J. S., Akhter, N., Rabaan, A. A., & Alqumber, M. A. A. (2023). Antimicrobial resistance: A growing serious threat for global public health. *Healthcare*, *11*(13), 1946. https://doi.org/10.3390/healthcare11131946

Sambaza, S., Maity, A., & Pillay, K. (2019). Enhanced degradation of BPA in water by PANI supported Ag/TiO_2 nanocomposite under UV and visible light. *Journal of Environmental Chemical Engineering*, *7*(1), 102880. https://doi.org/10.1016/j.jece.2019.102880

Sambaza, S. S., & Naicker, N. (2023). Contribution of wastewater to antimicrobial resistance: A review article. *Journal of Global Antimicrobial Resistance*, *34*, 23–29. https://doi.org/10.1016/j.jgar.2023.05.010

Samrot, A. V., Wilson, S., Sanjay Preeth, R. S., Prakash, P., Sathiyasree, M., Saigeetha, S., Shobana, N., Pachiyappan, S., & Rajesh, V. V. (2023). Sources of antibiotic contamination in wastewater and approaches to their removal—An overview. *Sustainability*, *15*, 12639. https://doi.org/10.3390/su151612639

Stanton, I. C., Bethel, A., Leonard, A. F. C., Gaze, W. H., & Garside, R. (2022). Existing evidence on antibiotic resistance exposure and transmission to humans from the environment: A systematic map. *Environmental Evidence*, *11*(1), 8. https://doi.org/10.1186/s13750-022-00262-2

Tadros, M., Goneau, L., Romaschin, A., Jarvis, M., & Matukas, L. (2018). Rapid detection of resistance to carbapenems and cephalosporins in *Enterobacteriaceae* using liquid chromatography tandem mass spectrometry. *PLoS One*, *13*(11), e0206842 https://doi.org/10.1371/journal.pone.0206842

Tenover, F. C. (2006). Mechanisms of antimicrobial resistance in bacteria. *American Journal of Medicine*, *119*, S3–S10. https://doi.org/10.1016/j.amjmed.2006.03.011

Tong, C., Hu, H., Chen, G., Li, Z., Li, A., & Zhang, J. (2021). Disinfectant resistance in bacteria: Mechanisms, spread, and resolution strategies. *Environmental Research*, *195*, 110897. https://doi.org/10.1016/j.envres.2021.110897

Ugoeze, K., Alalor, C., Ibezim, C., Chinko, B., Owonaro, P., Anie, C., Okoronkwo, N., Mgbahurike, A., Ofomata, C., Alfred-Ugbenbo, D., & Ndukwu, G. (2024). Environmental and human health impact of antibiotics waste mismanagement: A review. *Advances in Environmental and Engineering Research*, *5*(1), 005. https://doi.org/10.21926/aeer.2401005

Uluseker, C., Kaster, K. M., Thorsen, K., Basiry, D., Shobana, S., Jain, M., Kumar, G., Kommedal, R., & Pala-Ozkok, I. (2021). A review on occurrence and spread of antibiotic resistance in wastewaters and in wastewater treatment plants: Mechanisms and perspectives. *Frontiers in Microbiology*, *12*, 1–19. https://doi.org/10.3389/fmicb.2021.717809

Wijesinghe, H. G. S., Hare, D. J., Mohamed, A., Shah, A. K., Harris, P. N. A., & Hill, M. M. (2021). Detecting antimicrobial resistance in: *Escherichia coli* using benchtop attenuated total reflectance-Fourier transform infrared spectroscopy and machine learning. *Analyst*, *146*(20), 6211–6219. https://doi.org/10.1039/d1an00546d

Zhang, X., Zhang, T., & Fang, H. H. P. (2009). Antibiotic resistance genes in water environment. *Applied Microbiology and Biotechnology*, *82*(3), 397–414. https://doi.org/10.1007/s00253-008-1829-z

8 Biogenic Synthesis of CuO Nanoparticle in Resistance to Disperse Dyes and Microbes in Wastewater by *Strychnos potatorum* Seed

Ganesh Nagarathinam, Babu Selvarajan, Prakash Pandiyan, Ambikapathy Varatharaju, Gomathi Selvam, and Panneerselvam Annamalai

8.1 INTRODUCTION

Modernization and creative industrial developments have had both positive and negative effects on human society, with a major negative impact on water resources (Elhassadi, 2008). Wastewater effluents from homes and businesses can pose a major risk to aquatic life and human health (Pandey, 2002). Hazardous compounds like insecticides and organic dyes (such as unprocessed colored dyes) may be present in industrial wastewater. Colored chemical dyes harm aquatic bodies by reducing dissolved oxygen concentrations because they block sunlight (Farré et al., 2008). Numerous industries, such as plastics, paper, leather tanning, textile, and pharmaceutical sectors, use more than 10,000 industrial dyes. It was predicted that about 7×105 metric tonnes, or 15% of their expected annual production rate, were wasted during industrial operations that could contaminate natural water bodies (Lavanya, 2014). An estimated 3.7 million liters of water are polluted daily worldwide by the materials used in the textile industry. Additionally, due to its complex structure and refractory nature, industrial dye remediation was regarded as a significant challenge in wastewater treatment (Buthiyappan et al., 2016; Nezamzadeh-Ejhieh and Moazzeni, 2013; Senobari and Nezamzadeh-Ejhieh, 2018a). Because these harmful chemicals can cause skin allergies and irritation or cancer, they must be eliminated from water systems to protect aquatic life and people (Carmen and Daniel, 2012. For this reason, a number of water treatment techniques have been studied for the treatment of industrial wastewater effluents, such as electrolysis, flocculation, precipitation, coagulation, electron beam treatment, activated carbon, oxidation and reduction reactions, froth flotation, and photoelectrochemical treatment (Latif et al., 2010). Nevertheless, these methods often create secondary pollutants, such as carcinogenic ones, or move contaminants from one phase to another. Additionally, these traditional techniques, which require large amounts of photocatalysts, generate a lot of sludge (Bahrami and Nezamzadeh-Ejhieh, 2015; Nezamzadeh-Ejhieh and Khorsandi, 2010). This issue has been investigated using advanced oxidation processes (AOPs), among which semiconductor-based AOPs were among the most efficient and reasonably priced methods of handling hazardous dyes. Due to their high bandgap compatibility, AOPs can potentially employ the entire spectrum of the sun in photocatalytic reactions (Ajoudanian and Nezamzadeh-Ejhieh, 2015; Esmaili-Hafshejani and Nezamzadeh-Ejhieh, 2016; Jafari and Nezamzadeh-Ejhieh, 2017). To thoroughly break down electron-rich organic dyes into

 DOI: 10.1201/9781003269823-8

harmless products (such as CO_2 and H_2O) through mineralization, AOPs generate extremely reactive free radicals (Deng and Zhao, 2015; Esplugas et al., 2002; Garcia-Segura and Brillas, 2017; Zhou and He, 2007). For the breakdown of organic water contaminants, metal oxide nanoparticles (NPs) work best when hydrogen peroxide was present or when exposed to ultraviolet (UV) light (Lachheb et al., 2002; Raghav et al., 2016; Wong and Chu, 2003). However, many of the target pollutants that the industries release were either nonbiodegradable or hazardous. Reactive Yellow 160 (RY160) and colored dyes such as Nile Blue (NB) were examples of this. Methods that rely on hydrogen peroxide oxidation and UV radiation were insufficient to break down these dyes since they were so persistent (Qiao et al., 2005). Moreover, using hydrogen peroxide and UV radiation was not always safe or feasible. Humans can suffer severe consequences from hydrogen peroxide consumption, such as lipid peroxidation, damage to the central nervous system, and gastrointestinal tract irritation that may end in vomiting (Watt et al., 2004). Additionally, cancer was caused by UV rays (Matsumura and Anantha swamy, 2004).

The use of direct solar radiation was a notable and highly effective method in photocatalysis. It can be the best alternative to lessen the demand for expensive light sources. Since visible light accounts for the bulk of solar energy (usually consisting of 45% visible light and 5% UV irradiation), it has an effective effect on the photocatalytic process (Nezamzadeh-Ejhieh and Banan, 2012). Therefore, even without UV irradiation and an additional oxidizing agent, a photocatalyst with enhanced dye degradation capabilities was highly desirable for wastewater treatment. Additionally, using green practices could help increase the applicability of these technologies.

8.2 MATERIALS AND METHODS

8.2.1 Materials

The seeds were gathered from the Poninji Andavar holy groves in Sendurai, Ariyalur, Tamil Nadu, India (Lat. 11.2208550, Long. 79.236170). The NB/RY160 dye and copper acetate have been acquired. Every stage of the experiment has made use of double-distilled water (DDW). Unless otherwise noted, the remaining compounds were used. The characteristics of RY160 and NB were demonstrated.

8.2.2 Method

8.2.2.1 Preparation of Seed Extracts

The poor abundance of *Strychnos potatorum* L. suggests that it was uncommon and not widely distributed. The best way to preserve the phytocompounds in the sample was to thoroughly shade-dry the *S. potatorum* seeds for a week before grinding them into fine granules (Hassan et al., 2009; Shaygannia et al., 2015). 10 g of seed powder was added to 100 mL of DDW, and it was heated to 80°C for 3 h. After cooling to ambient temperature, the extract was filtered through Whatman No. 1 filter paper and then refrigerated to ensure that all of the biological components found in the plants were removed.

8.2.3 Synthesis of CuO NPs

For the large-scale production of NPs, plant extract was the most favored source of NP synthesis (Singh et al., 2019a; Fagier, 2021; Ansari and Alzohairy, 2018; Agarwal et al., 2017). Laser vaporization, sol-gel, microwave radiation, ultrasonic treatment, solid-state thermal degradation, and hydrothermal processes were some of the physical and chemical methods used in the synthesis. Energy, space, and expensive equipment were all needed for NPs (Hasnidawani et al., 2016). Green biosynthesis of NPs provides a safe, economical, and environmentally beneficial substitute for chemical and physical methods (Ansari and Alzohairy, 2018). The plant components were washed with well-running water to get rid of muck and suspended impurities after exotic materials and broken parts had been removed, and they were carefully separated from other plants. After that, the plant parts were left in the sun to dry for three or four days. Using a blender, the dried sample mash with less than 10% moisture was turned into a simplified powder.

The plant bits were heated to 60°C (Figure 8.1) after being cooked in various solvents (ethanol, water) to create the extract (Agarwal et al., 2017). The *S. potatorum* seed extract was mixed with 0.5 mm of copper acetate monohydrate to create a metal precursor. The mixture turned pale white or yellow, which was obvious evidence that the NPs were developing, and the phytochemical was produced when it was boiled for the required length of time and temperature. Heat treatment was used to create CuO NPs. CuO NPs were separated using a centrifuge and then cleaned with water and 100% ethanol. The precipitate was calcined for 3 h at a high temperature of 500°C to encourage the crystallization of CuO NPs and achieve crystal size (Figure 8.2). Agarwal et al. (2017); Hameed et al. (2011); Ahmad et al. (2019) showed the CuO NP green synthesis studies using different parts of plant extract. Proteins, alkaloids, carbohydrates, phenolic acids, terpenoids, and polyphenols

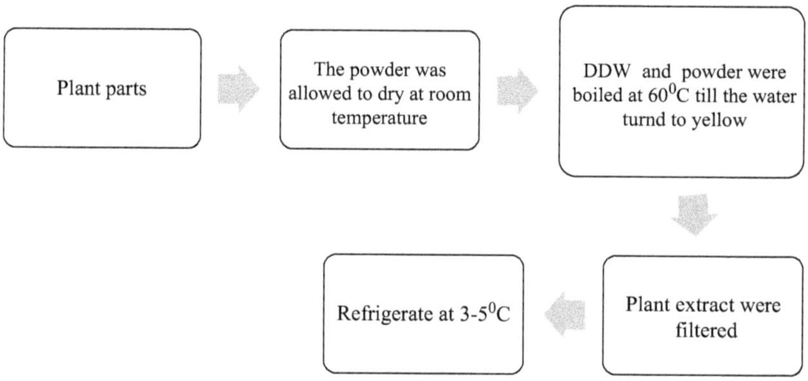

FIGURE 8.1 Extraction of plant based on the description in previous studies.

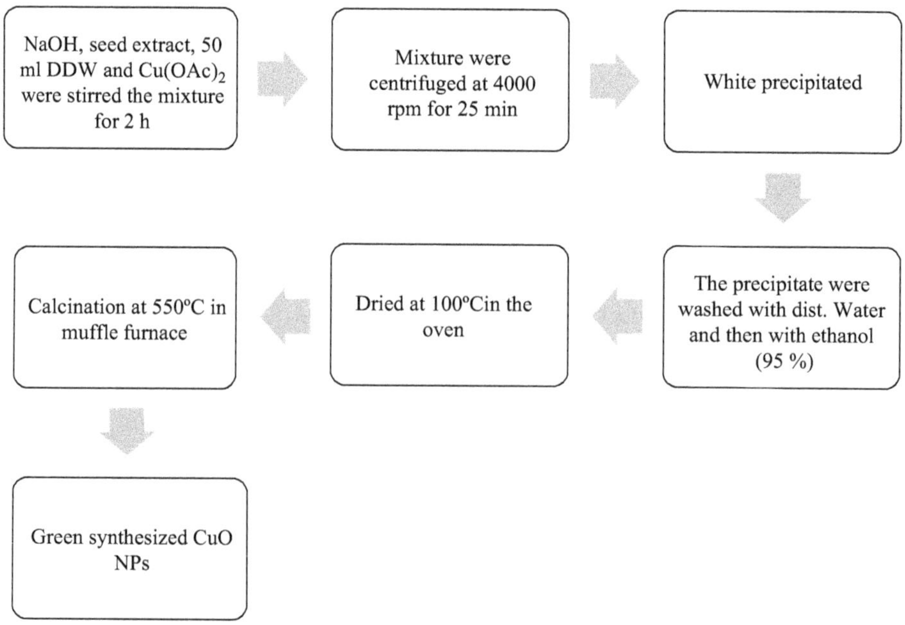

FIGURE 8.2 Preparations of green CuO NP synthesis (based on the description in previous studies).

were among the physiologically active ingredients of plant extracts. These substances were essential for stabilizing and initially reducing metal ions (Shamaila et al., 2016). The nucleophiles aromatic ring of phenols serves as an antioxidant and metal chelating agent. Flavonoids were the most significant biological reducing agents in plant extracts because they frequently give hydrogen atoms and electrons to other compounds. By providing metal ions with electrons, the protein acts as a reducing agent, stabilizing the process and causing NPs to form (Ovais et al., 2018). Previous studies have shown that various plants can be used to produce CuO NPs in an environmentally friendly manner.

$$Cu(CH_3COO)_2 + \text{plant extract} \rightarrow Cu(OH)_2$$

$$Cu(OH)_2 + \text{calcination} \rightarrow CuO + H_2O$$

8.2.4 Anti-Microbial Activity

8.2.4.1 Antibacterial Activity

An antibacterial test using the previously published agar disc diffusion technique was performed on plant extract. Gram-negative (*Klebsiella pneumonia*, *Escherichia coli*, and *Pseudomonas aeruginosa*) and Gram-positive (*Staphylococcus aureus* and *Bacillus subtilis*) bacteria were gathered from the Culture Collection Institute as test organisms. The test organism's concentration was maintained at the 0.5 McFarland standard (106^{-7} CFU/mL) while the test bacteria were cultivated in broth cultures. One hundred microliters of inoculum were put to NA plates. Three identical sterile standard discs were filled. CuO-synthesized *S. potatorum* seeds were introduced to each well at varying concentrations (25, 50, 75, and 100 µg/mL), and the plates were subsequently incubated at 33°C. After a day, the zone of inhibition (ZOI) was measured with a ruler. The standard was kept as the positive control.

8.2.5 Antifungal Activity

In compliance with the stated methodology, the antifungal test was conducted using the agar disc diffusion technique. *Aspergillus niger*, *Candida dublinienisis*, *Candida tropicalis*, and *Cyptococcus neoformans* were the test organisms used. In potato dextrose broth, a fungal broth culture was created, and it was cultured for the entire night. One hundred microliters of the microbial pathogen (0.5 McFarland) were spread out on PDA plates, and a sterile standard disc containing CuO synthesis was added. *S. potatorum* seeds were introduced to each disc at several concentrations (25, 50, 75, and 100 µg/mL) and incubated at 37°C. The ZOI was measured using a ruler 48 h later. Cycloheximide (20 mg/mL) was maintained as the positive control.

8.2.6 Photocatalytic Activity

To assume absorption–desorption equilibrium, 10 mg of biogenic CuO NPs were added to the experiment and swirled with 20 mL of 1.3×10^{-5} mol/L of NB dye and 4.8×10^{-5} mol/L of RY160 dye (40 ppm each). The experiment was conducted in the dark for 60 min. Changes in the dye concentration and absorption level in the dark were observed. After that, the resultant samples were placed directly in the sun for the photocatalytic experiment. A pyranometer was used to measure the solar flux, which came out to be 637 W/m^2. A few milliliters of the sample were extracted during the 20-min interval and centrifuged for 10 min at 10,000 rpm. A UV-visible spectrophotometer was used to analyze the supernatant. NB and RY160 have respective λ_{maxes} of 630 and 425 nm. After measuring the amount of dye left over from the photocatalytic experiment, the absorbance of the concentrated dye was calculated. The relationship between absorbance and solution concentration was determined (Swinehart, 1962). A UV–visible spectrophotometer was employed to assess the control samples and evaluate the photocatalytic degradation efficiency (*R*) of CuO nanoparticles (NPs) (Meena et al., 2016).

$$R = \{(A_0 - A_t)/A_0\} \times 100$$

A_0 and A_t denote the absorbance of a dye.

Experiments were conducted simultaneously under similar experimental conditions in three distinct reaction containers for NB and RY160 degradation. For both dyes, the degradation period was 120 min. Both dyes' computerized values in triplicate were determined by the relative standard error. Calculations were made to determine the limits of detection and quantification (Shrivastava et al., 2011).

$$\text{LOD} = X_{b1} + 3S_{b1}$$

$$\text{LOQ} = X_{b1} + 10S_{b1}$$

where X_{b1} was the mean concentration of the blank and S_{b1} was the standard deviation of the blank.

The LOD values for NB and RY160 were 3 and 8 mg/L, and the LOQ values were 7 and 16 mg/L, respectively.

8.2.7 Zeta Potential Analysis

Ten milligrams of CuO NPs were taken in 50 mL of DDW, and then the solutions were adjusted with HCl and NaOH.

8.2.8 Characterization Techniques

A UV-visible spectrophotometer was used to measure the UV-visible spectra in a quartz cuvette between 200 and 600 nm. With a sensitivity of 15×10^{-6} V/(W/m^2) and an operational temperature range of −49 to +80°C, pyranometers were employed to measure the solar flux. A Bruker Alpha instrument was used to analyze and view CuO NPs from 4,000 to 400 cm^{-1} in FTIR. FE-SEM and TEM were used to assess the morphology and size of the synthesized NPs at a voltage acceleration of 300 kV. XRD spectroscopy was used to elaborate on the composition of the CuO NPs. Using an X-ray diffractometer from $2\theta = 10°–80°$, crystallographic studies of biogenic CuO NPs were performed using CuO α radiation at $\lambda = 0.154$ nm. The XRD's scanning speed in continuous scan mode was 0.041°/min (step time: 25.19, step size: 0.017). Using N_2 adsorption desorption isotherms, the generated sample's BET (Brunauer–Emmett–Teller) specific surface area was determined. The zeta potential of the biogenic CuO NPs was investigated at various pH values (2–10) using the Malvern Zeta Sizer Nano ZS90.

8.2.9 Result and Discussion

8.2.9.1 Optical Study

The UV-V spectrum was used to analyze and display the CuO NPs that were synthesized from *S. potatorum* seeds. The existence of CuO NPs was confirmed by the strong band at 282 nm that was visible in the colloidal substance's final product. According to previous studies, the band absorption at 282 nm was associated with the copper metal interband transition of the core electrons found in

CuO NPs. The chemicals that caused the reduction of copper metal to CuO NPs were present, and their band edge was located at 480 nm). Utilizing the Tauc plot (Viezbicke et al., 2015) (αhv)1/2 technique, the gap between the synthesized CuO NPs was examined.

The energy of the light was hv, and the absorption coefficient was α in the Tauc plot. Planck's constant was 6.626×10–34 Js, and light had a frequency of v. The bandgap of CuO NPs was computed and contrasted with the value previously reported at 1.8 eV (Dhineshbabu et al., 2016). It was inferior to it. (Azimi and Nezamzadeh-Ejhieh, 2015; Derikvandi and Nezamzadeh-Ejhieh, 2017b; Gu et al., 2004; Karimi-Shamsabadi and Nezamzadeh-Ejhieh, 2016) cited the particle size of CuO NPs with varying bandgaps.

8.2.10 Morphological and Elemental Analysis

SEM and TEM were used to assess the morphological characterization of the synthesized CuO NPs. Agglomerated CuO NPs reveal spherical, morphologically rough agglomerates, and the XRD analysis of the sample shows 111,200 and 002,111 of Cu and O, respectively. The majority of them were eliminated when elements and molecules were present during the synthesis of CuO NPs using techniques including washing and calcination. CuO NPs were spherical in shape and homogeneous in size, but the inclusion of carbon in the sample caused an uneven and irregular size distribution.

8.2.11 Antimicrobial Activity of CuO-Synthesized *S. potatorum*

8.2.11.1 Antibacterial Activity of CuO-Synthesized *S. potatorum*

The current study reports that CuO NPs made from *S. potatorum* seed via disc diffusion were used to inoculate and screen the pathogen. *Staphylococcus aureus* exhibits the most inhibition at a concentration of 100 μL, while *E. coli* exhibits the lowest inhibition at a concentration of 25 μL.

8.2.12 Antifungal Activity of CuO-Synthesized *S. potatorum*

The current study reports that CuO NPs made from *S. potatorum* seed via disc diffusion were used to inoculate and screen the pathogen. *Candida dublinienisis* has the lowest inhibition at a concentration of 25 μL, while *Candida tropicalis* exhibits the highest inhibition at a concentration of 100 μL.

8.2.13 Photocatalytic Activity Performance of CuO NPs

By directly exposing the dyes NB and RY160 to sunlight and measuring their absorbance at 632 and 425 nm, the effectiveness of the synthesized CuO NPs was investigated. Following the addition of CuO NPs and exposure to sunlight, both dyes were examined for a decrease in absorbance band intensity over a predetermined period of time. In the NB dye sample, a somewhat full degradation was completed in 120 min, whereas RY160 required 140 min, and it might be degraded within the time interval. For NB and RY160, the percentages that could be used to calculate the deterioration were 95% and 80%. RY160 has a complicated structure, a slower rate of photocatalytic degradation, and a reduced degradation potential because of its larger molecular weight of 818.13 g/mol. However, because of the quick diffusion toward the photocatalytic process, the leading surface of NB with a lower molecular weight (353.84 g/mol) rapidly degraded (Lavanya, 2014).

$$CuO + h\nu \rightarrow CuO(e^- + h^+)$$

CuO NPs' electron and hole interactions with light caused deterioration in the photochemical reaction. The hydroxyl radical (OH) was created when the hole water was oxidized by the interaction of the molecular oxygen with the photogenerated electron in the band hole photogenerated reaction.

Primarily responsible for the production of reactive oxygen species (ROS), such as hydroxyl radicals (•OH–) and superoxide anions $\left(\bullet O_2^-\right)$. ROS were the molecules that caused the dye degradation.

$$CuO + h\nu \rightarrow CuO(e^-) + CuO(h^+)$$

$$CuO(e^-)O_2 \rightarrow O_2^-$$

$$CuO(h^+) + H_2O \rightarrow OH^-$$

$$O_2^- \text{ or } OH^- + \text{Dye} \rightarrow \text{degraded product}$$

This article describes the catalyst experiment that was conducted to control the presence and absence of the CuO NPs in the dark. The results eliminated the possibility of self-degradation, dye degradation, and catalysis activity. In the absence of dye degradation by CuO NPs in the dark, the experiment's results were minimal, indicating no discernible degradation effect. On the other hand, both NB and RY160 were more or less totally broken down when exposed to sunlight for the catalyst. The photocatalytic reaction provided a clear demonstration of the dye degradation experiment. The calculation of dye degradation kinetics was followed by Langmuir–Hinshelwood (L-H) (Nezamzadeh-Ejhieh and Ghanbari-Mobarakeh, 2015; Shams-Ghahfarokhi and Nezamzadeh-Ejhieh, 2015) (Figures 8.3 and 8.4).

$$r = -\frac{dc}{dt} = k\theta = k\left(\frac{KC}{1+C}\right)$$

where *r*, *K*, *C*, *k*, *c*, and *t* stand for the dye solution concentration (mg/L), the adsorption rate constant (L/mg), the reaction rate (mg/L/min), and the Langmuir constant related to the adsorption energy. The concentration of dye solution and the photocatalysis (mg/L/min) at every given time (min) is shown. The reaction was governed by zero-order kinetics (k_0=zero-order rate constant) as soon as KC≫1, which means that the reaction rate was independent of concentration when $r=k_0$.

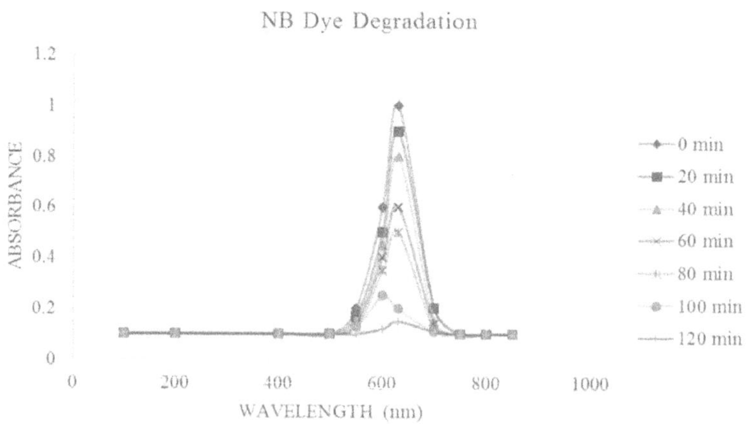

FIGURE 8.3 UV–visible spectra analyses of NB dye degradation.

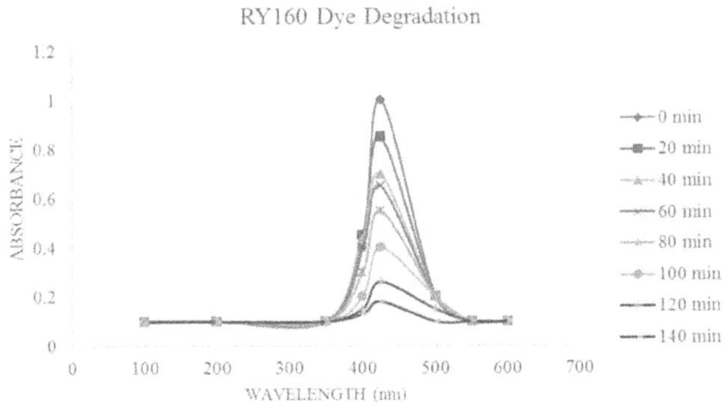

FIGURE 8.4 UV–visible spectra analyses of RY160 dye degradation.

Conversely, the rate of response varies proportionately or linearly with concentration when KC≥1 (for example, when the sample was diluted). (i.e., $r = -\frac{dc}{dt} = kKC = k_1C$ or in $\frac{C_t}{C_o} = -k_1t$).

Plotting the decline in dye concentration (C_t/C_o) as a function of time by the first-order rate constant indicates the first-order rate law reaction. The regression coefficient (R_2) and dye degradation (K′app) were determined using the pseudo-first-order linear fit as the rate constant: RY160: K′app=0.010/min and R_2=0.989; NB: K′app=0.019 and R_2=0.979.

CuO NPs from NB and RY160 demonstrated degradation efficiencies of 93% and 81%, respectively; the percentage was determined using the equation $R = \{(A_0 - A_t)/A_0\} \times 100$.

For NB and RY160, the RSE of the developed photocatalytic dye degradation methods was 1.09% and 0.73%, respectively. The zeta potential catalyst differed from the pH value that was given. According to the study, the aggregation of particles implies that the zeta potential of CuO NPs was roughly 0 at pH 6.1. As a model chemical for the catalytic potential, 4-nitrophenol was used to validate CuO NPs (Peng et al., 2017). The results showed that 4-nitrophenol was helpful in effectively degrading the catalytic property of CuO NP's extraordinary catalytic potential.

A study on dye degradation in tap water was conducted; however, the efficiency was lower than that of the DDW. TDS and numerous other dissolved species were found in tap water (Saeed et al., 2017). In any case, CuO NPs exhibit remarkable photocatalytic performance. The results of earlier research on dye degradation using various techniques included UV irradiation, costly equipment (electrochemical electrodes), and composites—mixtures of two or three different materials. CuO NPs can degrade dyes photocatalytically with just sunlight, which is an excellent technique for the dye in question. NB and RY160 CuO NPs were used cost-effective and efficient photocatalysts to degrade textile dye without the need for expensive electrodes or an external artificial light source. In our earlier study, we detailed the mechanism of biogenic CuO NP-based photocatalytic dye degradation: catalyst, target dye, dye concentration, irradiation/time (min) & Reference: Amisha et al. (2008): ZnO/Sepiolite, Reactive blue 4, 200 mg/120 min UV irradiation & Bansal and Sud (2011): TiO_2 NPs, Reactive red 141 (RR 141), 10 mg/L, 120 min UV irradiation & Amalraj and Anitha Pius (2014): Reactive Yellow 105 (RY105), 10 mg/L, 120 min UV irradiation & Annadurai et al. (2000): ZnO NPs, Reactive black 5, 496 mg/L, 60 min solar irradiation & Amisha et al. (2008): Reactive orange 4(RO4), 32 mg/L, 70 min, UV irradiation & Gonçalves et al. (2005): Reactive orange 4(RO4), 25 mg/L, 30 min UV irradiation & Vijayabalan et al. (2013): CuO NPs Coomassie brilliant blue R-250, 10 mg/L, 90 min solar Irradiation & Sankar et al. (2014): Congo red, 10 mg/L, 60 min solar irradiation & Aminuzzaman et al. (2017): Methyl orange, 10–5M, 60 min UV irradiation: SnO_2 NPs, Reactive Yellow 186, 40 ppm, 180 min solar irradiation & Kumar et al. (2018): C/PbO_2-electrodes Gaber et al. (2013) and Reactive yellow 145, 50 ppm, 60 min: Pb+Sn/PbO_2+SnO_2-electrodes & Gaber et al. (2013) and Reactive Yellow 145, 50 ppm, 60 min (Figures 8.5 and 8.6).

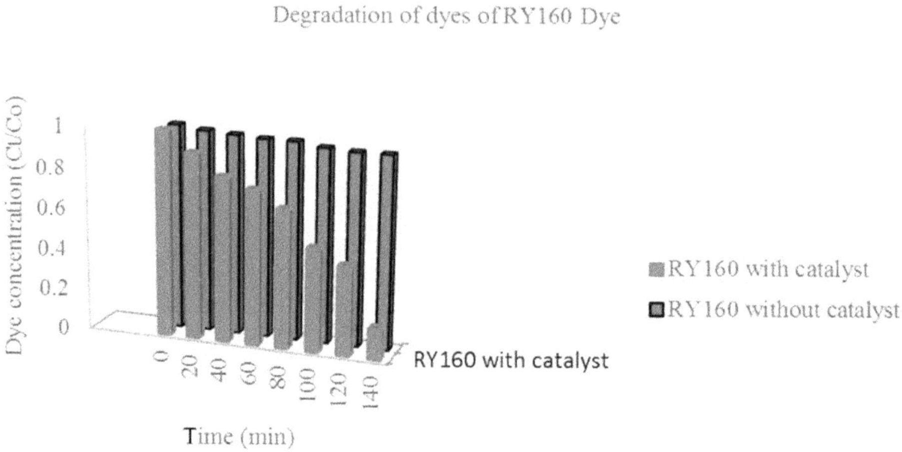

FIGURE 8.5 Degradation of dyes in the absence and presence of catalyst of Ry160 dye.

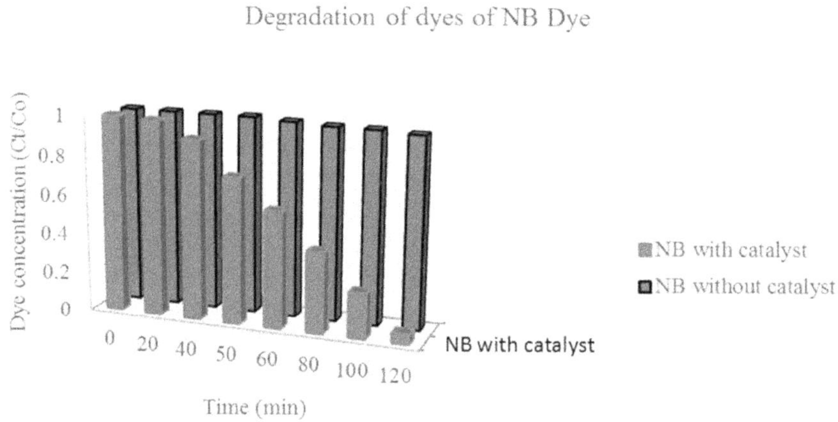

FIGURE 8.6 Degradation of dyes in the absence and presence of catalyst of NB dye.

8.3 CONCLUSION

With the help of *S. potatorum* seed extract, the synthesized CuO NPs were successfully recapitulated in a green, inexpensive, and straightforward manner. The characterization showed that CuO NPs formed a spherical monoclinic structure with a consistent size distribution of 2–6 nm. Effective photocatalytic potential with a 2.5 eV optical bandgap when exposed to sunlight. Here, we used copper acetate monohydrate and *S. potatorum* seed extract to create an efficient CuO NP-based photocatalyst for the degradation of NB and RY160 dyes under solar light irradiation. To evaluate the generated photocatalytic performance in the presence of solar light, "hard to degrade" industrial dyes like NB and RY160 were used in contrast to methylene blue and Congo red. Because of their slow diffusion rates and huge molecular weights, these dyes were difficult to degrade. To the best of our knowledge, this was the first study to use CuO NP-based green synthesis, or production from plant extract, for the photocatalytic degradation of RY160 and NB dyes. The catalysts were recognized for their easy recovery and reusability in addition to their remarkable dye degradation efficiency.

In this study, we demonstrate that synthetic CuO NPs may effectively break down dyes without the need for an artificial light source. As a result, our research offers fresh insights into affordable, ecofriendly photocatalysts that could successfully eliminate toxic hues from industrially tainted water without endangering the environment. It was found that the targeted dye was difficult to break down by the effective CuO NPs. Both DDW and actual tap water were used to test the photocatalytic activity of the synthesized CuO NPs. Additionally, even at high electrolyte concentrations, the enhanced stability of the generated CuO NPs was shown by the analysis of aggregation kinetics. Because of their high stability and massive photocatalytic capacity, the current study recommends creating green technique-based photocatalysts and applying them successfully in wastewater treatment to eliminate hazardous dyes.

According to the study, the pathogens found in wastewater were immune to both chemicals and microorganisms. They were tested using CuO-synthesized NPs, which demonstrated antimicrobial qualities in both antifungal and antibacterial activities. Through a photocatalytic process that uses sunlight, the synthesized CuO NP breaks down azo dye and bacteria in wastewater and purifies it. CuO NP was added to the process to stop it and break down the dye and microbes.

REFERENCES

Agarwal, H., Venkat Kumar, S., Rajeshkumar, S. 2017. A review on green synthesis of zinc oxide nanoparticles – an eco-friendly approach. *Resour. Technol.* 3 (4), 406–413, https://doi.org/10.1016/j.reffit.2017.03.002.

Ahmad, F., Ashraf, N., Ashraf, T., Bin Zhou, R., Yin, D.C. 2019. Biological synthesis of metallic nanoparticles (MNPs) by plants and microbes: their cellular uptake, biocompatibility, and biomedical applications. *Appl. Microbiol. Biotechnol.* 103 (7), 2913–2935, https://doi.org/10.1007/s00253-019-09675-5.

Ajoudanian, N., Nezamzadeh-Ejhieh, A. 2015. Enhanced photocatalytic activity of nickel oxide supported on clinoptilolite nanoparticles for the photodegradation of aqueous cephalexin. *Mater. Sci. Semicond. Process.* 36, 162–169, https://doi.org/10.1016/j.mssp.2015.03.042.

Amalraj, A., Anitha Pius, P. 2014. Photodegradation of reactive red 141 and reactive yellow 105 dyes using prepared TiO_2 nanoparticles. *Mater. Sci. Forum* 807, 65–79, doi:10.4028/www.scientific.net/MSF.807.65.

Aminuzzaman, M., Kei, L.M., Liang, W.H. 2017. Green synthesis of copper oxide (CuO) nanoparticles using banana peel extract and their photocatalytic activities. *AIP Conf. Proc.* 1828, 020016, https://doi.org/10.1063/1.4979387.020016.

Amisha, S., Selvam, K., Sobana, N., Swaminathan, M. 2008. Photomineralisation of reactive black 5 with ZnO using solar and UV-A light. *J. Korean Chem. Soc.* 52, 66–72.

Annadurai, G., Sivakumar, T., Rajesh Babu, S. 2000. Photocatalytic decolorization of Congo red over ZnO powder using Box–Behnken design of experiments. *Bioprocess Eng.* 23, 167–173, https://doi.org/10.1007/PL00009120.

Ansari, M.A., Alzohairy, M.A. 2018, One-pot facile green synthesis of silver nanoparticles using seed extract of *Phoenix dactylifera* and their bactericidal potential against MRSA. *Evid. Based Complement. Alternat. Med.* 2018, 1860280, https://doi.org/10.1155/2018/1860280.

Azimi, S., Nezamzadeh-Ejhieh, A. 2015. Enhanced activity of clinoptilolite-supported hybridized PbS-CdS semiconductors for the photocatalytic degradation of a mixture of tetracycline and cephalexin aqueous solution. *J. Mol. Catal. A Chem.* 408, 152–160, https://doi.org/10.1016/j.molcata.2015.07.017.

Bahrami, M., Nezamzadeh-Ejhieh, A. 2015. Effect of the supported ZnO on clinoptilolite nano-particles in the photodecolorization of semi-real sample bromothymol blue aqueous solution. *Mater. Sci. Semicond. Process.* 30, 275–284, https://doi.org/10.1016/j.mssp.2014.10.006.

Bansal, P., Sud, D. 2011. Photodegradation of commercial dye, Procion Blue HERD from real textile wastewater using nanocatalysts. *Desalination* 267, 244–249, https://doi.org/10.1016/j.desal.2010.09.034.

Buthiyappan, A., Abdul Aziz, A.R., Wan Daud, W.M.A. 2016. Recent advances and prospects of catalytic advanced oxidation process in treating textile effluents. *Rev. Chem. Eng.* 32 (1), 1–47, https://doi.org/10.1515/revce-2015-0034.

Carmen, Z., Daniel, S. 2012. Textile organic dyes – characteristics, polluting effects and separation/elimination procedures from industrial effluents – a critical overview. In: Tomasz, P., Aleksandra, M.S. (Eds) *Organic Pollutants Ten Years after the Stockholm Convention - Environmental and Analytical Update*, pp. 55–86. Rijeka: IntechOpen, https://doi.org/10.5772/32373.

Deng, Y., Zhao, R. 2015. Advanced oxidation processes (AOPs) in wastewater treatment. *J.Ind. Eng. Chem.* 1, 167–176, https://doi.org/10.1007/s40726-015-0015-z.

Derikvandi, H., Nezamzadeh-Ejhieh, A. 2017b. Increased photocatalytic activity of NiO and ZnO in photodegradation of a model drug aqueous solution: effect of coupling, supporting, particles size and calcination temperature. *J. Hazard Mater.* 321, 629–638, https://doi. org/10.1016/j.jhazmat.2016.09.056.

Dhineshbabu, N.R., Rajendran, V., Nithyavathy, N., Vetumperumal, R. 2016. Study of structural and optical properties of cupric oxide nanoparticles. *Appl. Nanosci.* 6, 933–939, https://doi.org/10.1007/s13204-015-0499-2.

Elhassadi, A. 2008. Pollution of water resources from industrial effluents: a case study - Benghazi, Libya. *Desalination* 222, 286–293, https://doi.org/10.1016/j.desal.2007.05.030.

Esmaili-Hafshejani, J., Nezamzadeh-Ejhieh, A. 2016. Increased photocatalytic activity of Zn(II)/Cu(II) oxides and sulfides by coupling and supporting them onto clinoptilolite nanoparticles in the degradation of benzophenone aqueous solution. *J. Hazard Mater.* 316, 194–203, https://doi.org/10.1016/j.jhazmat.2016.05.006.

Esplugas, S., Giménez, J., Contreras, S., Pascual, E., Rodríguez, M. 2002. Comparison of different advanced oxidation processes for phenol degradation. *Water Res.* 36, 1034–1042, https://doi.org/10.1016/S0043-1354(01)00301-3.

Fagier, M.A. 2021. Plant-mediated biosynthesis and photocatalysis activities of zinc oxide nanoparticles: a prospect towards dyes mineralization. *J. Nanotechnol.* 2021, 1–15, https://doi.org/10.1155/2021/6629180.

Farré, M.L., Pérez, S., Kantiani, L., Barceló, D. 2008. Fate and toxicity of emerging pollutants, their metabolites and transformation products in the aquatic environment. *TrAC Trends* Anal. *Chem. (Reference Ed.)* 27, 991–1007, https://doi.org/10.1016/j.trac.2008.09.010.

Gaber, M., Ghalwa, N.A., Khedr, A.M., Salem, M.F. 2013. Electrochemical degradation of reactive yellow 160 dye in real wastewater using C/PbO_{2-}, Pb + Sn/PbO_2+SnO_2 - and Pb/PbO_2 modified electrodes. *J. Chem.* 2013, 9, https://doi.org/10.1155/2013/691763.

Garcia-Segura, S., Brillas, E. 2017. Applied photoelectrocatalysis on the degradation of organic pollutants in wastewaters. *J. Photochem. Photobiol. C Photochem. Rev.* 31, 1–35, https://doi.org/10.1016/j.jphotochemrev.2017.01.005.

Gonçalves, M.S.T., Pinto, E.M.S., Nkeonye, P., Oliveira-Campos, A.M.F. 2005. Degradation of C.I. reactive orange 4 and its simulated dyebath wastewater by heterogeneous photocatalysis. *Dyes Pigments* 64, 135–139, https://doi.org/10.1016/j.dyepig.2004.05.004.

Gu, Y., Kuskovsky, I.L., Yin, M., O'Brien, S., Neumark, G.F. 2004. Quantum confinement in ZnO nanorods. *Appl. Phys. Lett.* 85, 3833–3835, https://doi.org/10.1063/1.1811797.

Hameed, B.H., Akpan, U.G., Wee, K.P. 2011. Photocatalytic degradation of acid red 1 dye using ZnO catalyst in the presence and absence of silver. *Desalin. Water Treat.* 27 (1–3), 204–209, https://doi.org/10.5004/dwt.2011.1388.

Hasnidawani, J.N., Azlina, H.N., Norita, H., Bonnia, N.N., Ratim, S., Ali, E.S. 2016. Synthesis of ZnO nanostructures using sol-gel method. *Procedia Chem.* 19, 211–216, https://doi.org/10.1016/j.proche.2016.03.095.

Hassan, S.W., Umar, R.A., Maishanu, H.M., Matazu, I.K., Faruk, U.Z., Sani, A.A. 2009. The effect of drying method on the nutrients and non-nutrients composition of leaves of gynandropsis gynandra (capparaceae). *Asian J. Biochem.* 2, 349–353, https://doi.org/10.3923/ajb.2007.349.353.

Karimi-Shamsabadi, M., Nezamzadeh-Ejhieh, A. 2016. Comparative study on the increased photoactivity of coupled and supported manganese-silver oxides onto a natural zeolite nano-particles. *J. Mol. Catal. A Chem.* 418–419, 103–114, https://doi.org/10.1016/j.molcata.2016.03.034.

Kumar, M., Mehta, A., Mishra, A., Singh, J., Rawat, M., Basu, S. 2018. Biosynthesis of tin oxide nanoparticles using Psidium Guajava leave extract for photocatalytic dye degradation under sunlight. *Mater. Lett.* 215, 121–124, https://doi.org/10.1016/j.matlet.2017.12.074.

Lachheb, H., Puzenat, E., Houas, A., Ksibi, M., Elaloui, E., Guillard, C., Herrmann, J.M. 2002. Photocatalytic degradation of various types of dyes (alizarin S, crocein orange G, methyl red, Congo red, methylene blue) in water by UV-irradiated Titania. *Appl. Catal. B Environ.* 39, 75–90, https://doi.org/10.1016/S0926-3373(02)00078-4.

Latif, A., Noor, S., Sharif, Q.M., Najeebullah, M. 2010. Different techniques recently used for the treatment of textile dyeing effluents: a review. *J. Chem. Soc. Pakistan.* 32 (1), 115–124.

Lavanya, C. 2014. Review article degradation of toxic dyes: a review. *Int. J. Curr.Microbiol. Applied Sci.* 3, 189–199.

Matsumura, Y., Anantha swamy, H.N. 2004. Toxic effects of ultraviolet radiation on the skin. *Toxicol. Appl. Pharmacol.* 195 (3), 298–308, https://doi.org/10.1016/j.taap.2003.08.019.

Meena, S., Vaya, D., Das, B.K. 2016. Photocatalytic degradation of Malachite Green dye by modified ZnO nanomaterial. *Bull. Mater. Sci.* 39, 1735–1743, https://doi.org/10.1007/s12034-016-1318-4.

Nezamzadeh-Ejhieh, A. 2017. Supporting of coupled silver halides onto clinoptilolite nanoparticles as simple method for increasing their photocatalytic activity in heterogeneous photodegradation of mixture of 4-methoxy aniline and 4-chloro-3- nitro aniline. *J. Colloid Interface Sci.*, 490, 478–487, https://doi.org/10.1016/j.jcis.2016.11.087.

Nezamzadeh-Ejhieh, A., Banan, Z. 2012. Sunlight assisted photodecolorization of crystal violet catalyzed by CdS nanoparticles embedded on zeolite A. *Desalination*, 284, 157–166, https://doi.org/10.1016/j.desal.2011.08.050.

Nezamzadeh-Ejhieh, A., Ghanbari-Mobarakeh, Z. 2015. Heterogeneous photodegradation of 2,4-dichlorophenol using FeO doped onto nano-particles of zeolite P. *J. Ind. Eng. Chem.* 21, 668–676, https://doi.org/10.1016/j.jiec.2014.03.035.

Nezamzadeh-Ejhieh, A., Khorsandi, M. 2010. Heterogeneous photodecolorization of eriochrome black T using Ni/P zeolite catalyst. *Desalination*, 262, 79–85, https://doi.org/10.1016/j.desal.2010.05.047.

Nezamzadeh-Ejhieh, A., Moazzeni, N. 2013. Sunlight photodecolorization of a mixture of methyl orange and bromocresol green by CuS incorporated in a clinoptilolite zeolite as a heterogeneous catalyst. *J. Ind. Eng. Chem*, 19, 1433–1442, https://doi.org/10.1016/j.jiec.2013.01.006.

Ovais, M. et al. 2018. Role of plant phytochemicals and microbial enzymes in biosynthesis of metallic nanoparticles. *Appl. Microbiol. Biotechnol.* 102 (16) 6799–6814, https://doi.org/10.1007/s00253-018-9146-7.

Pandey, S. 2002. Water pollution and health. *Kathmandu Univ. Med. J.* 4, 128–134.

Peng, C., Shen, C., Zheng, S., Yang, W., Hu, H., Liu, J., Shi, J. 2017. Transformation of CuO nanoparticles in the aquatic environment: influence of pH, electrolytes and natural organic matter. *Nanomaterials* 7, 326, https://doi.org/10.3390/nano7100326.

Qiao, R.P., Li, N., Qi, X.H., Wang, Q.S., Zhuang, Y.Y. 2005. Degradation of microcystin- RR by UV radiation in the presence of hydrogen peroxide. *Toxicon* 45, 745–752, https://doi.org/10.1016/j.toxicon.2005.01.012.

Raghav, R., Aggarwal, P., Srivastava, S. 2016. Tailoring oxides of copper-Cu2O and CuO nanoparticles and evaluation of organic dyes degradation. *AIP Conf. Proc.* 1724, 020078, https://doi.org/10.1063/1.4945198.

Saeed, K., Khan, I., Gul, T., Sadiq, M. 2017. Efficient photodegradation of methyl violet dye using TiO_2/Pt and TiO_2/Pd photocatalysts. *Appl. Water Sci.* 7, 3841–3848, https://doi.org/10.1007/s13201-017-0535-3.

Sankar, R., Manikandan, P., Malarvizhi, V., Fathima, T., Shivashangari, K.S., Ravikumar, V. 2014. Green synthesis of colloidal copper oxide nanoparticles using Carica papaya and its application in photocatalytic dye degradation. *Spectrochim. Acta Part A Mol. Biomol. Spectrosc.* 121, 746–750, https://doi.org/10.1016/j.saa.2013.12.020.

Senobari, S., Nezamzadeh-Ejhieh, A. 2018a. A comprehensive study on the photocatalytic activity of coupled copper oxide-cadmium sulfide nanoparticles. *Spectrochim. Acta Part A Mol. Biomol. Spectrosc.* 196, 334–343, https://doi.org/10.1016/j.saa.2018.02.043.

Shamaila, S. et al. 2016, Advancements in nanoparticle fabrication by hazard free ecofriendly green routes. *Appl. Mater. Today* 5, 150–199, https://doi.org/10.1016/j.apmt.2016.09.009.

Shams-Ghahfarokhi, Z., Nezamzadeh-Ejhieh, A. 2015. As-synthesized ZSM-5 zeolite as a suitable support for increasing the photoactivity of semiconductors in a typical photodegradation process. *Mater. Sci. Semicond. Process.* 39, 265–275, https://doi.org/10.1016/j.mssp.2015.05.022.

Sharma, J.K., Akhtar, M.S., Ameen, S., Srivastava, P., Singh, G. 2015. Green synthesis of CuO nanoparticles with leaf extract of *Calotropis gigantea* and its dye-sensitized solar cells applications. *J. Alloy. Comp.* 632,, 321–325, https://doi.org/10.1016/j.jallcom.2015.01.172.

Sharma, V.K., Feng, M. 2017. Water depollution using metal-organic frameworks-catalyzed advanced oxidation processes: a review. *J. Hazard Mater.* 372,, 3–16, https://doi.org/10.1016/j.jhazmat.2017.09.043.

Shaygannia, E., Bahmani, M., Zamanzad, B., Rafieian-Kopaei, M. 2015. A review study on *Punica granatum* L. *J. Evidence-Based Complement. Altern. Med.* 21,, 221–227, https://doi.org/10.1177/2156587215598039.

Shrivastava, A., Gupta, V., Article, R. 2011. Methods for the determination of limit of detection and limit of quantitation of the analytical methods. *Chronicles Young Sci.* 2,, 21–25, https://doi.org/10.4103/2229-5186.79345.

Singh, J. et al. 2019a. Biogenic synthesis of copper oxide nanoparticles using plant extract and its prodigious potential for photocatalytic degradation of dyes. *Environ. Res.* 177, 108569.

Singh, J., Kaur, S., Kaur, G., Basu, S., Rawat, M. 2019b. Biogenic ZnO nanoparticles: a study of blueshift of optical band gap and photocatalytic degradation of reactive yellow 186 dye under direct sunlight. *Green Processes Synth.* 8 (1) 272–280, https://doi.org/10.1515/gps-2018-0084.

Vijayabalan, A., Selvam, K., Krishnakumar, B., Swaminathan, M. 2013. Photocatalytic degradation of Reactive Orange 4 by surface fluorinated TiO_2 Wackherr under UV-A light. *Separ. Purif. Technol.* 108, 51–56, https://doi.org/10.1016/j.seppur.2013.01.025.

Watt, B.E., Proudfoot, A.T., Vale, J.A. 2004. Hydrogen peroxide poisoning. *Toxicol. Rev.* 23, 51–57, https://doi.org/10.2165/00139709-200423010-00006.

Wong, C.C., Chu, W. 2003. The hydrogen peroxide-assisted photocatalytic degradation of alachlor in TiO_2 suspensions. *Environ. Sci. Technol.* 37, 2310–2316, https://doi.org/10.1021/es020898n.

Zhou, M., He, J. 2007. Degradation of azo dye by three clean advanced oxidation processes: wet oxidation, electrochemical oxidation and wet electrochemical oxidation-A comparative study. *Electrochim. Act.* 53 (4), 1902–1910.

9 Tracking Antimicrobial Resistance in Wastewater

Emerging Threats and Mitigation Strategies

Romana Mirdhe and Huskur Chennarayappa Lakshman

9.1 INTRODUCTION

Risking the effectiveness of medical treatments and raising the incidence of infectious illnesses, antimicrobial resistance (AMR) is a serious global health concern (World Health Organization, 2020). Because of its possible involvement in environmental persistence and development of resistance, the extensive presence of antibiotic-resistant bacteria (ARB) and antibiotic resistance genes (ARGs) in wastewater systems is now a major cause for worry (Riquelme et al., 2021). Wastewater from hospitals, pharmaceutical industries, and households serves as a major reservoir for ARB and ARGs, contributing to their accumulation in aquatic ecosystems (Manaia et al., 2018). Furthermore, horizontal gene transfer (HGT) of resistance genes among microbial communities worsens the spread of AMR, posing risks to both humans and environmental health (Karkman et al., 2018). Given these challenges, understanding the mechanisms driving AMR persistence and dissemination in wastewater is crucial for developing effectual extenuation approaches for reducing spread of resistant strains (Zhuang et al., 2021).

Municipal wastewater treatment plays a crucial part in safeguarding public health and preserving aquatic ecosystems by removing contaminants before effluent discharge (Michael et al., 2013). A conventional WWTP integrates physical, chemical, and biological treatment methods to eliminate pollutants, including organic carbon, nutrients, and emerging micropollutants, including pharma and personal care products (Verlicchi et al., 2012). Biological treatment, in particular, relies on a diverse community of microorganisms to degrade organic matter and improve water quality (Ju et al., 2019). The microbial composition within WWTPs is highly dynamic and influenced by various factors, such as influent wastewater characteristics, operational conditions, and environmental variables (Pal et al., 2016). Over time, these microbial communities evolve and adapt, ensuring the sustained efficiency of the treatment process while responding to ecological and technological shifts (Zhang et al., 2017). However, conventional WWTPs are not specifically made for eliminating ARB and ARGs that can persist in effluent and sludge, potentially contributing to their dissemination in the environment (Manaia et al., 2018). As a result, there is a growing necessity for exploring advanced therapeutic approaches to enhance AMR removal and mitigate its spread through wastewater systems (Riquelme et al., 2021).

Emergence of bacteria and other microbes that are resistant to antibiotics is a major concern for world health. Since wastewater from homes, pharmaceutical companies, and hospitals acts as a reservoir and channel for such resistant strains, the rising prevalence of ARB and ARGs in wastewater systems is a major cause for worry. Physical, chemical, and biological processes are all included in the layout of conventional wastewater treatment plants (WWTPs) to eliminate pollutants. However, their effectiveness in eliminating ARB and ARGs remains a significant challenge. The persistence of resistance genes in treated effluents and sludge suggests that WWTPs may act as hotspots for

DOI: 10.1201/9781003269823-9

AMR dissemination, potentially reintroducing resistant strains into the environment. Additionally, HGT among microbial communities within WWTPs further exacerbates the spread of AMR. Understanding the mechanisms driving AMR persistence, transformation, and dissemination in wastewater treatment systems is crucial for developing effective mitigation strategies. By improving wastewater treatment processes and implementing advanced treatment technologies, it may be possible to curb the spread of AMR and safeguard both public health and environmental integrity.

The aim of our review paper is offering comprehensive analysis of AMR in wastewater treatment systems, with an emphasis upon persistence, dissemination, and potential mitigation strategies of ARB and ARGs.

By analyzing existing research and highlighting effective intervention strategies, the purpose of this review is to help improve wastewater management strategies aimed at mitigating the global threat of AMR.

9.2 EMERGING THREATS OF ANTIMICROBIAL RESISTANCE IN WASTEWATER SYSTEMS

9.2.1 Environmental Contamination

Antimicrobial resistance genes (ARGs) present in wastewater can withstand conventional treatment processes, allowing them to persist and eventually be discharged into natural water bodies like rivers, lakes, and coastal environments. Studies have shown that WWTPs, while effective at reducing microbial loads, are not entirely efficient in removing antibiotics and ARGs (Michael et al., 2013; Riquelme Breazeal et al., 2013; Kümmerer et al., 2021; Wang et al., 2022). As a result, these contaminants can accumulate in aquatic ecosystems, creating reservoirs of AMR Zhang et al. (2022). Once released into the environment, antibiotics exert selective pressure on native microbial communities, promoting survival and proliferation of resistant bacteria. This selective pressure fosters the emergence of new resistant strains while also facilitating HGT of ARGs among diverse bacterial species (Marti et al., 2014; Berendonk et al., 2015; Liu et al., 2021; Sun et al., 2023). Consequently, resistance genes can spread beyond wastewater sources, reaching drinking water supplies, agricultural irrigation systems, and even human and animal populations, thereby exacerbating the global AMR crisis (Chen et al., 2021).

Additionally, heavy metals, biocides, and antibiotic residues may all interact with one another to co-select for AMR (Pal et al., 2017). Continual monitoring, strict restrictions, and improved wastewater treatment technologies that are urgently needed to limit environmental development of AMR are underscored by the persistence of these toxins in aquatic ecosystems (Feng et al., 2022; Xie et al., 2023).

9.2.2 Urban and Industrial Sources of Antimicrobial Resistance in Wastewater

Wastewater originating from urban and industrial sources serves as a significant conduit for antibiotics, ARGs, and resistant microorganisms, posing a major challenge to wastewater treatment systems and public health (Rizzo et al., 2013). Among these sources, domestic sewage, hospital effluents, and industrial discharges contribute substantial loads of antimicrobial contaminants, thereby exacerbating the AMR burden in WWTPs.

9.2.3 Domestic Wastewater (Households)

A large portion of the drug residues and bacteria that are immune to antibiotics found in municipal wastewater originates from households. The majority of antibiotics that people ingest do not undergo full metabolism before being eliminated in feces and urine, which eventually make their way to WWTPs (Yuan et al., 2021; Li et al., 2022a). Moreover, the improper disposal of unused medications, such as flushing antibiotics down drains or toilets, exacerbates the antibiotic burden in wastewater systems (Zhao et al., 2021). This ongoing introduction of antibiotics into wastewater exerts selective pressure on bacterial communities, promoting the proliferation of ARB and facilitating HGT among microbial populations (Feng et al., 2022; Liu et al., 2023).

9.2.4 Hospital Wastewater

Hospitals are considered hotspots for AMR dissemination due to the high consumption of antibiotics for treating infections, surgical procedures, and prophylactic applications. Studies have shown that hospital wastewater has substantially high concentrations of antibiotics, ARGs, and multidrug-resistant (MDR) bacteria compared to municipal wastewater (Kümmerer, 2009). These resistant microorganisms can persist through conventional wastewater treatment processes and enter natural water bodies, increasing the risk of environmental and human exposure (Manaia et al., 2018). Furthermore, pathogenic bacteria, like *Escherichia coli* and *Pseudomonas aeruginosa*, carrying ESBL and carbapenemase genes, have been frequently detected in hospital effluents, raising concerns about the spread of untreatable infections (Hocquet et al., 2016).

9.2.5 Industrial Wastewater

Industrial wastewater, particularly from pharmaceutical manufacturing facilities, represents another major source of antibiotics and AMR determinants in the environment. Pharmaceutical plants release high concentrations of active pharmaceutical ingredients into wastewater, especially in regions with inadequate treatment infrastructure (Larsson, 2014). Research has shown that pharma factory wastewater may include antibiotic concentrations, which are many orders of magnitude higher than what is often observed in medical facilities. This creates a perfect situation where bacteria resistance to antibiotics might be selected and propagated (Fick et al., 2009). Other industries, such as agriculture, food processing, and biotechnology, also contribute to AMR contamination by releasing biocides, heavy metals, and antibiotic residues into wastewater, which can co-select for AMR (Pal et al., 2017).

9.2.6 Contribution to the AMR Burden in Wastewater

The combined influence of urban and industrial sources leads to a complex mixture of antibiotics, ARGs, and resistant pathogens in wastewater, posing significant challenges to efficiency of conventional treatment systems. Persistence of AMR determinants in treated effluents and sludge highlights the urgent need for enhanced surveillance, advanced treatment technologies, and stricter regulatory measures to curb the environmental dissemination of AMR (Wang et al., 2022; Xie et al., 2023). Strengthening policies on antibiotic disposal, upgrading wastewater treatment infrastructure, and promoting antibiotic stewardship programs are essential strategies to mitigate the AMR burden in wastewater systems (Zhang et al., 2022; Chen et al., 2023).

9.3 CLIMATE AND SEASONAL VARIATIONS IN THE SPREAD OF AMR IN WASTEWATER SYSTEMS

Climate change and seasonal variations have a substantial part in influencing spread, persistence, and evolution of AMR in wastewater systems. These environmental factors impact wastewater treatment efficiency, microbial community dynamics, and the dissemination of ARB and ARGs into natural ecosystems. Changes in temperature, precipitation, and extreme weather events contribute to the mobilization and distribution of AMR determinants, exacerbating the global AMR crisis (Eckert et al., 2020; Adegoke et al., 2021; Li et al., 2022b). Rising global temperatures have been linked to increased HGT and the proliferation of resistant bacterial strains in WWTPs, further complicating efforts to control AMR (MacFadden et al., 2022). Additionally, flooding and heavy rainfall events can lead to the overflow of untreated sewage, facilitating the uncontrolled spread of ARGs and antibiotic residues into surface waters and agricultural soils (Wu et al., 2023; Tiedje et al., 2023). To reduce dangers to human and environmental health posed by AMR, our results highlight the pressing need for wastewater treatment solutions that are climate-adaptive.

9.3.1 Impact of Rainfall and Extreme Weather Events

Increased rainfall and extreme weather events, such as hurricanes and floods, can result in the overflow of wastewater systems, allowing untreated or partially treated wastewater to be discharged into surface water bodies, soils, and groundwater (Wu et al., 2019). These discharges contain high concentrations of antibiotics, ARB, and ARGs, facilitating their spread in environment and increasing the risk of human and animal exposure (Ahmed et al., 2021). Moreover, storm water runoff can transport AMR contaminants from urban and agricultural areas into natural waterways, further exacerbating environmental contamination (Gouliouris et al., 2019).

9.3.2 Seasonal Temperature Variations and Microbial Dynamics

Temperature fluctuations throughout the year influence composition and activities of microbial communities in WWTPs and receiving water bodies. Higher temperatures in summer promote microbial activity, potentially accelerating HGT of ARGs among bacteria (Lamba and Ahammad, 2017). This increased gene exchange can enhance the development and persistence of new resistant strains in wastewater environments. Conversely, lower temperatures in winter may slow down the degradation of antibiotics and antimicrobial residues, prolonging their presence in wastewater and increasing selective pressure on microbial communities (Hendriksen et al., 2019).

9.3.3 Seasonal Variations in Antibiotic Use and AMR Trends

Seasonal fluctuations in human and veterinary antibiotic usage also contribute to variations in AMR prevalence in wastewater. For instance, winter months often see a surge in antibiotic consumption due to increased incidences of respiratory infections, leading to higher concentrations of antibiotic residues and ARGs in municipal wastewater (Khan et al., 2020). In contrast, agricultural antibiotic use may peak during certain seasons due to livestock treatment cycles, further influencing the AMR burden in wastewater systems (Berendsen et al., 2015).

9.3.4 Climate Change and Long-Term AMR Evolution

Conditions that encourage the selection and spreading of resistant bacteria in effluent and natural bodies of water could be created by long-term impacts of climate change, which include increased global temperatures and changing precipitation patterns. Higher bacterial reproduction rates and environmental stability of resistant strains are linked to warmer temperatures (MacFadden et al., 2018). Additionally, climate change-induced alterations in water quality, such as lower dissolved oxygen levels and increased organic matter loads, can affect microbial communities and resistance gene dynamics in wastewater ecosystems (Rocha et al., 2022).

9.4 IMPLICATIONS FOR WASTEWATER MANAGEMENT AND AMR MITIGATION

The influence of climate and seasonal variations on AMR highlights the need for adaptive wastewater management strategies. Implementing real-time monitoring of wastewater for AMR indicators, improving stormwater management infrastructure, and enhancing wastewater treatment

TABLE 9.1
Sources and Contributions of Antimicrobial Resistance in Wastewater

Source	Key Contributions to AMR in Wastewater	Implications
Domestic wastewater (households)	• Unmetabolized antibiotics excreted through urine and feces. • Improper disposal of medications (flushing antibiotics down the drain). • Antibiotic residues in gray water from laundry and dishwashing.	• Promotes bacterial resistance due to continuous exposure to sublethal antibiotic concentrations. • Horizontal gene transfer (HGT) enhances AMR spread.
Hospital wastewater	• High concentrations of antibiotics from patient treatments. • Presence of MDR bacteria such as *E. coli*, *K. pneumoniae*, and *P. aeruginosa*. • Waste from surgical procedures and infection control.	• Resistant bacteria persist through treatment plants. • AMR pathogens can enter surface waters and drinking water sources.
Pharmaceutical industry wastewater	• Direct discharge of active pharmaceutical ingredients (APIs). • Presence of highly concentrated antibiotics in untreated effluents. • Heavy metals and disinfectants that co-select for resistance genes.	• Creates hotspots for AMR evolution and selection pressure. • Antibiotic residues contaminate local water bodies and soil.
Agricultural and livestock wastewater	• Usage of antibiotics in livestock for growth and illness resistivity. • Manure runoff containing antibiotic residues and resistant bacteria. • Contaminated irrigation water.	• Transmission of AMR from animals to humans via food and water. • Contributes to the spread of resistance in soil microbiomes.

technologies to withstand climate variability are critical steps toward mitigating AMR spread (Wellington et al., 2013). Additionally, integrating climate change considerations into AMR surveillance and wastewater treatment policies can help develop more resilient and sustainable approaches to combating the environmental dissemination of AMR (Table 9.1).

9.5 DETECTION AND MONITORING OF ANTIMICROBIAL RESISTANCE IN WASTEWATER

Rapid emergence and spread of AMR in wastewater systems necessitate robust detection and monitoring strategies to assess the occurrence of ARGs and resilient bacteria. Among the various molecular and genomic approaches available, Next-generation sequencing (NGS) has revolutionized the study of AMR by providing high-throughput, detailed insights into the genetic composition of microbial communities in wastewater. This technology enables researchers to track the evolution, dissemination, and potential risks associated with ARGs, facilitating the development of targeted intervention strategies (Zhang et al., 2021).

9.6 NEXT-GENERATION SEQUENCING

NGS technologies, particularly whole-genome sequencing and targeted sequencing methods, have revolutionized the ability to detect ARGs in complex environmental samples such as wastewater. These high-throughput sequencing platforms provide a comprehensive approach to characterizing the genetic makeup of microbial populations and identifying resistance genes, even in low-abundance organisms (Zong et al., 2020). NGS allows for the identification of known and novel ARGs by comparing sequences to large databases, offering perceptions in diversity and evolution of resistivity traits. Technology is not only capable of identifying point mutations responsible for resistance but also can track gene transfer events between microbes, which is essential for understanding the spread of AMR (Couto et al., 2021).

Moreover, NGS enables the monitoring of AMR evolution by identifying mutations in bacterial populations that may indicate the development of resistance to commonly used antibiotics. For example, research conducted by Lupo et al. (2019) has shown how NGS can track the emergence of MDR pathogens in wastewater over time, providing an early warning system for public health threats.

9.7 METAGENOMICS

Metagenomics, particularly shotgun metagenomics, allows for a comprehensive analysis of the microbial communities in wastewater without the need for culturing individual pathogens. By sequencing the entire DNA content of a sample, shotgun metagenomics provides a snapshot of the microbial ecosystem, including both bacteria and other microorganisms, such as viruses and fungi. This method enables the detection of novel resistance genes and recognition of new mechanisms of resistivity that may not yet be present in clinical isolates (Köhler et al., 2018). Through metagenomic analysis, researchers can detect the presence of ARGs in both pathogenic and nonpathogenic bacteria, which is crucial for understanding potential reservoirs of resistance genes in environment. A study by Burgess et al. (2020) highlighted how metagenomics helped identify previously unrecognized ARGs in wastewater, which could be linked to human or animal waste contamination. These findings stress the essentiality of wastewater like a reservoir and vector for AMR, underscoring the need for continuous surveillance.

9.8 REGULAR SURVEILLANCE

Continuous surveillance of wastewater for antibiotics, ARGs, and resistant bacterial strains is critical for detecting and managing the spread of AMR. The dynamic nature of wastewater, which integrates the effluent from hospitals, households, industries, and agriculture, offers a suitable matrix for monitoring the presence and spread of resistant bacteria and ARGs. By tracking the concentration of specific antibiotics and resistance markers, early warnings of emerging AMR hotspots can be provided, allowing for timely interventions and control measures. Regular surveillance helps identify trends in AMR prevalence and can assist in pinpointing regions where intervention is most urgently needed. For instance, Stokes et al. (2022) demonstrated that wastewater surveillance could be used to track the seasonal and geographic distribution of antibiotic resistance, providing valuable data on areas at greater risk for outbreaks of resistant infections.

Further, regular monitoring of wastewater has shown its potential to monitor the effectiveness of public health strategies aimed at controlling AMR. In a study by Rizzo et al. (2021), wastewater surveillance data was used to evaluate the success of antibiotic stewardship programs, offering a real-time feedback mechanism for improving resistance management.

9.9 MITIGATION STRATEGIES

Mitigating AMR in wastewater is critical for reducing its environmental impact and protecting public health. Several strategies can be implemented to curb the spread of resistance, ranging from technological innovations to regulatory frameworks and public education. Below is an expanded discussion on these strategies with additional references.

9.9.1 Enhanced Wastewater Treatment

Advanced wastewater treatment technologies have emerged as a key solution for reducing the volume of antibiotics and ARGs in treated effluent. Membrane bioreactors (MBRs) and advanced oxidation processes (AOPs) are two promising technologies that have shown success in this regard.

- Membrane bioreactors (MBRs) integrate biological therapy using membrane filtration, providing an improved means of eliminating contaminants, such as ARGs and antibiotics. MBRs provide high-efficiency filtration that can remove both suspended solids and dissolved contaminants. Research by Kümmerer et al. (2019) observed that MBRs significantly reduce the concentration of antibiotic-resistant bacteria and ARGs, making them a viable technology for WWTPs.
- AOPs utilize strong oxidants, like ozone or hydrogen peroxide, for breaking down organic pollutants, including antibiotics. A study by Niemeyer et al. (2020) found that AOPs could effectively degrade numerous kinds of antibiotics and reduce the prevalence of ARGs in wastewater effluent. These processes can also act as a complementary treatment step to enhance the overall removal efficiency when combined with traditional treatment methods.

9.9.2 Regulatory Measures

Developing and enforcing regulations for the safe disposal of antimicrobial waste and wastewater effluent management is essential to controlling AMR in the environment. Regulatory measures can ensure that WWTPs adopt the best practices for managing antimicrobial contaminants and ARGs.

- EU and United States EPA have already started to develop guidelines for monitoring and controlling the discharge of antibiotics into the environment (Börjesson et al., 2021). These regulations could mandate the implementation of advanced treatment technologies and encourage WWTPs to adopt strategies for removing ARGs from treated effluent.
- Additionally, regulations could focus on the disposal of pharmaceutical waste, particularly unused or expired antibiotics, which are often improperly disposed of and end up in wastewater systems. Implementing take-back programs and encouraging proper disposal methods, such as returning antibiotics to pharmacies, can help reduce the introduction of antimicrobial agents into wastewater (Chahine et al., 2019).

9.9.3 Public Awareness and Education

Raising public awareness about AMR and its environmental impact is another important strategy for mitigating the problem. Educating the public about the proper disposal of medications can help reduce the amount of antibiotics entering wastewater systems.

- Public education campaigns can target improper disposal practices, such as flushing unused antibiotics down the toilet, which contributes to the contamination of water systems. In Pereira et al. (2021), the authors highlighted the effectiveness of public awareness campaigns in encouraging responsible antibiotic disposal and reducing the environmental load of antibiotics.
- Additionally, educating the public on the dangers of overusing antibiotics, both in humans and in animals, can lead to more responsible consumption. Misuse and mismanagement of antibiotics are major contributors to the development of AMR, and initiatives aimed at reducing unnecessary prescriptions and promoting adherence to prescribed treatments can help curb this trend (Williams et al., 2020).

9.9.4 One Health Approach

Humans, animals, and ecological health are all seen as interdependent under One Health paradigm. Addressing AMR through this comprehensive framework is essential for tackling the issue in a holistic manner.

- The One Health approach integrates surveillance systems across humans, animals, and environmental sectors, creating a collaborative effort to monitor and manage AMR. World Health Organization (WHO) has advocated for the adoption of the One Health approach to combat AMR by promoting cross-sectoral coordination and information sharing (WHO, 2021).
- Research by López-Vidal et al. (2020) demonstrated that a One Health approach enhances AMR mitigation efforts by addressing resources and routes via which resistance is able to propagate. As instance, antimicrobial use in agriculture is a significant driver of resistance, with runoff from farms often contaminating water sources. By reducing antibiotic use in livestock farming and improving waste management practices, the One Health approach could decrease the overall burden of AMR in environment (Tables 9.2–9.4).

TABLE 9.2
Detection and Monitoring Technologies for AMR in Wastewater

Technology	Methodology	Applications and Advantages	Limitations
Next-generation sequencing (NGS)	Whole-genome sequencing and targeted sequencing of microbial DNA to identify ARGs.	• Provides high-throughput, detailed genetic information. • Tracks AMR evolution and emerging resistance genes.	• High cost and technical expertise required.
Shotgun metagenomics	Sequencing all DNA in a sample to analyze microbial communities and resistance genes.	• Identifies known and novel resistance genes. • Finds bacteria that are resistant in both cultured and uncultured conditions.	• Requires large computational resources.
Quantitative PCR (qPCR)	Amplification of specific ARG sequences for detection and quantification.	• Highly sensitive for detecting target ARGs. • Rapid and cost-effective.	• Limited to known resistance genes.
Wastewater surveillance	Routine sampling and analysis of wastewater for antibiotic residues and ARGs.	• Provides early warnings of AMR hotspots. • Helps in tracking resistance trends.	• Requires standardized protocols for consistent monitoring.

TABLE 9.3
Emerging Threats of Antimicrobial Resistance in Wastewater

Threat	Cause and Mechanism	Potential Impact
Persistence of ARGs in treated effluent	Traditional WWTPs are not effective enough in removing ARGs.	• Resistant bacteria enter natural water bodies and drinking water sources. • Increases the likelihood of HGT among bacteria.
Urban and industrial wastewater as AMR hotspots	High loads of antibiotics and resistant bacteria from hospitals, pharmaceutical plants, and households.	• Continuous selection pressure which leads in evolution of MDR pathogens. • Industrial waste can introduce high antibiotic concentrations into rivers and lakes.
Climate and seasonal variations	Increased temperatures accelerate microbial activity and HGT. Flooding and heavy rainfall events spread untreated wastewater.	• AMR-contaminated floodwaters reach drinking water sources. • Extreme weather increases wastewater overflow, worsening resistance spread.
Co-selection with heavy metals and biocides	Wastewater containing disinfectants, metals (e.g., copper, zinc), and antibiotics.	• Selection pressure leads to bacteria acquiring multiple resistance mechanisms. • Reduced effectiveness of antimicrobial treatment in healthcare settings.

TABLE 9.4
Mitigation Strategies for AMR in Wastewater

Strategy	Key Actions	Effectiveness and Challenges
Enhanced wastewater treatment	• Use of membrane bioreactors (MBRs) for filtering bacteria and ARGs. • Advanced oxidation processes (AOPs), such as ozone and hydrogen peroxide, to degrade antibiotics.	• MBRs effectively remove ARGs but require high operational costs. • AOPs degrade antibiotics but need optimization for large-scale applications.
Regulatory measures	• Establish policies for pharmaceutical and hospital waste disposal. • Set permissible limits for antibiotic residues in wastewater.	• Stringent regulations can reduce AMR contamination at the source. • Requires strong enforcement and compliance mechanisms.
Public awareness and education	• Promote responsible antibiotic disposal through pharmacy take-back programs. • Educate healthcare professionals and farmers on antibiotic stewardship.	• Reduces unnecessary antibiotic use. • Requires long-term commitment and widespread public participation.
One Health approach	• Incorporate environmental, animal, and human health monitoring systems. • Foster cooperation among environmental, agricultural, and healthcare organizations.	• Provides a holistic strategy for AMR control. • Requires coordination at the national and global levels.

9.10 CONCLUSION

The monitoring and mitigation of AMR in wastewater is critical for controlling the spread of resistance and safeguarding public health. As AMR continues to pose significant global challenges, wastewater serves as both a reservoir and a pathway for the dissemination of resistance genes. Tracking the occurrence of antibiotics and ARGs through technologies like NGS and metagenomics has provided valuable insights into the dynamics of AMR in wastewater systems. These technologies enable the early detection of emerging threats, offering opportunities for targeted intervention and management. In addition to detection, effective mitigation strategies such as advanced wastewater treatment technologies, including MBRs and AOPs, have an important part in reducing the volume of antibiotics and ARGs in treated effluent. Regulatory measures to control antimicrobial waste disposal, combined with public education campaigns upon responsible usage and disposal of antibiotics, further contribute to minimizing the environmental load of antimicrobial contaminants. The One Health approach offers a comprehensive framework for tackling AMR, recognizing the interconnectedness between humans, animals, and environmental health. By integrating surveillance systems across these sectors, a collaborative effort can be made to address the root causes of AMR and prevent its further spread.

In conclusion, the review highlights the need for ongoing surveillance, research, and the implementation of multipronged mitigation strategies to manage the growing threat of AMR in wastewater. By combining technological innovation, regulatory action, public awareness, and cross-sector collaboration, we can work toward reducing the environmental impact of AMR and protecting both human and environmental health,

REFERENCES

Adegoke, A. A., A. I. Okoh, and O. O. Babalola. 2021. Climate Change and the Environmental Microbiome: Implications for the Dissemination of Antibiotic Resistance Genes and Pathogens. *Microorganisms* 9 (3): 528.

Ahmed, W., A. Bivins, P. M. Bertsch, K. Bibby, K. Farkas, P. Gyawali, and J. Mueller. 2021. Surveillance of Wastewater for Early Epidemic Prediction: A Review. *Science of the Total Environment* 761: 144–216.

Berendonk, T. U., C. M. Manaia, C. Merlin, D. Fatta-Kassinos, E. Cytryn, F. Walsh, and J. L. Martinez. 2015. Tackling Antibiotic Resistance: The Environmental Framework. *Nature Reviews Microbiology* 13 (5): 310–317.

Berendsen, B. J., R. S. Wegh, J. Memelink, T. Zuidema, and L. A. Stolker. 2015. The Analysis of Animal Faeces as a Tool to Monitor Antibiotic Usage. *Talanta* 132: 258–268.

Börjesson, M., M. Häger, and E. Kristiansson. 2021. Regulatory Frameworks for Controlling Antibiotic Pollution in Wastewater. *Environmental Pollution* 268: 115607.

Burgess, T., C. Marshall, and D. O'Reilly. 2020. Identification of Novel ARGs in Wastewater via Shotgun Metagenomics. *Environmental Microbiology Reports* 12 (1): 105–113.

Chahine, M., V. Pichon, and G. Marc. 2019. Waste Management Strategies for Pharmaceutical Waste: Addressing the Environmental Burden of Unused Antibiotics. *Journal of Environmental Management* 243: 1–8.

Chen, H., C. Liu, and L. Zhao. 2021. The Environmental Spread of Antimicrobial Resistance from Wastewater Effluent and Its Global Impact. *Environmental Pollution* 277: 116848.

Chen, L., P. Sun, and X. Wang. 2023. The Role of Wastewater Treatment Plants in Mitigating Antimicrobial Resistance: Current Challenges and Future Directions. *Environmental Science and Technology* 57 (8): 4513–4528.

Couto, N., M. Oliveira, and A. Fonseca. 2021. The Role of NGS in Monitoring Antimicrobial Resistance in Wastewater Environments. *Journal of Antimicrobial Chemotherapy* 76 (5): 1239–1248.

Eckert, E. M., A. Di Cesare, M. T. Kettner, M. Arias-Andres, D. Fontaneto, and H. P. Grossart. 2020a. Microplastics Increase the Spread of Antibiotic Resistance in the Environment. *Microbiome* 8: 40.

Eckert, E. M., A. Di Cesare, B. Stenzel, and G. Corno. 2020b. Impact of Seasonal Variations on the Spread of Antibiotic Resistance in Municipal Wastewater Treatment Plants. *Environmental Pollution* 260: 114033.

Feng, J., C. Tan, H. Wang, and Z. Li. 2022. Co-Selection of Antibiotics and Heavy Metals in Wastewater Systems: Implications for Antimicrobial Resistance Spread. *Science of the Total Environment* 800: 149555.

Fick, J., H. Soderstrom, R. H. Lindberg, C. Phan, M. Tysklind, and D. G. J. Larsson. 2009. Contamination of Surface, Ground, and Drinking Water from Pharmaceutical Production. *Environmental Toxicology and Chemistry* 28 (12): 2522–2527.

Gouliouris, T., F. Coll, C. Ludden, B. Blane, P. Naydenova, C. Crawley, and S. J. Peacock. 2019. Population Genomics of *Enterococcus faecium* Shows No Evidence for a Link Between Wastewater Isolates and Clinical Infection. *Microbial Genomics* 5 (11): 45–56.

Hendriksen, R. S., P. Munk, P. Njage, B. Van Bunnik, L. McNally, O. Lukjancenko, and F. M. Aarestrup. 2019. Global Monitoring of Antimicrobial Resistance Based on Metagenomics Analyses of Urban Sewage. *Nature Communications* 10 (1): 1124.

Hocquet, D., A. Muller, and X. Bertrand. 2016. What Happens in Hospitals Does Not Stay in Hospitals: The Environmental Impact of Antibiotic Resistance. *Nature Reviews Microbiology* 14 (8): 460–473.

Ju, F., K. Beck, X. Yin, A. Maccagnan, C. S. McArdell, H. P. Singer, and T. Zhang. 2019. Wastewater Treatment Plant Resistomes Are Shaped by Bacterial Composition, Genetic Exchange, and Upregulated Expression in the Effluent Microbiomes. *The ISME Journal* 13 (2): 346–360.

Karkman, A., K. Pärnänen, and D. J. Larsson. 2018. Fecal Pollution Can Explain Antibiotic Resistance Gene Abundances in Anthropogenically Impacted Environments. *Nature Communications* 9 (1): 1–10.

Khan, N. A., M. Shahnawaz, S. U. Khan, and A. Malik. 2020. Seasonal Variations in Antibiotic Resistance Pattern of *Escherichia coli* Isolates in Wastewater and Surface Water. *Environmental Science and Pollution Research* 27: 36745–36756.

Köhler, T., S. H. Lee, and R. Braun. 2018. Shotgun Metagenomics for the Detection of Antibiotic Resistance Genes in Environmental Samples. *Microbial Ecology* 76 (1): 14–27.

Kümmerer, K. 2009. Antibiotics in the Aquatic Environment–A Review–Part I. *Chemosphere* 75 (4): 417–434.

Kümmerer, K., A. Götz, and S. Decker. 2019. Effectiveness of Membrane Bioreactors in Wastewater Treatment and Antibiotic Resistance Removal. *Environmental Science and Pollution Research* 26 (3): 2490–2499.

Kümmerer, K., H. Holzinger, and A. Haider. 2021. Persistent Antibiotics in Wastewater: Implications for Antimicrobial Resistance. *Environmental International* 147: 106390.

Lamba, M., and S. Z. Ahammad. 2017. Sewage Treatment Effluents in Delhi: A Key Contributor of β-Lactam Resistant Bacteria and Genes to the Environment. *Chemosphere* 188: 249–256.

Larsson, D. G. J. 2014. Pollution from Drug Manufacturing: Review and Perspectives. *Philosophical Transactions of the Royal Society B: Biological Sciences* 369 (1656): 20130571.

Li, S., M. Gao, and X. Wang. 2022a. Climate Change and Antibiotic Resistance: Impacts on Wastewater Treatment Efficiency and Public Health Risks. *Science of the Total Environment* 814: 152691.

Li, S., X. Zhang, and S. Chen. 2022b. Occurrence and Fate of Antibiotics in Urban Wastewater: From Households to Wastewater Treatment Plants. *Science of the Total Environment* 814: 152739.

Liu, Y., Z. Li, and M. Wang. 2021. Horizontal Gene Transfer in Wastewater Treatment Plants and Its Implications for Antimicrobial Resistance Spread. *Science of the Total Environment* 789: 148113.

Liu, Y., Z. Li, and M. Wang. 2023. Horizontal Gene Transfer in Wastewater Treatment Plants and Its Implications for Antimicrobial Resistance Spread. *Science of the Total Environment* 789: 148113.

López-Vidal, Y., F. González-González, and H. Rodríguez-Villalobos. 2020. The One Health Approach to Antimicrobial Resistance. *Science of the Total Environment* 726: 138–150.

Lupo, A., L. M. Schouls, and A. Haenen. 2019. Tracking the Emergence of Multidrug-Resistant Pathogens in Wastewater via NGS. *Science of the Total Environment* 678: 344–351.

MacFadden, D. R., S. F. McGough, D. Fisman, M. Santillana, and J. S. Brownstein. 2018. Antibiotic Resistance Increases with Local Temperature. *Nature Climate Change* 8 (6): 510–514.

MacFadden, D. R., S. F. McGough, D. Fisman, M. Santillana, and J. S. Brownstein. 2022. Antibiotic Resistance Increases with Local Temperature. *Nature Climate Change* 12 (1): 89–97.

Manaia, C. M., G. Macedo, D. Fatta-Kassinos, and O. C. Nunes. 2018. Antibiotic Resistance in Urban Aquatic Environments: Can It Be Controlled? *Applied Microbiology and Biotechnology* 102 (5): 2073–2087.

Manaia, C. M., J. Rocha, N. Scaccia, R. B. M. Marano, E. Radu, F. Biancullo, and D. Fatta-Kassinos. 2018. Antibiotic Resistance in Wastewater Treatment Plants: Tackling the Black Box. *Environment International* 115: 312–324.

Marti, E., J. Jofre, and J. L. Balcázar. 2014. Prevalence of Antibiotic Resistance Genes and Bacterial Community Composition in a River Influenced by a Wastewater Treatment Plant. *PLoS One* 8 (10): e78906.

Michael, I., L. Rizzo, C. S. McArdell, C. M. Manaia, C. Merlin, T. Schwartz, C. Dagot, and D. Fatta-Kassinos. 2013. Urban Wastewater Treatment Plants as Hotspots for Antibiotic-Resistant Bacteria and Genes Spread into the Environment: A Review. *Science of the Total Environment* 447: 345–360.

Niemeyer, K., S. Huber, and M. Wenzel. 2020. Advanced Oxidation Processes for the Removal of Antibiotics and ARGs in Wastewater Treatment. *Chemosphere* 245: 125562.

Pal, C., J. Bengtsson-Palme, E. Kristiansson, and D. G. J. Larsson. 2017. Co-Occurrence of Resistance Genes to Antibiotics, Biocides, and Metals Reveals Novel Insights into Their Co-Selection Potential. *BMC Genomics* 16: 964.

Pal, C., J. Bengtsson-Palme, C. Rensing, E. Kristiansson, and D. G. J. Larsson. 2016. BacMet: Antibacterial Biocide and Metal Resistance Genes Database. *Nucleic Acids Research* 42 (D1): D737–D743.

Pereira, A. R., J. Sousa, and R. Silva. 2021. Public Awareness Campaigns for the Proper Disposal of Antibiotics. *Public Health Reports* 136 (2): 257–263.

Riquelme Breazeal, M. V., J. G. Frye, T. A. Bigelow, M. Wiedmann, and H. F. Oliver. 2013. Mechanisms of Antimicrobial Resistance in *Salmonella enterica*. *Microbiology Spectrum* 1 (2): 1–24.

Riquelme, M. V., C. Capozzi, C. Marcoux, Y. Gizard, T. U. Berendonk, and C. Dagot. 2021. Impact of Wastewater Treatment Processes on Antibiotic Resistance Genes: A Metagenomic Approach. *Water Research* 188: 116519.

Rizzo, L., C. Manaia, C. Merlin, T. Schwartz, C. Dagot, M. C. Ploy, and D. Fatta-Kassinos. 2013. Urban Wastewater Treatment Plants as Hotspots for Antibiotic-Resistant Bacteria and Genes Spread into the Environment: A Review. *Science of the Total Environment* 447: 345–360.

Rizzo, L., C. McArdell, and C. Dagot. 2021. Wastewater Surveillance as a Tool for Monitoring the Effectiveness of AMR Control Strategies. *Lancet Infectious Diseases* 21 (8): 1063–1070.

Rocha, J., D. Cacace, I. Kampouris, H. Guilloteau, T. Jäger, and C. M. Manaia. 2022. The Role of Water Reuse and Treatment on the Dissemination of Antibiotic Resistance in Europe: A Meta-Analysis. *Environmental International* 158: 106998.

Stokes, H. W., R. Brown, and A. O'Neill. 2022. Wastewater Surveillance of Antimicrobial Resistance: Temporal and Geographic Trends. *Science of the Total Environment* 804: 150188.

Sun, Q., S. Zhao, and W. Zhang. 2023. The Role of Wastewater Effluent in the Spread of Antimicrobial Resistance in Natural Water Bodies. *Environmental Microbiology* 25 (1): 162–173.

Tiedje, J. M., B. Aslam, and W. M. de Vos. 2023. The Role of Extreme Weather Events in Accelerating the Spread of Antibiotic Resistance. *Nature Reviews Microbiology* 21 (2): 112–124.

Verlicchi, P., M. Al Aukidy, and E. Zambello. 2012. Occurrence of Pharmaceutical Compounds in Urban Wastewater: Removal, Mass Load, and Environmental Risk After a Secondary Treatment—A Review. *Science of the Total Environment* 429: 123–155.

Wang, S., J. Zhang, and Y. Liu. 2022. Ineffectiveness of Conventional Wastewater Treatment in Removing Antibiotic Resistance Genes: A Systematic Review and Meta-Analysis. *Environmental Pollution* 297: 118712.

Wellington, E. M., A. B. Boxall, P. Cross, E. J. Feil, W. H. Gaze, P. M. Hawkey, and A. P. Williams. 2013. The Role of the Natural Environment in the Emergence of Antibiotic Resistance in Gram-Negative Bacteria. *The Lancet Infectious Diseases* 13 (2): 155–165.
World Health Organization (WHO). 2020. Antimicrobial Resistance. Retrieved from https://www.who.int/.
World Health Organization (WHO). 2021. One Health and Its Role in Tackling Antimicrobial Resistance. World Health Organization. Retrieved from https://www.who.int/news-room/fact-sheets/detail/one-health.
Williams, S. R., D. Harvey, and G. Linder. 2020. Reducing the Overuse of Antibiotics through Education and Policy. *Lancet Infectious Diseases* 20 (6): 671–672.
Wu, J., Y. Zhang, and H. Li. 2023. Flood-Induced Dissemination of Antibiotic Resistance Genes from Wastewater Treatment Plants to Natural Water Bodies. *Water Research* 245: 120319.
Wu, N., Y. Zhang, C. Li, D. Li, J. Luo, and X. Pan. 2019. Seasonal Variation of Antibiotic Resistance Genes and Microbial Communities in Municipal Wastewater Treatment Plants. *Science of the Total Environment* 697: 134–147.
Xie, B., S. Wang, and Q. Ma. 2023. Current Challenges and Future Perspectives for Advanced Wastewater Treatment Technologies to Combat Antimicrobial Resistance. *Environmental Science and Technology* 57 (2): 908–918.
Yuan, Z., X. Li, and H. Zhang. 2021. Environmental Fate of Pharmaceutical Residues from Household Sources: A Global Review on Wastewater Contamination and Treatment. *Environmental Pollution* 274: 115675.
Zhang, T., Y. Hu, and X. Tang. 2022. Accumulation and Fate of Antibiotics and Antibiotic Resistance Genes in Wastewater Treatment Systems. *Science of the Total Environment* 837: 155633.
Zhang, T., X. X. Zhang, and L. Ye. 2017. Plasmid Metagenome Reveals High Levels of Antibiotic Resistance Genes and Mobile Genetic Elements in Activated Sludge. *PLoS One* 6 (10): e26041.
Zhao, L., Q. Sun, and W. Zhang. 2021. Improper Disposal of Antibiotics in Household Wastewater and Its Environmental Impacts: A Review. *Environmental Pollution* 268: 115622.
Zhuang, M., Y. Achmon, Y. Cao, X. Liang, Y. Tian, and Y. Jiang. 2021. Antibiotic Resistance Genes in Drinking Water: A Review and Bibliometric Analysis. *Science of the Total Environment* 807: 151042.
Zong, Z., M. Liu, and H. Wang. 2020. Emerging Antimicrobial Resistance in Wastewater Treatment Plants: The Role of NGS Technology. *Environmental International* 135: 105351.

10 Monitoring Antimicrobial Resistance in Wastewater

An Escalating Environmental Concern

Kalpana, Sweta, Ruchi Singh, Richa Chandra, Prakash Chandra Gupta, Shilpa Deshpande Kaistha, and Nisha Sharma

10.1 INTRODUCTION

Antimicrobials, widely utilized in medicine and agriculture, are acknowledged as one of the most efficacious kinds of chemotherapy. They have preserved countless lives and contributed to the management of infectious diseases, which were the principal causes of mortality and morbidity before the advent of antibiotics. Antimicrobial resistance (AMR) denotes the capability of infections and microbes to evade the effects of antibiotics, enabling their proliferation. AMR is inevitable as microbes undergo genetic alterations to mitigate the detrimental impacts of antibiotic treatment. A report in India indicated a concerning rate of child mortality from antibiotic-resistant bacterial infections, resulting in approximately 50,000 newborn fatalities due to sepsis. The World Health Organization (WHO) stated that AMR results in 700,000 fatalities globally annually, with an anticipated 20 million deaths by 2050, incurring over $2.9 trillion in healthcare expenses. Methicillin-resistant *Staphylococcus aureus* (MRSA) constitutes a threat, accounting for 74% of all *S. aureus* infections worldwide. In 2019, mortality entirely owing to MRSA surpassed 100,000 (Yamin et al., 2023).

With current patterns of impervious infections in human beings as well as livestock suggesting a probable post-antibiotic age, AMR is generally agreed as one of the biggest 10 threats to public health. Although frightening projections indicate that resistant bacterial infections may become the main cause of mortality worldwide by 2025 comprising both genes and organisms, AMR can spread among humans, animals, and environmental media like air and water, therefore highlighting the need of the "One Health approach" in addressing the interrelations of humans, animals, and environmental health (Liguori et al., 2022). In places with limited access to water, sanitation, and hygiene specifically, the environmental spread of AMR is alarming. While AMR surveillance has attracted interest in both clinical and veterinary areas, studies on ambient AMR tracking/surveillance are still rare (Alfahl et al., 2024).

Wastewater surveillance (WWS), or the monitoring of urban sewage discharge, serves as an economical method in favor of forecasting the prevalence of antibiotic-resistant bacteria (ARB) and their associated genes within a community. Municipal sewage is a valuable resource for monitoring ARB due to its bacterial biomass, which includes pathogens from the entire community population. This biomass contributes to various infection stages through biological samples such as feces, urine, nasal secretions, dermal excretions, and respiratory mucus collected from domestic settings, healthcare facilities, and long-term care institutions (Tiwari et al., 2022). Unregulated industrial wastewater is a significant source of toxic compounds, resulting in health issues such

DOI: 10.1201/9781003269823-10

as cancer, immunological dysfunction, and respiratory ailments. The Environmental Protection State Inspectorate (PIOS) indicated that around 60% of liquid effluent endangers public health and the environment. Risk assessments, including hazard identification, exposure assessment, and risk characterization, are crucial for ensuring safer wastewater management. Pharmaceutical firms, utilizing 99% of water for drug production, discharge harmful substances into the environment. Treatment techniques such as microfiltration, ultrafiltration (RO), agglomeration, and coalescence are employed to diminish the concentration of deleterious substances in effluent (Sathya et al., 2022). AMR is a multifaceted issue influenced by various factors, including travel, trade, climate change, and demographic shifts. Current monitoring concentrates on select pathogens and passively disseminates laboratory findings. Urban sewage serves as an appealing monitoring technique because of its extensive, predominantly healthy demographic and without informed consent. An increasing percentage of the world population resides in metropolitan areas and is linked to sewage systems. Sewage samples have few practical and logistical obstacles for collection. The majority of microbiological research on sewage has concentrated on the hazards associated with inadequately treated sewage discharge or overflow from heavy rains; however, some assessments have also been performed on pathogen monitoring. Sewage has demonstrated utility in worldwide polio eradication initiatives. Therefore, monitoring of AMR is very crucial to control the hazardous effects of sewage and wastewater resulting in global health impacts affecting the socioeconomic growth of world (Hendriksen et al., 2019).

10.1.1 Antibiotics, Antibiotic-Resistant Bacteria (ARB), and Antibiotic Resistance Genes (ARGs) in Water, Wastewater, and Sewage Sludge

The regulation of ARG release into the environment is contingent upon wastewater treatment (WWT). A wastewater treatment plant (WWTP) comprises different chemical constituents, entailing antiseptics, metallic elements (such as copper and zinc), medicaments (such as antibiotics), and various chemical constituents. Chemical attributes are among the various selection forces that impact the transmission, expression, or mobilization of ARGs, as well as the emergence, occurrence, and propagation of ARB. From human and animal microbiomes, wastewater (sewage) frequently transports ARGs, ARB, and a vast array of benign and infectious bacteria into WWTPs (Nguyen et al., 2021).

In aquatic surroundings, ARB can disseminate ARGs, which are enduring and can be assimilated by other microbes, hence augmenting antibiotic resistance. Antibiotics elevate aquatic AMR genes. The dissemination and co-contamination of drugs and ARG present significant health hazards, attracting scrutiny (Wang et al., 2021).

10.1.2 Development and Proliferation of ARG

The antibiotic resistome comprises ARGs that act as vectors for the dissemination of ARGs in this environment, potentially augmenting the prevalence and diversity of pathogens and commensal microbes (Hazra et al., 2022).

The biological treatment process in WWTPs facilitates the origin and spread of ARGs, resulting in the creation of novel ARB. WWTPs are substantial producers of ARB and their associated genes, contaminating many ecosystems. Research indicates a rise in effluent ARGs relative to influent water concentrations. Resistance genes have been detected in aerosols above wastewater treatment facilities (Gao et al., 2023).

10.1.3 Relationships between ARG and ARB Pathogens

WWT aims to reduce waterborne illnesses and regulate the discharge of Antimicrobial Respiratory Species into the environment. WWTPs contain chemical constituents like disinfectants, metals, and medications, which influence the transmission, expression, and mobilization of ARGs, facilitating their formation, persistence, and proliferation. Chemical factors play a crucial role in WWT (Guo et al., 2015; Li et al., 2019; Zhang et al., 2017). Urban sludge introduces ARGs, ARB, and various bacteria into WWTPs. These genes often cluster on transposable elements, creating complex resistance regions. The acquisition of an extrachromosomal DNA molecule can confer multidrug resistance (MDR) to the host bacteria, particularly when it emerges on extrachromosomal DNA or a vector with a high virulence gene factor (Venturini et al., 2010, 2013; Mangat et al., 2017). WWT aims to eliminate pathogens before effluent release, necessitating re-evaluation in this circumstance of AMR. Noninfectious microbes can act as pools for ARGs, and WWTPs could potentially regulate the spread of ARGs. The rise in infections due to ARB is portrayed as a critical challenge for the healthcare system in the 21st century (Cadman and Marinez, 2014). Between 2014 and 2016, nearly one million individuals died from ARBs due to ineffective treatment or inherent untreatability (Humphreys and Fleck, 2016). The development of antibiotic resistance and its associated gene transfer events is primarily due to the excessive and often improper use of antibiotics in human and veterinary medicine (Holmes et al., 2016). The mobilization of ARGs complicates the effective eradication of ARB once transmission begins. WWTPs process wastewater containing ARB and ARGs, as well as residual antibiotics and metals. Understanding factors influencing their proliferation and dissipation during treatment and identifying barriers to their spread in reuse is crucial. Dense and active microbial communities in WWTPs may serve as optimal environments for ARB selection and horizontal gene transfer (HGT) (Hong et al., 2018).

Antibiotic resistance is a critical human health hazard, with rising levels of ARB and associated ARGs in aquatic surroundings linked to anthropogenic contributions (World Health Organization, 2014). Shortly following the first clinical case reports, ARGs linked to serious diseases and antibiotic resistance have been found in WWTPs or the environment (Kieffer et al., 2016). Human-affected settings have shown observations of proliferation, bioaccumulation, ARB, and ARG transportation suggesting possible human exposure and the necessity of assessing any health hazards (Ashbolt et al., 2013). WWTPs handle wastewater including ARB, ARGs, residual antibiotics, metals, and other selective agents, hence requiring knowledge of parameters controlling proliferation and dissipation during treatment and identification of reuse barriers (Manaia, 2017). The dense and dynamic microbial communities in a WWTP may be ideal for selecting ARB and ARGs in treated effluent, as bacteriophages are detected in high proportions in secondary sewage effluent, facilitating HGT (Huang et al., 2016), while they interact in addition to infect dense microbial colonization's at elevated contagious rates (Muniesa and Jofre, 2004). The proliferation of active bacteria promotes swift virulent phages and substantial per-cell viral yield (Fuhrman, 1999), creating optimal conditions for lateral gene transfer via transduction, where bacteriophages inadvertently acquire and transfer bacterial DNA fragments, including ARGs (Miller et al., 1992). Concerns related to ARB and ARGs may hinder future attempts to recycle the treated wastewater.

10.1.4 Risks Related to ARGs and ARB in Wastewater

A methodology for risk assessment of AMR in WWT detects several sources of antibiotic resistance, analyzes paths and fates of determinants, evaluates exposure and risk quantification, and points forth mitigating options. A few research studies have looked at the possible relationship between the use of recycled water and human health impacts related to ARB as well as antibiotic-resistant genes exposure. The most direct path for human contact is direct potable reuse, using treated wastewater for human consumption. Nonpotable reuse also has possible hazards, including breathing in aerosol generated during irrigation or cooling, and the results showed very low antibiotic resistance bacteria and antibiotic-resistance genes in treated wastewater that has undergone treatment (Garner et al., 2021b).

10.2 SOURCES OF AMR IN WASTEWATER

By allowing the mixing of germs from numerous environments, including people and animals, hence providing a medium for genetic transmission among many species, water has a vital ecological and evolutionary function in AMR. This is especially true in wastewater, given high nutrient concentrations and bacterial burdens as well as the existence of antibiotic drugs and their derivatives (Sambaza & Naicker, 2023).

The investigation of the impact of diverse pollution sources on the spread of AMR is presently the subject of research, with fecal pollution identified as a significant contributor to ARB and ARG. Alawi et al. (2022) reported that sewage systems, agricultural activity, and animal-based transmission routes all contribute to the frequent fecal pollution of water. Many studies support this idea since it is believed that the chief genesis of antibiotic contamination in the environment is the excretion of poorly digested antibiotics by humans and animals, hence maybe causing AMR by constant microorganism exposure (Flach et al., 2024).

10.2.1 Hospital Wastewater: Antibiotics and Resistant Organisms Abundant

Unlike residential wastewater, hospital wastewater (HWW) is hazardous and infectious. Harmful bacteria and viruses, dangerous organic pollutants, radioactive compounds, and pharmaceutical chemicals including antibiotics and psychiatric medications make up the effluent generated from radiology departments, operating rooms, diagnostic laboratories, and infectious wards. The large concentration of numerous harmful bacteria in HWW emphasizes the prospective public health concern connected with HWW release into water bodies. Hospital waste in China is a major source of ARG and variants. Particularly with enhanced amounts of drugs and antibiotic compounds, the COVID-19 epidemic has greatly worsened hospital waste, therefore increasing the risk to human wellness and the environmental surroundings (Yuan & Pian, 2023)

HGT increases antibiotic resistance in bacteria, putting organisms at various trophic levels at risk. Humans, at the apex of the food chain, are exposed to elevated levels of disinfectants, antibiotics, ARGs, and ARB. Environmental studies should focus on enrichment, bioaccumulation, and biomagnification of these substances. Protocols for disinfectants and antibiotics have been refined amid COVID-19 (Chen et al., 2021).

ARB, ARGs, and intestinal opportunistic pathogens are abundant in hospital effluent. Antibiotics are more commonly provided to hospitalized patients, which increases the quantities of these bacteria in the wastewater. Different healthcare institutions produce wastewater with water demand ranges in wealthier countries between 200 and 1,200 L/bed/day and in poorer countries between 200 and 400 L/bed/day. This produces 250–570 m^3/day of wastewater with high concentrations of traditional contaminants with chemical oxygen demand (COD), demand for oxygen in biochemical processes, nutrients, dissolved solids, total suspended solids (TSS), and chlorides, fecal coliforms, along with other microorganisms (Puljko et al., 2023; Werkneh & Islam, 2023).

By means of synergistic effects, antibiotic residues present in hospital effluent can be rather harmful to aquatic species, especially when found in low water table. This can be attributed to their great use in healthcare environments, and metronidazole, ciprofloxacin, chloramphenicol, and sulfamethoxazole are the most often found antibiotics at high concentrations in hospital effluents (Deguenon et al., 2022).

10.2.2 Pharmaceutical and Industrial Wastewater: Presence of Antibiotic Residues Resulting from Medication Production

Increased infiltration of untreated chemicals into sewage overflow, cattle farms, and rivers arising from pharmaceutical industrialization in hospitals has resulted from this. Furthermore, pharmaceutical by-products are introduced into the sewage systems. Eliminating pharmaceutical waste ranges from 10% to 60%; therefore, wastewater treatment facilities fall short in this regard. According to a recent study, sludge releases 70% of antibiotics and genetic molecules (Husain Khan et al., 2023).

Globally used pharmaceutically active compounds like antibiotics, antidepressants, and steroids create drug residue into aquatic environments and their residues are becoming water pollutants endangering aquatic ecosystems (Akhter et al., 2024).

10.2.3 Agricultural Runoff and Livestock Farming: The Role of Antibiotics in Animal Husbandry

Livestock farming operation is a notable source of AMR amid excessive use of antibiotics in food production and waste management methods. Because of the close proximity of the animals and the greater risk of infection transmission, confined animal feeding operations (CAFOs) still need intensive antibiotic utilization per animal (Meyer et al., 2024). Livestock farming is crucial for global food security and contributes to livelihoods and cultural practices. However, intensive production methods, such as CAFOs, frequently cause environmental pollution. Large amounts of manure created by cattle can cause nutrient runoff and contamination of water bodies with contaminants like nitrogen and phosphorus, contributing to eutrophication and algal blooms (Williams, 2024). Animal husbandry is essential for the European Community's agriculture, and getting the greatest results depending on the use of healthy and high-quality feed. Several active compounds, including antibiotics, are utilized to sustain animal physical state and improve breeding yields. Medicated nutrition is a homogeneous blend of foodstuff and veterinary pharmaceutical drugs, administered orally (Ghimpeţeanu et al., 2022). The misuse of antibiotics in the livestock and poultry industry is a significant global environmental issue, posing a significant research hotspot (Tian et al., 2021).

Because of their repercussions on human health and changes in the ecology, antibiotics expelled into the environment are generating questions. Unmetabolized antibiotics can be discharged into the environment via wastewater, agricultural organic manure, and inadvertent animal manure leaks due to world population increase and demand for cattle (Kim et al., 2024).

10.2.4 Municipal and Domestic Wastewater: Ongoing Release of Antibiotic-Contaminated Effluent from Residences

Black water and gray water are the two categories for domestic wastewater; black water comes from toilets while gray water comes from all other wastewater excluding toilets. Although some gray water is persistent, its volume is one to seven times greater but organic compound in this is lower. Residential buildings account for over 75% of domestic wastewater output; the unsettled percentage adds up from public facilities, offices, and businesses. Though their elements are the same, both forms of wastewater differ in content and volume (Widyarani et al., 2022).

Macrolide pollution in freshwater systems comes from home sewage, hospital effluent, and pharmaceutical manufacturing plants. Urban rivers in developed Chinese cities are clearly contaminated with antibiotics; however, research studies on highly populated cities in northwest China are few. The evaluation of antibiotic pollution depends on macrolide antibiotic distribution (Yang et al., 2024a).

Furthermore, the degree of knowledge regarding health consequences of consuming water polluted with antibiotics was assessed, which showed that only 35% ($n=158$; $p=0.003$) of households knew regarding the health hazards of making use of water polluted with antibiotics (Karimi et al., 2022).

Long-term wastewater irrigation in urban farming significantly raises residual antibiotic levels in soil two to three times greater than in nonirrigated soil. Along with this enrichment, metal cation with organic carbon in soil generates antibiotic-containing compounds to extend their lifetime. Furthermore, contributing to antibiotic pollution are offshore aquaculture projects in coastal areas. Furthermore, influencing soil vulnerability in urban green areas could be the air movement of antibiotics (Yang et al., 2024b).

10.3 MECHANISMS OF AMR DISSEMINATION IN WASTEWATER SYSTEMS

Antibiotic resistance is recognized by the WHO as a major worldwide public health hazard of the modern era. Thus, especially in sewage treatment facility pollutants, urgent studies on environmental causes of antibiotic resistance spread are required to minimize detrimental health repercussions. Wastewater systems, especially urban runoff and hospitals, disseminate AMR and call for knowledge to help control its repercussions on the environment and public health (Leiva et al., 2021).

10.3.1 Antibiotic Residue Selection Pressure: Fostering the Viability and Proliferation of Resistant Bacteria

AMR evolution is being driven by selection pressure in the environment. When an antimicrobial agent is present in a microbe, those whose mutations confer resistance or acquired resistance genes have a viability preference and multiply; susceptible ones get destroyed, hence establishing the predominance of resistant strains among the population. The environment functions as a reserve for AMR; hospital effluent and agricultural trash facilitate the dispersal of resistant microbes and their ARG in soil, water, and air. Travel, sanitation, and infection control policies all influence the way selection pressure influences AMR evolution (Oliveira et al., 2024).

Moreover, actively contributing to the growth of AMR is the incorrect use of antibiotics, specifically broad-spectrum antibiotics that kill both hazardous and good bacteria inside the body. By disturbing the typical bacterial flora and giving resistant germs an opportunity to grow and spread, such approaches generate an environment that encourages their favorable development. The utilization and misutilization of antimicrobial compounds have been associated with enhanced AMR in foodborne bacteria (Endale et al., 2023).

Currently, one of the biggest dangers to human health is antibiotic resistance. Antibiotic resistance bacterium and ARG can be selectively pressured by antibiotic residues and their extensive usage in humans, animals, and the surroundings, therefore hastening the flow of residues of antibiotics. The burden of AMR in humans rises as ARG permeates the population, hence possibly having negative health consequences for individuals (Giacometti et al., 2021; Ding et al., 2023).

Concerns about the health of human's risks of antibiotic resistance connected to environmental residues of antibiotics include the possibility of altering the microbial community in humans and encouraging the emergence and option of antimicrobial-resistant bacteria and the risk of producing selection pressure on the environmental microorganisms leading to environmental resistance to antibiotics (Ahmad et al., 2023).

10.3.2 Horizontal Gene Transfer (HGT)

With an eye toward the effects of complex mixes of possible HGT in municipal and hospital wastewater, a study explored the horizontal transfer of ARG between bacteria in wastewater. Exposed to many sterile-filtered wastewaters, an *E. coli* recipient with a GFP tag was allowed to mate with a diverse wastewater bacterial population. An indication of HGT came from the recipient strain acquiring sulfamethoxazole resistance (Hutinel et al., 2021).

The mainstay of biological WWT, activated sludge, helps eliminate pathogens, fecal bacteria, and contaminants connected to sewage. Given residual antibiotics in sewage, HGT has been hypothesized to help ARGs propagate within WWTPs. Little data, however, points to sewage-associated antibiotics selecting for resistance in WWTPs either through HGT or elsewhere (Brown et al., 2024).

Because of their high microbial density and presence of residual antibiotics and other contaminants, WWTPs significantly help remove and mitigate pollutants, but they cannot totally stop the spread of ARGs. ARGs can be transferred to symbiotic bacteria and pathogens when ARG hosts come into contact with people. Horizontal and vertical gene transfer have been documented to help ARGs proliferate among bacteria in the contaminated water system (Ma et al., 2024).

10.3.3 Biofilm Formation: Safeguarding Resistant Microorganisms in Wastewater Treatment Facilities (WWTPs)

One possible method that has become somewhat well-known recently is biofilm-mediated reduction of contaminants from wastewater. Dependent on quorum sensing, this approach entails bacterial cells exchanging signaling chemicals, genetic materials, and metabolites. Understanding the bioremediation of contaminated water depends on an awareness of bacterial diversity in biofilm by the use of metagenomic techniques such as 16S rDNA (Saini et al., 2023). While little research has looked at the co-occurrence of biofilm-forming tendency in MDR clinical bacterial isolates in pharmaceutical industry wastewater (PIWW). Recent studies have examined the widespread occurrence of ARB communities in PIWW. The study estimated the antibiotic resistance, morphological, and genotypic characterization of biofilm-forming MDR-identified clinical bacterial strains by evaluating physicochemical characteristics, heavy metal concentrations, organic pollutants, and antibiotic detection in PIWW samples (Jyoti et al., 2024).

Inconsistent and inadequate WWT can cause microbially driven corrosion, environmental contamination, and microbial proliferation (Shivaram et al., 2023). Strong biofilm-forming potential of HWW could be a reservoir of MDR *A. baumannii* strains, endangering the public health (Eze et al., 2021).

Four different kinds of plastics high-density polyethylene, polyvinyl chloride, polyethylene terephthalate, and polystyrene, which are main polymer types found in urban wastewater, were employed in the colonizing experiment. The potential of WWTPs depends on enough treatment, disposal, and policies for managing hazardous waste (Bydalek et al., 2023).

10.3.4 Microbial Adaptation and Evolution: The Emergence of Multidrug-Resistant (MDR) Infections

A worldwide public health issue generating major death and disease is MDR. High mortality, morbidity, and treatment expenses are related to it (He et al., 2024). The capacity of some bacterial strains to create biofilms, which account for 65%–80% of human infections, aggravates the growth of MDR and extensively drug-resistant microorganisms (Mba & Nweze, 2021). Previously controllable infections have become life-threatening as traditional antibiotics have failed to combat pathogens such as *Staphylococcus aureus*, *E. coli*, and *P. aeruginosa*. This change has raised death rates and severely taxed systems of health care all over. With forecasts showing MDR infections could cause more deaths than cancer by 2025, the WHO ranks antibiotic resistance as one of the top ten public health concerns (Srivastav et al., 2024).

10.4 MONITORING AND SURVEILLANCE TECHNIQUES FOR AMR IN WASTEWATER

Especially at the human population level, wastewater-based surveillance (WBS) is a useful instrument for tracking AMR/ARB infections. Excreted from several sources including feces, urine, nasal mucus, skin, and sputum, ARB pathogens and their genetic markers are expelled shortly after the start of infections and during several phases (Tiwari et al., 2024). At a local level, a culture-based strategy can be more practical; yet, the availability of clinical data limits WWS of ARB validation efforts (Tiwari et al., 2022).

Knowledge on assessing the extent and frequency of antibiotic resistance in WWTPs nowadays depends on whole-genome sequencing (WGS) and quantitative polymerase chain reaction (qPCR). While WGS provides high-resolution quantification of designated sequences inside an array, qPCR lets one detect and quantify certain ARGs in ambient samples. But because of the breadth of a resistome, primers' selectivity for qPCR experiments limits ARG detection (Drane et al., 2024).

Conventional AMR surveillance compiles one patient at a time from time-consuming, culture-based techniques. Coupling "pooled" WBS samples with developing NGS technology provides a perhaps more quick and strong complement for AMR surveillance at the community level, though (Harrington et al., 2022).

10.4.1 Next-Generation Sequencing

By offering highly throughput and reasonably priced DNA and RNA molecule analysis, next-generation sequencing (NGS) has transformed genomics research. Rapid sequencing of millions of DNA fragments made possible by this technology has shed light on the structural framework of the genome, genetic variants, patterns of gene expression, and epigenetic changes (Satam et al., 2023). Research focusing on uncommon genetic illnesses, the genomics of cancer, microbiome studies, infectious disease analysis, as well as the genetics of human populations have been made possible by NGS platforms, hence broadening the field of genomics study. Targeting treatments, precision medicine approaches, and better diagnosis techniques all of which NGS has made possible have become important. Its adoption in the water industry has been restricted, nevertheless, by cost, specialist knowledge, data analysis and interpretation difficulties, lack of standardized approaches, and fast technological change. Efforts are required to remove these challenges and enable NGS technology to be available to water professionals so that our knowledge of complicated microbial populations in water and wastewater may grow (Garner et al., 2021a).

10.4.2 Metagenomic Sequencing

In water and wastewater, metagenomic sequencing also called shotgun metagenomic sequencing is a technique used to examine microbial populations and their functional capacity (Garner et al., 2021a). This method has expanded quickly in usage to investigate the functional capacity of microbial populations. Currently used for taxonomic profiling, shotgun metagenome sequencing enables parallel sequencing of DNA from every species-level detecting organism with great coverage. After that, the resulting sequencing reads are matched to a reference database to identify taxa, therefore enabling wide taxonomic coverage, accurate functional profiling, and the possibility of finding hitherto unidentified species and strains of bacteria (Cason et al., 2022).

NGS applied by shotgun metagenomics sequences DNA collected from microbial populations living in the studied environment. One can characterize the resistome (collective ARGs transported across a microbial population) by means of the ensuing metagenome. The sulfonamide resistance gene was the most often occurring ARG; succeeded by the qacE gene, which imparts resistance to quaternary ammonium compounds. The most often occurring ARGs against each antibiotic were the genes associated with aminoglycoside resistance, namely ant(3")-I and aminoglycoside kinase, as well as the genes conferring resistance to macrolides, ABC-F type ribosomal protection protein Msr(E) and macrolide phosphotransferase gene are of significant interest (Sekizuka et al., 2022).

Another work investigated utilizing shotgun metagenomics and culture-based techniques, the makeup of bacterial populations, the resistome, and mobilome at a WWTP. Four sampling sites of a WWTP in Antioquia, Colombia, gathered wastewater samples; twenty-four metagenomes were investigated. Mostly found in WWTP influent and effluent samples were bacterial families of clinical relevance including Moraxellaceae, Aeromonadaceae, and Enterobacteriaceae (Rodríguez et al., 2021). The study employed metagenomic NGS on various samples obtained from untreated wastewater, and wastewater sourced from an activated sludge bioreactor, river water collected both upstream and downstream of the wastewater discharge point, and upper respiratory tract swabs taken from the WWTP revealed the resistome profile and mobility of genes after analysis of the taxonomic composition of microorganisms (Zieliński et al., 2022).

10.4.3 Whole-Genome Sequencing (WGS)

A potent genomic analytic tool, WGS, offers a comprehensive record of a person's genetic composition, therefore allowing the discovery of genetic variants (Ma et al., 2022). Research on plants and animals as well as cancer, unusual genetic illnesses, and population genetics all find application for it. Mostly prokaryotes, WGS uses two kinds of sequencing techniques: small WGS for smaller genomes and large WGS for bigger genomes. For the purpose of genome assemblies, researchers typically opt for long-read sequencing, and for mutation calling short-read sequencing is recommended. Accurate genome assembly without a reference sequence has been effectively achieved and new genomes sequenced by utilizing a combination of short-read and long-read sequencing methodologies (Satam et al., 2023).

Molecular characterization of non-selective *E. coli* isolates revealed variations in phylogenetic group and host source distribution between WWTP and Lake Biwa samples in a study looking at possible antibiotic resistance pollution by *E. coli* host sources in aquatic environments. Similar patterns shown by AMP-resistant E. coli isolates from WWTP discharge and Lake Biwa indicated a clear correlation between the resistance of antibiotics disseminated via WWTP effluents and Lake Biwa (Ma et al., 2022).

One of the studies demonstrated that worldwide AMR gene diversity and prevalence differ by location, and enhancing sanitation and health may potentially mitigate the global impact of AMR. It advocates for a metagenomic study of sewage to facilitate ongoing worldwide surveillance and forecasting (Hendriksen et al., 2019).

10.4.4 Qualitative Polymerase Chain Reaction

A labor-efficient and all-encompassing approach for identifying ARGs in wastewater is qPCR. Conventional qPCR does, however, only examine four targets in a single run, which is a constraint. Targeting several genes in numerous samples, high-throughput qPCR (HT-qPCR) technology can help lower expenses, time, labor, and reagent needs (Shrestha et al., 2024).

Extra attention is needed to minimize contamination, including field blanks in sample collection. Among its various benefits are fast screening of big samples, high-throughput analysis, and detection and quantification of particular ARGs in complex environmental samples (Srathongneam et al., 2024).

The work sought to simultaneously monitor protozoa, bacteria, viruses, and ARG targets in composite influent wastewater samples gathered over a 24-hour timeframe, specifically for the surveillance of infectious diseases using a mix of molecular and microbiological techniques. Aliquots obtained from centrifuged pellets from wastewater along with ultrafiltered supernatant drove optimization for the concentration of biomass as well as nucleic acid purification techniques. Implementing a dual approach that incorporates (reverse transcription (RT))-qPCR as well as HT-qPCR techniques, the obtained nucleic acids were investigated to identify the frequency and spatial and temporal patterns of harmful bacteria, protozoa, and viruses in Finnish wastewater. The reproducibility as well as detection capability of custom-quantitative PCR and high-throughput qPCR techniques were assessed as a proof of concept for simultaneous performance of WBS for pathogens and ARGs. The reproducibility and susceptibility of in-house qPCR and HT-qPCR techniques were assessed as a proof of concept for simultaneous performance of WBS for pathogens and ARGs (Sarekoski et al., 2024).

10.4.5 Culture-Based Approaches

For the identification of particular bacteria, culture-based techniques are fast, cheap, and cost-effective; but they lack resolution, neglect unculturable bacteria, and have low sensitivity. They are most suited for microbes that grow at a slow pace as well as viable but nonculturable (VBNC) as well as for bacteria that multiply effectively in rich media within 24h (Clarke et al., 2024). WGS, antimicrobial susceptibility testing, and gene frequency analysis all benefit from culture-based approaches. They cannot, however, monitor live but unculturable bacteria and are time-consuming. Regardless of target viability, non-culture-based approaches, i.e., metagenomics are more helpful for ARG detection instead of culture-based approaches (Baba et al., 2022). Potential bias in quantification makes culture-based methods favored over PCR-based methods. Important opportunistic pathogens present in wastewater like *Acinetobacter* spp., *Aeromonas* spp., and *Pseudomonas* spp. may flourish in both natural aquatic habitats and wastewater and gain genes imparting many drug resistance. AMR study in water environments in Japan should employ culture-independent techniques going forward (Singh et al., 2024a). Contemporary optoelectronic technologies, fiber optics, microfluidics, and indicator dyes can augment the sensitivity and efficacy of optical testing systems. FDA-approved automated AST systems, such as the MicroScan WalkAway 205, BD Phoenix Automated Microbiology System, Vitek 2 System, and Sensititre ARIS 2X, are essential in clinical laboratories. These devices possess several microdilution trays designed for monoderm bacteria, diderm bacteria, automatically evaluating turbidness in broth media from multiwell plates and employing a redox indicator to identify microbes. Certain technologies, like the Alfed 60 ASTTM system from Alifax, employ laser-light scattering technology that enables the detection of bacterial proliferation in broth media (Vasala et al., 2020).

10.5 ENVIRONMENTAL AND PUBLIC HEALTH IMPACTS OF AMR IN WASTEWATER

The use of antibiotics in farming, aquaculture, pharmaceutical manufacture, and anthropogenic activity has resulted in AMR found in environmental soils along with streams. Because of their amalgamation of microbes, antibiotics, and various contaminants including heavy metals, WWTPs are important artificial lakes as well as distribution routes for AMR. The environment is a main AMR reservoir and dissemination route since up to 90% of antimicrobial dosages can be eliminated as active chemicals or metabolites into the surroundings (Kaiser et al., 2022).

ARB and ARGs find a vital reservoir in the aquatic environment, which presents hazards for direct and indirect exposures to humans and animals resulting in forward transmission. Prior knowledge on certain impurities and AR factors found in urban sludge is necessary for the discovery of local solutions to prevent the spread of these pollutants (Asaduzzaman et al., 2022).

Public health and environmental safety are seriously threatened by the complicated interaction of human activities with microbial evolution. Inaccurate WWT techniques can cause ARB to build up and subsequently contaminate aquatic habitats, irrigation of agricultural land, water supplies, direct contact with water, and so spread (La Rosa et al., 2025).

Faced with antibiotic resistance, the battle calls for an approach that considers human health issues as well as environmental ones, therefore emphasizing environmental pollution of antibiotics. Anyone regardless of age and gender can be affected by AMR; it is not limited to any nation. A significant challenge within the realm of public health is its complexity throughout the world, and the evolution along with the spread of AMR is influenced by interdependent elements linked to health care, agriculture, pharmaceuticals, incorrect waste management, trade, and finance (Salam et al., 2023).

10.5.1 Impacts of AMR on Public Health

Once regarded as a major medical breakthrough, antibiotics' indiscriminate use has turned into a nightmare for human health care. Treatment only becomes feasible with last-resort antibiotics since resistance to antibiotic-resistant organisms can cause untenable diseases and mortality. Opportunistic pathogens including *E. coli*, *Klebsiella pneumoniae*, *E. faecalis*, and *E. faecium* call the gut microbiome a varied environment home. Antibiotic residue intake can change the human microbiome, therefore encouraging the improvement as well as selecting antibiotic resistance in humans. Escalation of virulence, pathogenicity, treatment failure, and disease outbreaks characterized by extended periods of morbidity and rising death rates follow from AMR. Zoonoses or vectors allow resistance from animals to humans to be conveyed, therefore influencing human health (Bobate et al., 2023).

10.5.2 Contamination of Natural Water Bodies

Natural microbial populations, which are vital for ecological processes and water quality maintenance, may suffer in rivers from antibiotics (Grenni, 2022). Though metagenomic tools have been utilized to overcome the noncultivability of most environmental bacteria, there are no set techniques for spotting both ARGs and antibiotics. Treatment techniques do not fully remove antibiotics found in wastewater, which fuels their presence in aquatic environments. Research indicates that growing numbers of ARB and developing resistance genes are connected to the addition of antibiotics in aquatic systems. In aquatic environments, subinhibitory quantities of antibiotics affect HGT and mutagenesis in bacteria, hence distributing resistance genes into the surroundings (Kusi et al., 2022). Antibiotics and antibiotic resistance have been found to potentially arise from surface waters (Osbiston et al., 2021). Sustained selection pressure resulting from the application of biosolids, manure, wastewater polluted with antibiotics, ARB harboring ARGs, and irrational use of antimicrobials in agricultural soil helps ARGs to arise and become rather common in soils. Urban watersheds, streams, ponds, lakes, and other water features also help ARGs to flourish (Reddy et al., 2022).

10.5.3 Risk to Potable Water Resources

Currently, there are no established regulations governing the management of ARG and mobile genetic elements, particularly those ascribed with plasmid-mediated antimicrobial resistance (PMR) in potable water. Similarly, wastewater can serve as a potential source of contaminants for potable water resources. Plasmids act as carriers for the transmission of AMR across various ecological niches underscoring the necessity for reliable detection techniques, which would facilitate the global tracking and analysis of AMR sources. Consequently, this would allow for the establishment of mitigation strategies aimed at lowering the threat of infections and diseases ascribed with AMR, ultimately extending the effectiveness of currently available antimicrobial agents for use in both veterinary and *Homo sapiens* (Alawi et al., 2022).

The microbial communities present in diverse water sources and distribution networks can significantly influence water quality, particularly in relation to potable water and recreational water activities. Furthermore, the composition of bacterial communities may be associated with the presence of infectious agents that are significant for human health. The focus of this analysis is on the key fecal-indicator microorganisms identified in drinking water, highlighting the limitations of their effectiveness as indicators in other water sources vital for *Homo sapiens* well-being. *E. coli* along with *Enterococci* serve as two prominent fecal indicator organisms that are extensively utilized for monitoring the quality of recreational waters. Since these bacteria are found in the feces of both humans and animals, they are designated as indicators of bacterial contamination in fecal matter, specifically concerning the presence of *Enterococcus* species may imply a likelihood of persistent fecal impurity in water bodies (Stec et al., 2022).

10.5.4 Influence on Aquatic Ecosystems

AMR is transmitted in aquatic settings mostly, so effective community-based intervention plans depend on the identification of main sources for human and animal exposure (Grenni, 2022). To guarantee animal health and food supply, antimicrobial compounds are applied in aquaculture, plant development, and animal care. From these industries, wastewater and runoff water can be considered diffuse sources of antimicrobial compounds with their TPs in aquatic habitats. Their breakdown and sorption to organic particles define the presence of antibiotics in aquatic surroundings. These processes are influenced by elements like half-life, water quality, meteorological variables, and soil properties. While certain antibiotics are more durable and can thus accumulate to larger amounts, others are readily broken down (Hanna et al., 2023). Mostly from household, industrial, and hospital wastewater, emerging pollutants seriously endanger the environment and individuals. Major categories from which these pollutants originate are pharmaceuticals, personal care items, and endocrine-disturbing compounds (Okoye et al., 2022).

10.6 APPROACHES FOR WASTEWATER TREATMENT FOR LOWERING AMR

Although AMR presents environmental and health hazards, the worldwide water scarcity calls for the reuse of treated wastewater for agriculture. Because of dangerous disposal and rising population, antibiotics found in personal care, medications, and human health products pose hazards to aquatic habitats. WWT seeks to clean the environment by eliminating toxins. More research should concentrate on removing antibiotics from the environment by means of efficient methods including combination ones, enhanced oxidation processes, adsorptive treatment techniques, and membrane processes (Kumar and Pal, 2018).

10.6.1 Traditional Wastewater Treatment Approaches

Customary WWT techniques encompass preliminary primary, secondary (biological), and tertiary treatment stages.

10.6.1.1 Primary Treatment Approach

This eliminates suspended particles and organic debris. This preliminary phase aims to eliminate coarse, suspended, and floating materials from untreated wastewater. It encompasses screening to capture solid items and sedimentation by gravity to eliminate floating solids. This physical solid/liquid separation is a mechanical process; however, chemicals may occasionally be employed to expedite sedimentation. This stage of the treatment decreases the biochemical oxygen demand (BOD) of the entering wastewater by 20%–30% and the total suspended particles by approximately 50%–60% (Alvim et al., 2020).

10.6.1.2 Secondary Treatment Approach

This employs biological methods alike activated sludge processes and membrane bioreactors. The WWT process entails the elimination of dissolved organic debris, which bacteria convert into carbon dioxide, water, and energy for their proliferation. Secondary treatment eliminates around 85% of suspended particles and Chemical oxygen demand (COD), segregating the biological sludge from the clarified water. This sludge can be utilized as a co-substrate alongside other waste materials in a biogas facility to produce biogas, a combination of CH_4 and CO_2, which provides heat and electricity for subsequent energy distribution.

The residual water is subjected to nitrification or denitrification, thereafter undergoing chlorine treatment. The treated water may still harbor impurities; therefore, it must undergo filtering and disinfection with sodium hypochlorite. The processed water is deemed safe for irrigation applications. Solid waste produced in the time of primary and secondary treatment operations undergoes additional processing in a gravity-thickening tank, subsequently in a centrifuge dewatering tank, and ultimately in a lime stabilization tank. Treated solid waste can be utilized for multiple purposes, including landfilling, fertilizers, and construction materials. Additional techniques employed in full-scale reactors encompass ponds, trickling filters, anaerobic processes such as up-flow anaerobic sludge blanket (UASB) reactors, artificial wetlands, microbial fuel cells, and methanogenic reactors. UASB reactors have been utilized for WWT for an extended period, with research indicating an average COD cutoff rate of 85% without the provision of external heat. In a separate experiment, inoculating UASB reactors with cow manure yielded reductions of 81%, 75%, and 76% in COD, TSS, and total sulfate, respectively (Raju et al., 2020).

10.6.1.3 Tertiary Treatment Approach

This approach is employed to eliminate contaminants from wastewater, achieving the removal of around 99% of impurities. This process entails the treatment of water either singularly or in conjunction with modern techniques, namely, ultrasonication, ultraviolet light, and exposure to ozone to eliminate bacterial and heavy metal contaminants. The processed water is subjected to ultrasonication, ultraviolet light treatment, and ozone chamber exposure for thorough elimination. The ultrasonication (US) method renders cells nonviable via free-radical assault and physical damage to cellular membranes. The synergistic application of US, ultraviolet (UV), and exposure to ozone triggers free radicals that cohere to the cell membranes of biological pollutants, facilitating the ingress of chemical oxidants inside cells to target interior structures. The US, independently or in conjunction with others, promotes the deagglomeration of germs and enhances the efficacy of various chemical disinfectants. A combined treatment approach for textile wastewater was evaluated, comparing the efficacy of ultrasound as a pretreatment in conjunction with ultraviolet radiation against various amalgamation of ultrasound, UV radiation, TiO_2 photocatalysis, and ozone to enhance the wastewater disinfection process.

Nonetheless, basic methods are energy-intensive and necessitate extensive infrastructure, rendering them inappropriate for densely populated regions. Primary remedies may not totally eliminate pollutants and require large space and costs (Shamshad and Rehman, 2025).

10.6.2 Methods of Wastewater Treatment

10.6.2.1 Physical Methods

Physical methods are still the foundation for most WWT process flow systems and apply all physical forces to eliminate pollutants (Cheremisinoff, 2019).

By utilizing physical forces such as gravity, filtration, sedimentation, flotation, and screening, this method separates solids, liquids, and gases. To clear suspended particles, oil and grease, and certain metals from wastewater, physical methods are often straightforward, low cost, and successful. Physical methods cannot, nonetheless, eliminate dissolved or colloidal contaminants including organic waste, nutrients, pathogens, and hazardous compounds. Although filtration helps to remove solid particles by running the wastewater via a porous material, like sand, gravel, or a membrane, while sedimentation is the process by which solid particles settle in a tank under gravitational force. In centrifugation, separation of phases is accomplished by centrifuging the wastewater at accelerated rpm, flotation helps in the separation of solid/liquid particles by injecting air bubbles into the wastewater and creates a floating layer can be taken off (Sravan et al., 2024).

10.6.2.2 Biological Methods

Although biological methods are environmentally beneficial and effective for treating wastewater, their use relies on the kind and makeup of the wastewater. The biological treatment process entails the biodegradation of organic materials in water by microbes, purification via oxidation, and transformation into innocuous compounds and CO_2. Ammonia is converted into nitrates and nitrogen, whereas dissolved organic molecules are converted into sedimentable compounds. The proliferation of microorganisms in sludge escalates following the degradation of organic substances, resulting in gas production and the aggregation of organic waste and microbes into flakes (Trifirò & Zanirato, 2024). Biological WWT is classified into aerobic and anaerobic procedures. Aerobic treatment necessitates oxygen for microbes to decompose organic matter, while anaerobic treatment works in the absence of oxygen. Aerobic treatment is prevalent and is categorized into suspended and attached growth systems. Anaerobic treatment is mostly employed for the digestion of high-strength wastewater or sludge, resulting in biogas as a by-product (Sarkar and Mohan, 2020).

Customary biological treatment methodologies encompass aerobic, anaerobic, anoxic, facultative procedures, bioremediation, phytoremediation, and mycoremediation (Martins et al., 2024). Aerobic techniques encompass activated sludge, trickling filters, and oxidation ponds, necessitating oxygen for the degradation of organic waste. Anaerobic techniques employ digesters and up-flow reactors devoid of oxygen, resulting in methane generation. Anoxic techniques employ denitrification filters and anoxic zones within activated sludge systems to convert nitrate into nitrogen gas. Phytoremediation employs wetlands, aquatic flora, and terrestrial vegetation to absorb, collect, or decompose contaminants in wastewater (Yeruva et al., 2020).

Using hyacinth and algae species, an investigation on river Ganga's water to raise its biological oxygen demand (BOD) was reported therefore providing a convenient and reasonably priced treatment. In another study using anaerobic filters and fluidized bed reactors, treatment of fish processing effluent was done with high BOD and COD levels to find that 80% of COD could be eliminated. Eliminating more than 95% COD, a multistage rotating biological contractor reactor was judged to be more suitable (Birol & Das, 2010).

10.6.2.3 Chemical Methods

Chemical water treatment methods utilize chemicals to get rid of contaminants and toxins from water (Ahmed et al., 2021). Coagulation and flocculation are two interconnected approaches employed in water treatment to eliminate suspended particles and contaminants. Coagulation entails the use of agents such as alum or ferric chloride to destabilize suspended particles, resulting in the creation of bigger aggregates known as "microflocs" (Ramli and Aziz, 2023).

Flocculation promotes the aggregation of bigger particles known as "macroflocs," which can be removed from the water by sand or an alternative filtering media (Jagaba et al., 2023). Disinfectants such as chlorine, chloramines, or ozone are introduced to water to eradicate or neutralize bacteria, viruses, and other pathogenic microorganisms. Continuous oversight of the disinfection procedure is essential to assure that the proper concentration is sustained all over the dispersal network (Gao et al., 2023).

Pre-chlorination entails the addition of a minimal quantity of chlorine to water to reduce algal growth, prevent smells, and eliminate organic waste. Coagulation and flocculation are succeeded by disinfection utilizing agents such as chlorine, chloramines, or ozone. Postchlorination introduces a residuary disinfectant that remains in the water as it traverses the supply system (Rana et al., 2024).

Chemical reactions in the disinfection process entail interactions between the disinfectant and diverse microorganisms present in the water. Chlorination, chloramination, ozone disinfection, and UV disinfection are among the most frequently employed disinfectants. Chemical precipitation is employed to eliminate contaminants, including suspended particles, heavy metals, and phosphorus.

Adsorbents are solid materials employed to banish pollutants from water via the adsorption mechanism. Activated carbon, silica gel, zeolites, and clay minerals are widely utilized adsorbents in water treatment processes. Ion exchange is an extensively utilized technique for the treatment of water and waste, entailing the transfer of ions between a liquid phase (water or waste) and a solid phase (resin). This technology aims to eliminate various pollutants, amid heavy metals, radioactive isotopes, and other organic hazardous substances (Mutegoa, 2024). Figure 10.1 illustrates the various methods of WWT.

10.6.3 Recent Advances in Technologies for Treatment of Wastewater

The growing spectrum of pollutants in wastewater calls for modern treatment techniques and creativity. Conventional techniques suffer with new pollutants, sludge generation, and energy use. Advanced technologies meant to eradicate harmful pollutants and toxins include ozone generation, electrocoagulation, hybrid technologies, nanotechnology, and membrane technologies. Though their energy consumption can result in significant expenses, these techniques employ physical,

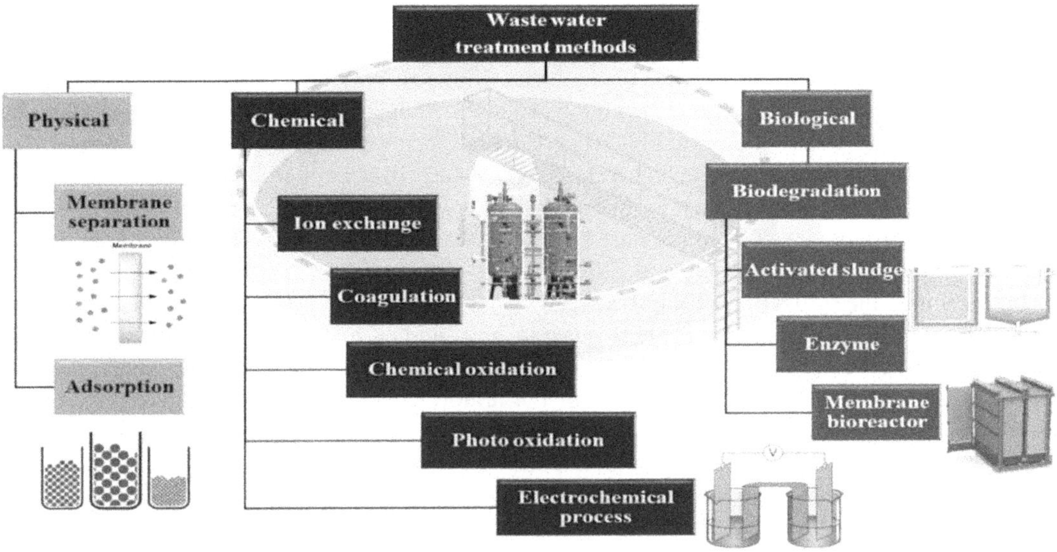

FIGURE 10.1 Wastewater treatment methods.

chemical, and biological approaches. The ultimate aim is to create wastewater treatment facilities able to totally eradicate newly arising pollutants. In this section, novel processes are discussed.

10.6.3.1 Application of 2D-Metal–Organic Frameworks for Wastewater Treatment

MXenes, identified in 2011, are two-dimensional (2D) materials characterized by the generic formula Mn+1XnTx. They are part of a group of transition metal carbides, nitrides, and carbonitrides. M signifies an early transition metal, X designates carbon or nitrogen, and T symbolizes surface termination groups. MXenes offer importance in environmental solutions amid monitoring the quality of water, WWT, and desalination (Dixit et al., 2022). Wang et al. (2018) formulated a composite, Co_3O_4@CNF/persulfate ($2KHSO_5 \cdot KHSO_4 \cdot K_2SO_4$), for water filtration, attaining removal efficiencies of 46.5% and 58.4% for organic contaminants (Wang et al., 2018). A composite of MXene and alginate was synthesized to enhance lead adsorption by occupying interlayer voids. MXenes act as adsorbents to remove Cr(VI) and methyl orange (MO) from water, with Ti_2C_2Tx-MXene nanosheets effectively eliminating copper ions. Mercury, a poisonous metal, is primarily found in coal and garbage combustion. The MoS_2–Ti_3C_2Tx nanocomposite demonstrated enhanced degrading efficiency, reducing mercury ions concentration from 50 to 0.01 μmol/L in 2 min (Pouramini et al., 2023).

A TiO_2/NH_2-MIL-125(Ti) nanorod array on carbon fiber cloth (CFc/TiO_2/Ti-MOF) was created for autonomous solar desalination. The membrane successfully eliminated 58.5% ofloxacin, 62.8% tetracycline, 94.1% methylene blue, and 76.4% Rhodamine B within 120 min of sunlight exposure. Metal–organic frameworks (MOFs) are proficient in WWT, productively assimilating on surface, heavy metals such as mercury, lead, and cadmium. Ammonia-functionalized MOFs have been investigated for the reduction of mercury concentrations to permissible levels in drinking water (Singh et al., 2024b).

A Zn-BTC-based metal–organic scaffold was employed for the surface assimilation of six aromatic insecticides from water samples. The MOFs exhibited significant chemical stability and durability owing to the connection among BTC and Zn. These MOFs may be advantageous for environmental pollutant separation and enrichment due to their notable durability (Lal et al., 2024).

10.6.3.2 Application of Biomass for Wastewater Treatment

Traditional WWT is incapable of directly recycling nutrients from wastewater; however, microalgae effectively assimilate nitrogen and phosphorus into their biomass, rendering them suitable for

WWT applications (Tena et al., 2024). The aim of algal-based WWT is to lower the cost of producing algae. High-rate algal ponds for wastewater treatment (WWTHRAPs) offer a potential solution capable of addressing these difficulties in conducive climates. WWTHRAPs possess the prospects for biofuel generation as a by-product, with the expenses of algae growing and harvesting offset by the WWT process. These shallow, paddlewheel-aerated open raceway ponds enhance natural biological treatment processes (Mehrabadi et al., 2015). A study illustrated the production of *Chlorella pyrenoidosa* and *Phormidium* algae in municipal wastewater via a 100 L outdoor biofilm system. Following six days of elevated temperature treatment, the resultant biomass exhibited stability and a substantial lipid content (Nguyen et al., 2022).

In one study, the utilization of unprocessed biomass from agricultural practices as adsorbents for water purification, employing waste to generate activated carbons via chemical activation with KOH, was done. The effectiveness of trash materials, including stubble, sawdust, fibers, bamboo blossoms, and other plants, was evaluated for the removal of methylene blue. Adsorption investigations were performed in a batch setup, assessing the effects of kinetics, pH, and temperature. The natural adsorbents demonstrated significant adsorption efficacy for MB, with *Adansonia digitata* L. exhibiting the highest removal capacity of 188.3 mg/g. The investigation finished with the development of slow filters including natural adsorbents or activated carbons. The greatest quantities of methylene blue removed via steady flux were diminished than those achieved with various unprocessed biomass category in a batch system; yet, these quantities were equivalent to findings of research (Cansado et al., 2024).

10.6.3.3 Application of Biochar for Wastewater Treatment

Biochar, a carbon-dense material derived from the combustion of biomass discards, represents a promising solution for mitigating pollutants and decreasing carbon emissions. By tailoring feedstock and pyrolysis parameters, biochar can be designed to address specific pollutants in industrial WWT methodologies. It escalates flocculation, dewatering, adsorption, and oxidation in municipal WWT, facilitating sludge management, odor reduction, and nutrient retrieval. Biochar mitigates stormwater effects by enhancing soil structure, erosion resistance, water retention capacity, and hydraulic conductivity. The viability of expanding biochar production necessitates the collaboration among multiple stakeholders to optimize environmental, societal, and economic advantages (He et al., 2022). Biochars derived from several biomasses have been utilized as sorbents for organic pollutants in water and wastewater. Switchgrass biochars efficiently eliminated methylene blue, orange G, and Congo red, but chicken manure biochar demonstrated potential for the removal of hazardous phenol and 2,4-dinitrophenol. The adsorption ability of biochar for heavy metals is contingent upon its properties and the specific nature of the target metal. Biochar's porous, biostable, and accessible characteristics have attracted attention as a substitute for traditional adsorbents such as activated carbon. The utilization of this material in anaerobic digestion is on the rise owing to its porous, bio-stable, and readily available characteristics. The potential of biochar as an effective alternative for heavy metal adsorption is increasing (Enaime et al., 2020).

10.6.3.4 Nanomaterials: Potential Use in Wastewater Treatment

Nanomaterials enrich water treatment and contamination management owing to their adsorption capacities, chemical attributes, mechanical qualities, cost-effectiveness, and energy efficiency. They eliminate 1–10 nm particles in wastewater, encompassing viruses and organic matter. Research is being conducted on nanoparticles for the treatment of sewage by eliminating organic materials, fungi, and viruses. They are employed in newer oxidation schemes for effective WWT (Mbarek et al., 2022).

10.6.3.4.1 Nanomembranes

Nanomembranes are membranes made of various nanofibers that exclude undesired tiny particles from aqueous solutions. They are applicable for ultrafiltration, reverse osmosis, and nanofiltration.

Nanomembranes are classified into three categories: inorganic, organic, and organic–inorganic hybrids. Organic nanomembranes are made up of polymeric materials such as cellulose acetates, polyethersulfone, cellulose nitrates, polyacrylonitrile, polyvinyl alcohol, and biomacromolecules. Inorganic nanomembranes consist of metals, metal oxides, carbon nanotubes (CNTs), and graphene. The introduction of inorganic material into a polymeric system results in the formation of inorganic–organic hybrid nanomembranes (Wei et al., 2017).

Nanomembrane utilizing the water channel protein aquaporin for water purification was engineered, resulting in enhanced water flux of 65% and a substantial decline in $MgCl_2$ and NaCl concentrations in water (Sun et al., 2013). Tijing et al. (2016) developed a hydrophobic nanomembrane utilizing polyvinylidene fluoride-co-hexafluoropropylene and CNTs as nanofillers, enhancing water flux by 33%–59% and achieving a salt eliminating capacity over 99.99% was developed (Tijing et al., 2016). A nanomembrane for desalination utilizing zirconium (VI) derived from a MOF was engineered, which effectively eliminated multivalent ions such as Al^{+3}, Ca^{+2}, and Mg^{+2} from water (Liu et al., 2015). Bolisetty et al. (2017) employed amyloid fibril-based membranes to extract arsenic from polluted water, with a removal efficiency of 99.6% (Bolisetty et al., 2017).

Nanoparticles can readily infiltrate human bodies through food chains, resulting in oxidative stress and other complications. Ultra-thin nanomembrane composed of cellulose nanofibrils was developed to eliminate nanoparticles from water, featuring pore sizes between 2.5 and 12 nm and a thickness of approximately 30 nm (Zhang et al., 2016). A CNT sponge-based nanomembrane was engineered that eliminated 100% of TiO_2 and AuNPs, about 80% of CdSNPs, rhodamine B, and MO molecules from water (Li et al., 2010).

Organic pollutants, including oils, hydrocarbons, medicines, pesticides, and phenols, endanger aquatic organisms and can disrupt the ecosystem. A multitude of tests have been conducted to eliminate organic contaminants from water utilizing nanomembranes. A composite nanomembrane utilizing cellulose nanocrystals within a chitosan (CS) matrix was designed (Karim et al., 2014), whereas in another study a nanomembrane was fabricated by vertically aligning CNTs on a steel mesh. Employing immobilized silver nanoclusters on aminated polyacrylonitrile nanofibrous membranes for the separation of oil from water was prepared (Lee and Baik, 2010).

Biological pollutants, including viruses, bacteria, plankton, and algae, are accountable for waterborne infections in general. TiO_2 nanowire-based ultrafiltration nanomembranes that have antibacterial, antifouling, photocatalytic oxidation, and simultaneous dispersive properties were developed. Highly selective, nanoporous membrane utilizing copolymers for viral filtration, including a pore diameter of around 17 nm and exhibiting substantial stability at 2 bars, were created. A nanomembrane was prepared by applying a covering of T-alumina and titania nanocrystals, resulting in the complete elimination of microbes and around 25% reduction of ions (Umar et al., 2022).

10.6.3.4.2 Nanoadsorbents

Nanoadsorbents have demonstrated efficacy in WWT by eliminating pollutants. Research has demonstrated their efficacy in eliminating heavy metal cations and dyes from polluted water. Mesoporous TiO_2 was employed to eliminate Reactive Red 195 from the liquid phase at ambient temperature and in the absence of light. Nanoadsorbents can be enhanced to augment their adsorption capabilities and efficacy in eliminating pollutants. Silica-shell coatings containing Fe_3O_4 particles show enhanced adsorption efficacy in improving contaminated water (Anoob et al., 2024). Carbon nano tubes (CNTs) embedded with silver nanoparticles efficiently inactivate bacteria, presenting several advantages over conventional adsorbents; nonetheless, their higher expense for large-scale WWT raises concerns. An investigation illustrated the efficacy of silver and titanium oxide-doped activated carbon as an adsorbent for the removal of total phenols from pharmaceutical effluents, employing a sustainable method produced by Shea butter leaf extracts (Pandey et al., 2022).

Magnetic hydroxyapatite served as an adsorbent for the extraction of Cd(II) and Zn(II) from wastewater. γ-Alumina nanofibers containing functional groups were utilized for the riddance of heavy metals. A poly-methacrylic acid grafted CS–bentonite nanocomposite was produced for the efficient riddance of Hg(II), Pb(II), and Cd(II) metal ions. Diamine-functionalized mesoporous silica on multiwalled carbon nanotubes (MWCNTs) was synthesized for the removal of heavy metals, exhibiting superior adsorption capacity compared to standard MWCNTs due to the synergistic adsorption properties of the two materials (Thekkudan et al., 2017). GO/aminated lignin aerogel was formulated for the adsorption of MG dye in wastewater, evaluating its adsorption capacity across different doses, pH levels, contact durations, and reaction temperatures. The GO/AL-AG attained 90% efficiency after 5 cycles (Chen et al., 2020). Lignin-derived nanomaterials exhibit considerable promise for the treatment of water and wastewater, especially in the reduction of dyes, nitroarenes, and heavy metals. Crosslinking β-cyclodextrin with polyethyleneimine magnetic nanoadsorbents (β-CD/PEI-Fe_3O_4) was investigated for the concurrent extraction of MO and lead ions ($Pb2^+$) from wastewater. The margins of β-CD, featuring oxygen-containing groups and free amino groups on PEI, served as functional sites for effective Pb_2^+ absorption. Cubic NaA zeolite, a microporous crystalline aluminosilicate, was produced from natural halloysite rocks for the adsorption of ammonium ions ($NH4^+$) from wastewater. The adsorbent system demonstrated reusability and efficacy in eliminating $NH4^+$ contaminants from wastewater (Jain et al., 2021).

10.7 POLICY STRATEGIES AND POTENTIAL DIRECTIONS

Antimicrobial resistance (AMR) is escalating despite increased awareness and media coverage. The use of human antimicrobials raised by 65% from 2000 to 2015, with a rise in antibiotics like oxazolidinones, carbapenems, polymyxins, and glycylcyclines. AMR causes annual deaths in the United States and Europe, affecting 35,000 individuals annually. The United Nations' 2030 Agenda for Sustainable Development in 2015 highlighted the importance of AMR, which could hinder the achievement of Sustainable Development Goals. A robust response requires multisectoral collaboration. Governments have implemented strategies to mitigate overconsumption and misuse of antibiotics, such as prohibiting growth boosters in all EU Member States since 2006. However, most nations have not implemented policies aimed at reducing antimicrobial consumption through population-wide initiatives. Legislative action is needed to define governmental responsibilities in the fight against AMR (Serwecińska, 2020; Jani and Kumari, 2024).

10.7.1 Pharmaceutical Waste Management Strategies

The National Pollution Disposal and Sewage Elimination (NPDES) Program is an essential project designed to guarantee the safe disposal of expired and unused pharmaceuticals that may otherwise be deposited in landfills and wastewater systems. It encompasses techniques including Hospital and Clinical Protocols, Extended Producer Responsibility, and Take-Back Programs, necessitating the preliminary dismissal of hospital wastewater prior to disposal. The initiative underscores the significance of collaboration among governmental entities, healthcare professionals, pharmaceutical manufacturers, and civil society organizations, with pharmacists serving a pivotal role in advising on the appropriate utilization and discard of medications. Public–private partnerships are being utilized to collaboratively manage water resources, improving water and wastewater management policies and practices. The NPDES program establishes a permit system for the utilization and disposal of sewage sludge and implements water quality control in accordance with C.F.R. Part 130. The nationwide pretreatment program seeks to reduce pollution from commercial and nondomestic wastewater sources while protecting Publicly Owned Treatment Works infrastructure (Shah, 2024).

10.7.2 Statutory Plans for AMR Monitoring in Wastewater

Monitoring of AMR is essential for diminishing infections, prioritizing initiatives, evaluating intervention outcomes, informing treatment guidelines, mitigating adverse effects, and developing novel antimicrobial agents. The WHO's Global AMR and use surveillance system (GLASS) enhances information and standardizes AMR monitoring worldwide. The European Union's EARS-Net and ESAC-Net encompass 30 European nations. WWS can forecast ARB and their associated genes at the population level. A comprehensive study was undertaken to ascertain prevalent methodologies for AMR monitoring, recognize implementation obstacles, and evaluate application constraints. The survey sought to condense the array of frequently utilized aims and techniques into a definitive list of validated methods. Monitoring activities are currently extending into adjacent regions utilizing the existing infrastructure of the EPA National Rivers and Streams Assessment (Manaia, 2023).

10.8 CONCLUSION

With the serious worries for public health and ecological systems, the monitoring of AMR in wastewater has emerged as a pivotal environmental concern. With the help of AMR identification and monitoring in wastewater, vital frameworks in resistance evolution and transmission are reported, therefore furnishing an ideal perspective on the transmission of resistant bacteria.

Recent developments in monitoring techniques include metagenomic sequencing, CRISPR-based detection, and advanced biosensors are revolutionizing the ability to precisely and effectively find and count resistant bacteria. These new approaches not only help better understand AMR dynamics in complex wastewater environments but also offer good tools for real-time monitoring and early warning systems.

Nonetheless, problems prevail over the preciseness of techniques, data uniformity assurance, and rectification of shortcomings in the global monitoring system. Consolidating innovative surveillance procedures with meticulous wastewater management systems, legislative actions, and international collaboration can aid in appropriately tackling the augmenting challenge of AMR. Minimizing the environmental consequences of AMR and guarding public health for successive generations rest on continual scientific investigations and breakthroughs in this discipline.

REFERENCES

Ahmad, N., Joji, R. M., & Shahid, M. (2023). Evolution and implementation of One Health to control the dissemination of antibiotic-resistant bacteria and resistance genes: A review. *Frontiers in Cellular and Infection Microbiology*, 12, 1065796.

Ahmed, S. F., Mofijur, M., Nuzhat, S., Chowdhury, A. T., Rafa, N., Uddin, M. A., & Show, P. L. (2021). Recent developments in physical, biological, chemical, and hybrid treatment techniques for removing emerging contaminants from wastewater. *Journal of Hazardous Materials*, 416, 125912.

Akhter, S., Bhat, M. A., Ahmed, S., & Siddiqui, W. A. (2024). Antibiotic residue contamination in the aquatic environment, sources and associated potential health risks. *Environmental Geochemistry and Health*, 46(10), 387.

Alawi, M., Torrijos, T. V., & Walsh, F. (2022). Plasmid-mediated antimicrobial resistance in drinking water. *Environmental Advances*, 8, 100191.

Alfahl, Z., Chueiri, A., Carolan, S., Darcy, G., Hussain, N., Cahill, N., & O'Connor, L. (2024). Antimicrobial resistance detection methods in water environments: A scoping review. *Sustainable Microbiology*, 1(1), qvae034.

Alvim, C. B., Bes-Piá, M. A., & Mendoza-Roca, J. A. (2020). Separation and identification of microplastics from primary and secondary effluents and activated sludge from wastewater treatment plants. *Chemical Engineering Journal*, 402, 126293.

Anoob, F., Arachchi, S., Azamathulla, H. M., Al-mahbashi, N., & Rathnayake, U. (2024). Nanoadsorbents as an effective wastewater treatment candidate for pharmaceutical contaminants; towards sustainable policy development. *Case Studies in Chemical and Environmental Engineering*, 9, 100639.

Asaduzzaman, M., Rousham, E., Unicomb, L., Islam, M. R., Amin, M. B., Rahman, M., & Islam, M. A. (2022). Spatiotemporal distribution of antimicrobial resistant organisms in different water environments in urban and rural settings of Bangladesh. *Science of the Total Environment*, 831, 154890.

Ashbolt, N. J., Amézquita, A., Backhaus, T., Borriello, P., Brandt, K. K., Collignon, P., & Topp, E. (2013). Human health risk assessment (HHRA) for environmental development and transfer of antibiotic resistance. *Environmental Health Perspectives*, 121(9), 993–1001.

Baba, H., Nishiyama, M., Watanabe, T., & Kanamori, H. (2022). Review of antimicrobial resistance in wastewater in Japan: Current challenges and future perspectives. *Antibiotics*, 11(7), 849.

Birol, E., & Das, S. (2010). Estimating the value of improved wastewater treatment: The case of River Ganga, India. *Journal of Environmental Management*, 91(11), 2163–2171.

Bobate, S., Mahalle, S., Dafale, N. A., & Bajaj, A. (2023). Emergence of environmental antibiotic resistance: Mechanism, monitoring and management. *Environmental Advances*, 13, 100409.

Bolisetty, S., Reinhold, N., Zeder, C., Orozco, M. N., & Mezzenga, R. (2017). Efficient purification of arsenic-contaminated water using amyloid–carbon hybrid membranes. *Chemical Communications*, 53(42), 5714–5717.

Brown, C. L., Maile-Moskowitz, A., Lopatkin, A. J., Xia, K., Logan, L. K., Davis, B. C., & Pruden, A. (2024). Selection and horizontal gene transfer underlie microdiversity-level heterogeneity in resistance gene fate during wastewater treatment. *Nature Communications*, 15(1), 5412.

Bydalek, F., Webster, G., Barden, R., Weightman, A. J., Kasprzyk-Hordern, B., & Wenk, J. (2023). Microplastic biofilm, associated pathogen and antimicrobial resistance dynamics through a wastewater treatment process incorporating a constructed wetland. *Water Research*, 235, 119936.

Cadman, H., & Marinez, L. (2014). *Antimicrobial Resistance Global Report on Surveillance*. World Health Organization, Geneva, Switzerland pp. 1–42.

Cansado, I. P. d. P., Geraldo, P. F., Mourão, P. A. M., Castanheiro, J. E., Carreiro, E. P., & Suhas. (2024). Utilization of biomass waste at water treatment. *Resources*, *13*(3), 37. https://doi.org/10.3390/resources13030037

Cason, C., D'Accolti, M., Soffritti, I., Mazzacane, S., Comar, M., & Caselli, E. (2022). Next-generation sequencing and PCR technologies in monitoring the hospital microbiome and its drug resistance. *Frontiers in Microbiology*, 13, 969863.

Chen, H., Liu, T., Meng, Y., Cheng, Y., Lu, J., & Wang, H. (2020). Novel graphene oxide/aminated lignin aerogels for enhanced adsorption of malachite green in wastewater. *Colloids and Surfaces A: Physicochemical and Engineering Aspects*, 603, 125281.

Chen, Z., Guo, J., Jiang, Y., & Shao, Y. (2021). High concentration and high dose of disinfectants and antibiotics used during the COVID-19 pandemic threaten human health. *Environmental Sciences Europe*, 33, 1–4.

Cheremisinoff, P. N. (2019). *Handbook of Water and Wastewater Treatment Technology*. Routledge, London.

Clarke, L. M., O'Brien, J. W., Murray, A. K., Gaze, W. H., & Thomas, K. V. (2024). A review of wastewater-based epidemiology for antimicrobial resistance surveillance. *Journal of Environmental Exposure Assessment*, *3*(1), 7.

Deguenon, E., Dougnon, V., Houssou, V. M. C., Gbotche, E., Ahoyo, R. A., Fabiyi, K., & Boko, M. (2022). Hospital effluents as sources of antibiotics residues, resistant bacteria and heavy metals in Benin. *SN Applied Sciences*, 4(8), 206.

Ding, D., Wang, B., Zhang, X., Zhang, J., Zhang, H., Liu, X., & Yu, Z. (2023). The spread of antibiotic resistance to humans and potential protection strategies. *Ecotoxicology and Environmental Safety*, 254, 114734.

Dixit, F., Zimmermann, K., Dutta, R., Prakash, N. J., Barbeau, B., Mohseni, M., & Kandasubramanian, B. (2022). Application of MXenes for water treatment and energy-efficient desalination: A review. *Journal of Hazardous Materials*, 423, 127050.

Drane, K., Sheehan, M., Whelan, A., Ariel, E., & Kinobe, R. (2024). The role of wastewater treatment plants in dissemination of antibiotic resistance: Source, measurement, removal and risk assessment. *Antibiotics*, 13(7), 668.

Enaime, G., Baçaoui, A., Yaacoubi, A., & Lübken, M. (2020). Biochar for wastewater treatment—Conversion technologies and applications. *Applied Sciences*, 10(10), 3492.

Endale, H., Mathewos, M., & Abdeta, D. (2023). Potential causes of spread of antimicrobial resistance and preventive measures in one health perspective-a review. *Infection and Drug Resistance*, 16, 7515–7545.

Eze, E. C., El Zowalaty, M. E., & Pillay, M. (2021). Antibiotic resistance and biofilm formation of acinetobacter baumannii isolated from high-risk effluent water in tertiary hospitals in South Africa. *Journal of Global Antimicrobial Resistance*, 27, 82–90.

Flach, K. A., Bones, U. A., Wolff, D. B., de Oliveira Silveira, A., da Rosa, G. M., Carissimi, E., & Silvestri, S. (2024). Antibiotic resistant bacteria and genes (ARB and ARG) in water and sewage treatment units: A review. *Environmental Nanotechnology, Monitoring & Management*, 21, 100941.

Fuhrman, J. A. (1999). Marine viruses and their biogeochemical and ecological effects. *Nature*, 399(6736), 541–548.

Gao, Y., Sun, Z., Guo, Y., Qiang, Z., & Ben, W. (2023). Virus inactivation by sequential ultraviolet-chlorine disinfection: Synergistic effect and mechanism. *Chemosphere*, 314, 137632.

Garner, E., Davis, B. C., Milligan, E., Blair, M. F., Keenum, I., Maile-Moskowitz, A., & Pruden, A. (2021a). Next generation sequencing approaches to evaluate water and wastewater quality. *Water Research*, 194, 116907.

Garner, E., Organiscak, M., Dieter, L., Shingleton, C., Haddix, M., Joshi, S., & Hamilton, K. A. (2021b). Towards risk assessment for antibiotic resistant pathogens in recycled water: A systematic review and summary of research needs. *Environmental Microbiology*, 23(12), 7355–7372.

Ghimpețeanu, O. M., Pogurschi, E. N., Popa, D. C., Dragomir, N., Drăgotoiu, T., Mihai, O. D., & Petcu, C. D. (2022). Antibiotic use in livestock and residues in food—A public health threat: A review. *Foods*, 11(10), 1430.

Giacometti, F., Shirzad-Aski, H., & Ferreira, S. (2021). Antimicrobials and food-related stresses as selective factors for antibiotic resistance along the farm to fork continuum. *Antibiotics*, 10(6), 671.

Grenni, P. (2022). Antimicrobial resistance in rivers: A review of the genes detected and new challenges. *Environmental Toxicology and Chemistry*, 41(3), 687–714.

Guo, M. T., Yuan, Q. B., & Yang, J. (2015). Distinguishing effects of ultraviolet exposure and chlorination on the horizontal transfer of antibiotic resistance genes in municipal wastewater. *Environmental Science & Technology*, 49(9), 5771–5778.

Hanna, N., Tamhankar, A. J., & Lundborg, C. S. (2023). Antibiotic concentrations and antibiotic resistance in aquatic environments of the WHO Western Pacific and South-East Asia regions: A systematic review and probabilistic environmental hazard assessment. *The Lancet Planetary Health*, 7(1), e45–e54.

Harrington, A., Vo, V., Papp, K., Tillett, R. L., Chang, C. L., Baker, H., & Oh, E. C. (2022). Urban monitoring of antimicrobial resistance during a COVID-19 surge through wastewater surveillance. *Science of the Total Environment*, 853, 158577.

Hazra, M., Joshi, H., Williams, J. B., & Watts, J. E. (2022). Antibiotics and antibiotic resistant bacteria/genes in urban wastewater: A comparison of their fate in conventional treatment systems and constructed wetlands. *Chemosphere*, 303, 135148.

He, B., Zhu, T. T., Liang, Y., Wei, H. J., Huang, Z. L., Liang, L. J., & Sun, J. (2024). Adaptive evolution in asymptomatic host confers MDR Salmonella with enhanced environmental persistence and virulence. *Science of the Total Environment*, 908, 168340.

He, M., Xu, Z., Hou, D., Gao, B., Cao, X., Ok, Y. S., & Tsang, D. C. (2022). Waste-derived biochar for water pollution control and sustainable development. *Nature Reviews Earth & Environment*, 3(7), 444–460.

Hendriksen, R. S., Munk, P., Njage, P., Van Bunnik, B., McNally, L., Lukjancenko, O., & Aarestrup, F. M. (2019). Global monitoring of antimicrobial resistance based on metagenomics analyses of urban sewage. *Nature Communications*, 10(1), 1124.

Holmes, A. H., Moore, L. S., Sundsfjord, A., Steinbakk, M., Regmi, S., Karkey, A., & Piddock, L. J. (2016). Understanding the mechanisms and drivers of antimicrobial resistance. *The Lancet*, 387(10014), 176–187.

Hong, P. Y., Julian, T. R., Pype, M. L., Jiang, S. C., Nelson, K. L., Graham, D., & Manaia, C. M. (2018). Reusing treated wastewater: Consideration of the safety aspects associated with antibiotic-resistant bacteria and antibiotic resistance genes. *Water*, 10(3), 244.

Huang, X., Zhao, Z., Hernandez, D., & Jiang, S. C. (2016). Near real-time flow cytometry monitoring of bacterial and viral removal efficiencies during water reclamation processes. *Water*, 8(10), 464.

Humphreys, G., & Fleck, F. (2016). United Nations meeting on antimicrobial resistance. *Bulletin of the World Health* Organization, 94(9), 638.

Hutinel, M., Fick, J., Larsson, D. J., & Flach, C. F. (2021). Investigating the effects of municipal and hospital wastewaters on horizontal gene transfer. *Environmental Pollution*, 276, 116733.

Jagaba, A. H., Birniwa, A. H., Usman, A. K., Mu'azu, N. D., Yaro, N. S. A., Soja, U. B., & Lawal, I. M. (2023). Trend and current practices of coagulation-based hybrid systems for pulp and paper mill effluent treatment: Mechanisms, optimization techniques and performance evaluation. *Journal of Cleaner Production*, 429, 139543.

Jain, K., Patel, A. S., Pardhi, V. P., & Flora, S. J. S. (2021). Nanotechnology in wastewater management: A new paradigm towards wastewater treatment. *Molecules*, 26(6), 1797.

Jani, M., & Kumari, S. (2024). Pharmaceutical wastewater and environmental contamination: Challenges, innovations, and future direction. *PEXACY International Journal of Pharmaceutical Science*, 3(1), 79–98.
Jyoti, K., Soni, K., & Chandra, R. (2024). Pharmaceutical industrial wastewater exhibiting the co-occurrence of biofilm-forming genes in the multidrug-resistant bacterial community poses a novel environmental threat. *Aquatic Toxicology*, 273, 107019.
Kaiser, R. A., Taing, L., & Bhatia, H. (2022). Antimicrobial resistance and environmental health: A water stewardship framework for global and national action. *Antibiotics*, 11(1), 63.
Karim, Z., Mathew, A. P., Grahn, M., Mouzon, J., & Oksman, K. (2014). Nanoporous membranes with cellulose nanocrystals as functional entity in chitosan: Removal of dyes from water. *Carbohydrate Polymers*, 112, 668–676.
Karimi, K. J., Ahmad, A., Duse, A., Mwanthi, M., & Ayah, R. (2022). Prevalence of antibiotic use and disposal at household level in informal settlements of Kisumu, Kenya. *International Journal of Environmental Research and Public Health*, 20(1), 287.
Khan, A. H., Aziz, H. A., Palaniandy, P., Naushad, M., Cevik, E., & Zahmatkesh, S. (2023). Pharmaceutical residues in the ecosystem: Antibiotic resistance, health impacts, and removal techniques. *Chemosphere*, 339, 139647.
Kieffer, N., Poirel, L., Bessa, L. J., Barbosa-Vasconcelos, A., Da Costa, P. M., & Nordmann, P. (2016). VIM-1, VIM-34, and IMP-8 carbapenemase-producing *Escherichia coli* strains recovered from a Portuguese river. *Antimicrobial Agents and Chemotherapy*, 60(4), 2585–2586.
Kim, J. W., Hong, Y. K., Kwon, O. K., & Kim, S. C. (2024). Difference of microbial community in the stream adjacent to the mixed antibiotic effluent source. *Toxics*, 12(2), 135.
Kumar, A., & Pal, D. (2018). Antibiotic resistance and wastewater: Correlation, impact and critical human health challenges. *Journal of Environmental Chemical Engineering*, 6(1), 52–58.
Kusi, J., Ojewole, C. O., Ojewole, A. E., & Nwi-Mozu, I. (2022). Antimicrobial resistance development pathways in surface waters and public health implications. *Antibiotics*, 11(6), 821.
La Rosa, M. C., Maugeri, A., Favara, G., La Mastra, C., Magnano San Lio, R., Barchitta, M., & Agodi, A. (2025). The impact of wastewater on antimicrobial resistance: A scoping review of transmission pathways and contributing factors. *Antibiotics*, 14(2), 131.
Lal, S., Singh, P., Singhal, A., Kumar, S., Gahlot, A. P. S., Gandhi, N., & Kumari, P. (2024). Advances in metal–organic frameworks for water remediation applications. *RSC Advances*, 14(5), 3413–3446.
Lee, C., & Baik, S. (2010). Vertically-aligned carbon nano-tube membrane filters with superhydrophobicity and superoleophilicity. *Carbon*, 48(8), 2192–2197.
Leiva, A. M., Piña, B., & Vidal, G. (2021). Antibiotic resistance dissemination in wastewater treatment plants: A challenge for the reuse of treated wastewater in agriculture. *Reviews in Environmental Science and Bio/Technology*, 20, 1043–1072.
Li, B., Qiu, Y., Song, Y., Lin, H., & Yin, H. (2019). Dissecting horizontal and vertical gene transfer of antibiotic resistance plasmid in bacterial community using microfluidics. *Environment International*, 131, 105007.
Li, H., Gui, X., Zhang, L., Wang, S., Ji, C., Wei, J., & Cao, A. (2010). Carbon nanotube sponge filters for trapping nanoparticles and dye molecules from water. *Chemical Communications*, 46(42), 7966–7968.
Liguori, K., Keenum, I., Davis, B. C., Calarco, J., Milligan, E., Harwood, V. J., & Pruden, A. (2022). Antimicrobial resistance monitoring of water environments: A framework for standardized methods and quality control. *Environmental Science & Technology*, 56(13), 9149–9160.
Liu, X., Demir, N. K., Wu, Z., & Li, K. (2015). Highly water-stable zirconium metal–organic framework UiO-66 membranes supported on alumina hollow fibers for desalination. *Journal of the American Chemical Society*, 137(22), 6999–7002.
Ma, C. Y., Ihara, M., Liu, S., Sugie, Y., & Tanaka, H. (2022). Tracking the source of antibiotic-resistant Escherichia coli in the aquatic environment in Shiga, Japan, through whole-genome sequencing. *Environmental Advances*, 8, 100185.
Ma, J., Sun, H., Li, B., Wu, B., Zhang, X., & Ye, L. (2024). Horizontal transfer potential of antibiotic resistance genes in wastewater treatment plants unraveled by microfluidic-based mini-metagenomics. *Journal of Hazardous Materials*, 465, 133493.
Manaia, C. M. (2017). Assessing the risk of antibiotic resistance transmission from the environment to humans: non-direct proportionality between abundance and risk. *Trends in Microbiology*, 25(3), 173–181.
Manaia, C. M. (2023). Framework for establishing regulatory guidelines to control antibiotic resistance in treated effluents. *Critical Reviews in Environmental Science and Technology*, 53(6), 754–779.
Mangat, C. S., Bekal, S., Irwin, R. J., & Mulvey, M. R. (2017). A novel hybrid plasmid carrying multiple antimicrobial resistance and virulence genes in *Salmonella enterica* serovar Dublin. *Antimicrobial Agents and Chemotherapy*, 61(6), 1110–1128.

Martins, R. A., Salgado, E. M., Gonçalves, A. L., Esteves, A. F., & Pires, J. C. (2024). Microalgae-based remediation of real textile wastewater: Assessing pollutant removal and biomass valorisation. *Bioengineering*, 11(1), 44.
Mba, I. E., & Nweze, E. I. (2021). Nanoparticles as therapeutic options for treating multidrug-resistant bacteria: Research progress, challenges, and prospects. *World Journal of Microbiology and Biotechnology*, 37, 1–30.
Mbarek, W. B., Escoda, L., Saurina, J., Pineda, E., Alminderej, F. M., Khitouni, M., & Suñol, J. J. (2022). Nanomaterials as a sustainable choice for treating wastewater: A review. *Materials*, 15(23), 8576.
Mehrabadi, A., Craggs, R., & Farid, M. M. (2015). Wastewater treatment high rate algal ponds (WWT HRAP) for low-cost biofuel production. *Bioresource Technology*, 184, 202–214.
Meyer, C., Price, S., & Ercumen, A. (2024). Do animal husbandry operations contaminate groundwater sources with antimicrobial resistance: Systematic review. *Environmental Science and Pollution Research*, 31(11), 16164–16176.
Miller, R. V., Ripp, S., Replicon, J., Ogunseitan, O. A., & Kokjohn, T. A. (1992). Virus-mediated gene transfer in freshwater environments. In *Gene Transfers and Environment: Proceedings of the Third European Meeting on Bacterial Genetics and Ecology (BAGECO-3), 20–22 November 1991, Villefranche-sur-Mer, France* (pp. 51–62). Springer, Berlin, Heidelberg.
Muniesa, M., & Jofre, J. (2004). Factors influencing the replication of somatic coliphages in the water environment. *Antonie Van Leeuwenhoek*, 86, 65–76.
Mutegoa, E. (2024). Efficient techniques and practices for wastewater treatment: An update. *Discover Water*, 4(1), 69.
Nguyen, A. Q., Vu, H. P., Nguyen, L. N., Wang, Q., Djordjevic, S. P., Donner, E., & Nghiem, L. D. (2021). Monitoring antibiotic resistance genes in wastewater treatment: Current strategies and future challenges. *Science of the Total Environment*, 783, 146964.
Nguyen, L. N., Aditya, L., Vu, H. P., Johir, A. H., Bennar, L., Ralph, P., & Nghiem, L. D. (2022). Nutrient removal by algae-based wastewater treatment. *Current Pollution Reports*, 8(4), 369–383.
Okoye, C. O., Nyaruaba, R., Ita, R. E., Okon, S. U., Addey, C. I., Ebido, C. C., & Chukwudozie, K. I. (2022). Antibiotic resistance in the aquatic environment: Analytical techniques and interactive impact of emerging contaminants. *Environmental Toxicology and Pharmacology*, 96, 103995.
Oliveira, M., Antunes, W., Mota, S., Madureira-Carvalho, Á., Dinis-Oliveira, R. J., & da Silva, D. D. (2024). An overview of the recent advances in antimicrobial resistance. *Microorganisms*, 12(9), 1920.
Osbiston, K., Oxbrough, A., & Fernández-Martínez, L. T. (2021). Antibiotic resistance levels in soils from urban and rural land uses in Great Britain. *Access Microbiology*, 3(1), 000181.
Pandey, P., Khan, F., Agarwal, S., & Singh, M. (2022). Nano adsorbents in wastewater treatment: A new paradigm in wastewater management. *Letters in Applied NanoBioScience*, 12, 125.
Pouramini, Z., Mousavi, S. M., Babapoor, A., Hashemi, S. A., Pynadathu Rumjit, N., Garg, S., & Chiang, W. H. (2023). Recent advances in MXene-based nanocomposites for wastewater purification and water treatment: A review. *Water*, 15(7), 1267.
Puljko, A., Rozman, S. D., Barišić, I., Maravić, A., Jelić, M., Babić, I., & Udiković-Kolić, N. (2023). Resistance to critically important antibiotics in hospital wastewater from the largest Croatian city. *Science of the Total Environment*, 870, 161805.
Raju, S., Carbery, M., Kuttykattil, A., Senthirajah, K., Lundmark, A., Rogers, Z., & Palanisami, T. (2020). Improved methodology to determine the fate and transport of microplastics in a secondary wastewater treatment plant. *Water Research*, 173, 115549.
Ramli, S. F., & Aziz, H. A. (2023). Potential use of tin tetrachloride and polyacrylamide as a coagulant-coagulant aid in the treatment of highly coloured and turbid matured landfill leachate. *Process Safety and Environmental Protection*, 170, 971–982.
Rana, V., Sheokand, M., Jain, K., Dhaka, S., Godara, S. K., Urmaliya, D. N., & Dhaka, R. K. (2024). Effective and affordable water purification technologies for rural development. In Sughosh, M., Arun, L. S., Sylvester, C. I., & Eric van, H. (Eds) *Water Resources Management for Rural Development* (pp. 91–106). Elsevier, Amsterdam, Netherlands.
Reddy, S., Kaur, K., Barathe, P., Shriram, V., Govarthanan, M., & Kumar, V. (2022). Antimicrobial resistance in urban river ecosystems. *Microbiological Research*, 263, 127135.
Rodríguez, E. A., Ramirez, D., Balcázar, J. L., & Jiménez, J. N. (2021). Metagenomic analysis of urban wastewater resistome and mobilome: A support for antimicrobial resistance surveillance in an endemic country. *Environmental Pollution*, 276, 116736.
Saini, S., Tewari, S., Dwivedi, J., & Sharma, V. (2023). Biofilm-mediated wastewater treatment: A comprehensive review. *Materials Advances*, 4(6), 1415–1443.

Salam, M. A., Al-Amin, M. Y., Salam, M. T., Pawar, J. S., Akhter, N., Rabaan, A. A., & Alqumber, M. A. (2023). Antimicrobial resistance: A growing serious threat for global public health. *Healthcare*, 11(13), 1946.

Sambaza, S. S., & Naicker, N. (2023). Contribution of wastewater to antimicrobial resistance: A review article. *Journal of Global Antimicrobial Resistance*, 34, 23–29.

Sarekoski, A., Lipponen, A., Hokajärvi, A. M., Räisänen, K., Tiwari, A., Paspaliari, D., & WastPan Study Group. (2024). Simultaneous biomass concentration and subsequent quantitation of multiple infectious disease agents and antimicrobial resistance genes from community wastewater. *Environment International*, 191, 108973.

Sarkar, O., & Mohan, S. V. (2020). Synergy of anoxic microenvironment and facultative anaerobes on acidogenic metabolism in a self-induced electrofermentation system. *Bioresource Technology*, 313, 123604.

Satam, H., Joshi, K., Mangrolia, U., Waghoo, S., Zaidi, G., Rawool, S., & Malonia, S. K. (2023). Next-generation sequencing technology: Current trends and advancements. *Biology*, 12(7), 997.

Sathya, K., Nagarajan, K., Carlin Geor Malar, G., Rajalakshmi, S., & Raja Lakshmi, P. (2022). A comprehensive review on comparison among effluent treatment methods and modern methods of treatment of industrial wastewater effluent from different sources. *Applied Water Science*, 12(4), 70.

Sekizuka, T., Itokawa, K., Tanaka, R., Hashino, M., Yatsu, K., & Kuroda, M. (2022). Metagenomic analysis of urban wastewater treatment plant effluents in Tokyo. *Infection and Drug Resistance*, 15, 4763–4777.

Serwecińska, L. (2020). Antimicrobials and antibiotic-resistant bacteria: A risk to the environment and to public health. *Water*, 12(12), 3313.

Shah, P. H. (2024). Current situation of industrialization and sustainable development with special reference to position of 'Vishwamitri'River in Vadodara. *Journal of Advanced Zoology*, 45(2), 206–221.

Shamshad, J., & Rehman, R. U. (2025). Innovative approaches to sustainable wastewater treatment: A comprehensive exploration of conventional and emerging technologies. *Environmental Science: Advances*, 4, 189–222.

Shivaram, K. B., Bhatt, P., Applegate, B., & Simsek, H. (2023). Bacteriophage-based biocontrol technology to enhance the efficiency of wastewater treatment and reduce targeted bacterial biofilms. *Science of the Total Environment*, 862, 1607.

Shrestha, S., Malla, B., & Haramoto, E. (2024). High-throughput microfluidic quantitative PCR system for the simultaneous detection of antibiotic resistance genes and bacterial and viral pathogens in wastewater. *Environmental Research*, 255, 119156.

Singh, S., Ahmed, A. I., Almansoori, S., Alameri, S., Adlan, A., Odivilas, G., & Elamin, W. (2024a). A narrative review of wastewater surveillance: Pathogens of concern, applications, detection methods, and challenges. *Frontiers in Public Health*, 12, 1445961.

Singh, S., Sivaram, N., Nath, B., Khan, N. A., Singh, J., & Ramamurthy, P. C. (2024b). Metal organic frameworks for wastewater treatment, renewable energy and circular economy contributions. *NPJ Clean Water*, 7(1), 124.

Srathongneam, T., Sresung, M., Paisantham, P., Ruksakul, P., Singer, A. C., Sukchawalit, R., & Sirikanchana, K. (2024). High throughput qPCR unveils shared antibiotic resistance genes in tropical wastewater and river water. *Science of the Total Environment*, 908, 167867.

Sravan, J. S., Matsakas, L., & Sarkar, O. (2024). Advances in biological wastewater treatment processes: Focus on low-carbon energy and resource recovery in biorefinery context. *Bioengineering*, 11(3), 281.

Srivastav, A. K., Das, P., Jaiswal, P., & Parashar, V. (2024). CRISPR-cas systems for targeted therapy in multidrug-resistant infections: A new Frontier in Medical Microbiology. *International Journal of Innovative Research in Science, Engineering and Technology*, 13(11), 18332–18340.

Stec, J., Kosikowska, U., Mendrycka, M., Stępień-Pyśniak, D., Niedźwiedzka-Rystwej, P., Bębnowska, D., & Grywalska, E. (2022). Opportunistic pathogens of recreational waters with emphasis on antimicrobial resistance—a possible subject of human health concern. *International Journal of Environmental Research and Public Health*, 19(12), 7308.

Sun, G., Chung, T. S., Jeyaseelan, K., & Armugam, A. (2013). Stabilization and immobilization of aquaporin reconstituted lipid vesicles for water purification. *Colloids and Surfaces B: Biointerfaces*, 102, 466–471.

Tena, F. O., Bickel, V., Steinweg, C., & Posten, C. (2024). Continuous microalgae cultivation for wastewater treatment—Development of a process strategy during day and night. *Science of the Total Environment*, 912, 169082.

Thekkudan, V. N., Vaidyanathan, V. K., Ponnusamy, S. K., Charles, C., Sundar, S., Vishnu, D., & Subramanian, S. (2017). Review on nanoadsorbents: A solution for heavy metal removal from wastewater. *IET Nanobiotechnology*, 11(3), 213–224.

Tian, M., He, X., Feng, Y., Wang, W., Chen, H., Gong, M., & van Eerde, A. (2021). Pollution by antibiotics and antimicrobial resistance in livestock and poultry manure in China, and countermeasures. *Antibiotics*, 10(5), 539.

Tijing, L. D., Woo, Y. C., Shim, W. G., He, T., Choi, J. S., Kim, S. H., & Shon, H. K. (2016). Superhydrophobic nanofiber membrane containing carbon nanotubes for high-performance direct contact membrane distillation. *Journal of Membrane Science*, 502, 158–170.

Tiwari, A., Krolicka, A., Tran, T. T., Räisänen, K., Ásmundsdóttir, Á. M., Wikmark, O. G., & Pitkänen, T. (2024). Antibiotic resistance monitoring in wastewater in the Nordic countries: A systematic review. *Environmental Research*, 246, 118052.

Tiwari, A., Kurittu, P., Al-Mustapha, A. I., Heljanko, V., Johansson, V., Thakali, O., & Heikinheimo, A. (2022). Wastewater surveillance of antibiotic-resistant bacterial pathogens: A systematic review. *Frontiers in Microbiology*, 13, 977106.

Trifirò, F., & Zanirato, P. (2024). The battle of the physic and chemistry against pathogenic microorganism. *Mathews Journal of Pharmaceutical Science*, 8(1), 10027. doi:10.30654/mjps.10027.

Umar, W., Ur Rehman, M. Z., Umair, M., Ayub, M. A., Naeem, A., Rizwan, M., & Karri, R. R. (2022). Use of nanotechnology for wastewater treatment: Potential applications, advantages, and limitations. In Janardhan, R. K., Rama, R. K., Nabisab, M. M., & Erick, R. B. (Eds) *Sustainable Nanotechnology for Environmental Remediation* (pp. 223–272). Elsevier, Amsterdam, Netherlands.

Vasala, A., Hytönen, V. P., & Laitinen, O. H. (2020). Modern tools for rapid diagnostics of antimicrobial resistance. *Frontiers in Cellular and Infection Microbiology*, 10, 308.

Venturini, C., Beatson, S. A., Djordjevic, S. P., & Walker, M. J. (2010). Multiple antibiotic resistance gene recruitment onto the enterohemorrhagic *Escherichia coli* virulence plasmid. *The FASEB Journal*, 24(4), 1160–1166.

Venturini, C., Hassan, K. A., Roy Chowdhury, P., Paulsen, I. T., Walker, M. J., & Djordjevic, S. P. (2013). Sequences of two related multiple antibiotic resistance virulence plasmids sharing a unique IS 26-related molecular signature isolated from different *Escherichia coli* pathotypes from different hosts. *PLoS One*, 8(11), e78862.

Wang, J., Zhang, Z., Zhang, Q., Liu, J., & Ma, J. (2018). Preparation and adsorption application of carbon nanofibers with large specific surface area. *Journal of Materials Science*, 53, 16466–16475.

Wang, R., Ji, M., Zhai, H., Guo, Y., & Liu, Y. (2021). Occurrence of antibiotics and antibiotic resistance genes in WWTP effluent-receiving water bodies and reclaimed wastewater treatment plants. *Science of the Total Environment*, 796, 148919.

Wei, G., Su, Z., Reynolds, N. P., Arosio, P., Hamley, I. W., Gazit, E., & Mezzenga, R. (2017). Self-assembling peptide and protein amyloids: From structure to tailored function in nanotechnology. *Chemical Society Reviews*, 46(15), 4661–4708.

Werkneh, A. A., & Islam, M. A. (2023). Post-treatment disinfection technologies for sustainable removal of antibiotic residues and antimicrobial resistance bacteria from hospital wastewater. *Heliyon*, 9(4), e15360.

Widyarani, W., D. R., Hamidah, U., Komarulzaman, A., Rosmalina, R. T., & Sintawardani, N. (2022). Domestic wastewater in Indonesia: Generation, characteristics and treatment. *Environmental Science and Pollution Research*, 29(22), 32397–32414.

Williams, J. (2024). Contribution of livestock farming to environmental pollution in China. *Indian Journal of Animal Health*, 4, 43–53.

World Health Organization (WHO). (2014) *Antimicrobial Resistance: Global Report on Surveillance*. WHO, Geneva, Switzerland.

Yamin, D., Uskoković, V., Wakil, A. M., Goni, M. D., Shamsuddin, S. H., Mustafa, F. H., & Yusof, N. Y. (2023). Current and future technologies for the detection of antibiotic-resistant bacteria. *Diagnostics*, 13(20), 3246.

Yang, C., Yan, S., Zhang, B., Yao, X., Mo, J., Rehman, F., & Guo, J. (2024a). Spatiotemporal distribution of the planktonic microbiome and antibiotic resistance genes in a typical urban river contaminated by macrolide antibiotics. *Environmental Research*, 262, 119808.

Yang, L., Zhao, F., Yen, H., Feng, Q., Li, M., Wang, X., & Chen, L. (2024b). Urbanization and land use regulate soil vulnerability to antibiotic contamination in urban green spaces. *Journal of Hazardous Materials*, 465, 133363.

Yeruva, D. K., Ranadheer, P., & Venkata Mohan, S. (2020). Eco-electrogenic engineered flow through wetland system for tertiary treatment of acidogenic effluents from biohydrogen production. *Journal of Hazardous, Toxic, and Radioactive Waste*, 24(3), 04020020.

Yuan, T., & Pian, Y. (2023). Hospital wastewater as hotspots for pathogenic microorganisms spread into aquatic environment: A review. *Frontiers in Environmental Science*, 10, 1091734.

Zhang, Q. G., Deng, C., Soyekwo, F., Liu, Q. L., & Zhu, A. M. (2016). Sub-10 nm wide cellulose nanofibers for ultrathin nanoporous membranes with high organic permeation. *Advanced Functional Materials*, 26(5), 792–800.

Zhang, Y., Gu, A. Z., He, M., Li, D., & Chen, J. (2017). Subinhibitory concentrations of disinfectants promote the horizontal transfer of multidrug resistance genes within and across genera. *Environmental Science & Technology*, 51(1), 570–580.

Zieliński, W., Hubeny, J., Buta-Hubeny, M., Rolbiecki, D., Harnisz, M., Paukszto, Ł., & Korzeniewska, E. (2022). Metagenomics analysis of probable transmission of determinants of antibiotic resistance from wastewater to the environment—A case study. *Science of the Total Environment*, 827, 154354.

11 From Waste to Resource
Mitigating Antimicrobial Resistance in Wastewater-Irrigated Agriculture

Nisha Singh, Vineet Kumar Maurya, and Deepak Singh

11.1 INTRODUCTION

Water is the lifeblood of agriculture, yet its availability is increasingly threatened by global water scarcity. Rapid population growth, urbanization, industrialization, and climate change have placed immense pressure on freshwater resources, particularly in arid and semi-arid regions (Singh et al., 2020a). According to the United Nations, over two billion people live in countries experiencing high water stress, and this number is expected to rise in the coming decades. Agriculture, which accounts for approximately 70% of global freshwater withdrawals, is at the forefront of this crisis (Alcamo et al., 2000; Singh and Goyal, 2018). To address water shortages, farmers and policymakers have turned to alternative water sources, with wastewater emerging as a critical resource for irrigation. The use of wastewater in agriculture is not a new practice. For centuries, farmers have recognized its dual benefits as a reliable water source and a nutrient-rich fertilizer (Gupta et al., 2023; Singh et al., 2020b). In many developing countries, untreated or partially treated wastewater is widely used to irrigate crops, particularly in urban and peri-urban areas where access to clean water is limited. This practice supports food security, reduces dependence on chemical fertilizers, and provides livelihoods for millions of smallholder farmers. However, the use of untreated or inadequately treated wastewater poses significant risks to human health and the environment, particularly through the spread of antimicrobial resistance (AMR) (Singh et al., 2021a).

AMR is one of the most pressing global health challenges of the 21st century. It occurs when microorganisms such as bacteria, viruses, fungi, and parasites develop the ability to resist the drugs designed to kill them. The overuse and misuse of antibiotics in human medicine, agriculture, and livestock production have accelerated the emergence and spread of AMR (Hosain et al., 2021). Wastewater, especially when untreated, serves as a reservoir for antibiotic-resistant bacteria (ARB) and antibiotic resistance genes (ARGs). These contaminants can enter the environment through irrigation, contaminating soil, crops, and water sources, and ultimately entering the food chain (Pant et al., 2024). This creates a direct pathway for AMR to impact human health, making wastewater irrigation a critical concern in the global fight against AMR. The connection between wastewater and AMR is multifaceted. Wastewater from hospitals, pharmaceutical manufacturing plants, and livestock farms often contains high levels of antimicrobial residues, ARB, and ARGs. When this wastewater is used for irrigation without adequate treatment, it introduces these contaminants into agricultural systems (Naghavi et al., 2024). Horizontal gene transfer (HGT) further exacerbates the problem by enabling the spread of ARGs among different bacterial species. The result is a growing prevalence of resistant strains that can cause infections that are difficult, and sometimes impossible, to treat (Michaelis and Grohmann, 2023). This chapter explores the transition from untreated to treated wastewater in irrigation, with a focus on addressing the risks of AMR. It begins by examining the global context of water scarcity and the role of wastewater in agriculture. It then provides an overview of AMR, its mechanisms, and its connection to wastewater. The chapter highlights the environmental and public health risks associated with the use

DOI: 10.1201/9781003269823-11

of untreated wastewater and evaluates the effectiveness of wastewater treatment technologies in mitigating these risks. Finally, it outlines strategies for promoting the safe and sustainable use of wastewater in agriculture, emphasizing the need for improved infrastructure, policy frameworks, and global collaboration. The objectives of this chapter are threefold: (1) to raise awareness of the risks posed by AMR in wastewater-irrigated agriculture, (2) to explore the potential of treated wastewater as a safer alternative, and (3) to provide actionable recommendations for policymakers, researchers, and stakeholders. By addressing these challenges, we can promote the safe reuse of wastewater, protect human health, and ensure a sustainable future for agriculture in the face of growing water scarcity.

11.2 WASTEWATER USE IN AGRICULTURE AND ITS ROLE IN AMR

11.2.1 Global Perspective on Wastewater Use

The use of wastewater in agriculture is a practice as old as civilization itself. Ancient societies, such as those in Mesopotamia and the Indus Valley, recognized the value of wastewater for irrigating crops and enhancing soil fertility (Lofrano and Brown, 2017). In modern times, wastewater irrigation has become a vital strategy for addressing water scarcity, particularly in regions where freshwater resources are limited (Desai et al., 2023; Singh et al., 2021b). Countries like Israel and Jordan have pioneered the use of treated wastewater for irrigation, setting an example for other water-scarce regions. In Sub-Saharan Africa, untreated wastewater is commonly used in urban agriculture, providing livelihoods for smallholder farmers but posing significant health risks. Similarly, in South Asia, countries like India and Pakistan rely on wastewater to irrigate crops in peri-urban areas, where the demand for fresh produce is high.

Today, wastewater is used to irrigate approximately 10% of the world's croplands, supporting food production for millions of people. However, the extent and methods of wastewater use vary significantly across regions, reflecting differences in water availability, infrastructure, and regulatory frameworks. Historically, wastewater irrigation was driven by necessity rather than by design. In many developing countries, the lack of adequate sanitation and wastewater treatment infrastructure has led to the direct use of untreated wastewater in agriculture. This practice continues today, particularly in urban and peri-urban areas where rapid population growth has outpaced the development of water and sanitation systems. In contrast, developed countries have increasingly adopted treated wastewater for irrigation, supported by advanced treatment technologies and stringent regulatory standards. Despite these differences, the common thread across all regions is the recognition of wastewater as a valuable resource for sustaining agricultural productivity (Tella et al., 2025).

The benefits of wastewater reuse in agriculture are manifold. First and foremost, it addresses water scarcity by providing a reliable alternative to freshwater sources. In regions where rainfall is erratic or groundwater reserves are depleted, wastewater ensures a consistent water supply for crops (Khan et al., 2022). Additionally, wastewater is rich in nutrients such as nitrogen, phosphorus, and potassium, which are essential for plant growth. By recycling these nutrients, wastewater irrigation reduces the need for chemical fertilizers, lowering production costs and minimizing environmental pollution. Furthermore, the use of wastewater in agriculture can contribute to climate change mitigation by reducing the energy and emissions associated with the production of chemical fertilizers. Despite its benefits, the use of wastewater in agriculture is not without challenges. The risks associated with untreated or inadequately treated wastewater, such as the spread of pathogens and contaminants, have raised concerns about food safety and public health. These risks are particularly pronounced in regions where regulatory oversight is weak, and treatment infrastructure is lacking. Nevertheless, the potential of wastewater to support sustainable agriculture has led to its widespread adoption in many parts of the world (Bangre et al., 2024; Singh et al., 2020a).

11.2.2 Sources of Antimicrobial Residues in Wastewater

11.2.2.1 Hospitals and Healthcare Facilities

Hospitals are major contributors to the antimicrobial load in wastewater. The widespread use of antibiotics in healthcare settings, coupled with the excretion of unmetabolized drugs by patients, results in the discharge of high concentrations of antimicrobial residues into sewage systems. Studies have shown that hospital effluents can contain a wide range of antibiotics, including beta-lactams, fluoroquinolones, and macrolides, as well as disinfectants and other chemicals that can promote the development of AMR. The presence of these residues in wastewater creates an environment conducive to the selection and proliferation of ARB (Hounmanou et al., 2025).

11.2.2.2 Pharmaceutical Manufacturing Plants

Pharmaceutical manufacturing is another significant source of antimicrobial residues in wastewater. In many countries, particularly in South Asia, the production of antibiotics is concentrated in industrial hubs where wastewater treatment infrastructure is often inadequate. Effluents from these plants can contain extremely high concentrations of active pharmaceutical ingredients, which are released into nearby water bodies (Hanna et al., 2023). Hyderabad, a major pharmaceutical manufacturing hub in India, has been the focus of several studies on AMR in wastewater. Research conducted in the city's Musi River, which receives effluents from pharmaceutical plants, hospitals, and households, has detected extremely high concentrations of antibiotics and ARGs. Wastewater from the Musi River is used to irrigate crops in nearby agricultural fields, leading to the contamination of soil and produce. Studies have found high levels of ARB in vegetables such as spinach and lettuce, posing a direct risk to consumer health (Hegde et al., 2025). In Beijing, untreated wastewater from livestock farms is commonly used for irrigation. A study conducted in the region found that wastewater-irrigated soils contained high levels of ARB and ARGs, including genes conferring resistance to tetracyclines, sulfonamides, and fluoroquinolones. The study also detected ARGs in crops grown in these soils, highlighting the potential for AMR to enter the food (Wu et al., 2024). In addition, Mexico City has a long history of using untreated wastewater for irrigation. Research in the Mezquital Valley, where wastewater from the city is used to irrigate crops, has revealed high levels of ARB and ARGs in soil and water samples. The study also found that farmers and their families had a higher prevalence of antibiotic-resistant infections compared to the general population, underscoring the public health risks associated with untreated wastewater irrigation (Chávez-Mejía et al., 2019).

11.2.2.3 Livestock and Aquaculture

The use of antibiotics in livestock farming and aquaculture is a major driver of AMR. Antibiotics are routinely administered to animals for growth promotion, disease prevention, and treatment, leading to the excretion of unmetabolized drugs in manure and urine. When this waste is washed into water bodies or used as a fertilizer, it introduces antimicrobial residues and ARB into the environment. In regions with intensive livestock farming, such as China and the United States, the impact of agricultural runoff on water quality and AMR is particularly pronounced.

11.2.2.4 Household and Municipal Wastewater

Household wastewater also contributes to the antimicrobial load in sewage systems. The disposal of unused or expired medications, as well as the excretion of antibiotics by individuals undergoing treatment, adds to the concentration of antimicrobial residues in municipal wastewater. While the concentrations of antibiotics in household wastewater are generally lower than those in hospital or industrial effluents, the cumulative impact of these residues on AMR cannot be overlooked (Figure 11.1).

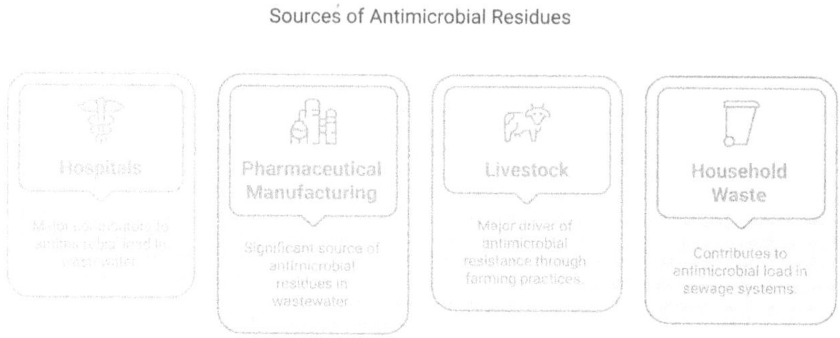

FIGURE 11.1 Illustrates various sources of antimicrobial residues in wastewater.

11.3 MECHANISMS OF AMR DEVELOPMENT AND SPREAD

11.3.1 Selective Pressure from Antimicrobial Residues

The presence of sublethal concentrations of antibiotics in wastewater exerts selective pressure on microbial communities, promoting the survival and proliferation of ARB. Even at low concentrations, antibiotics can inhibit the growth of susceptible bacteria while allowing resistant strains to thrive. This selective pressure is a key driver of AMR in untreated wastewater (Tucker, 2021).

11.3.2 Horizontal Gene Transfer (HGT)

HGT is a primary mechanism by which ARGs are disseminated among bacterial populations. In wastewater, HGT can occur through conjugation (transfer of plasmids), transformation (uptake of free DNA), or transduction (transfer of DNA by bacteriophages). The high density and diversity of microorganisms in wastewater create ideal conditions for HGT, enabling the rapid spread of resistance genes across different bacterial species.

11.3.3 Biofilms and Persistence of ARB

Biofilms, which are structured communities of bacteria embedded in a self-produced matrix, play a significant role in the persistence and spread of ARB in wastewater. Biofilms provide a protective environment for bacteria, shielding them from antibiotics and other stressors. Within biofilms, ARB can exchange genetic material and maintain resistance traits, even in the absence of selective pressure.

11.3.4 Environmental Reservoirs of AMR

Untreated wastewater serves as an environmental reservoir for ARB and ARGs. When this wastewater is used for irrigation, it introduces these contaminants into soil, crops, and water sources, creating a pathway for AMR to enter the food chain. The persistence of ARB and ARGs in the environment further exacerbates the problem, as they can be transferred to humans through direct contact, consumption of contaminated food, or exposure to contaminated water.

11.4 WASTEWATER TREATMENT AND ITS ROLE IN AMR MITIGATION

The transition from untreated to treated wastewater in agriculture represents a critical step toward mitigating the risks of AMR. Treated wastewater, when properly managed, can provide a safer alternative for irrigation, reducing the introduction of ARB and ARGs into the environment.

However, the effectiveness of wastewater treatment technologies in removing AMR-related contaminants varies, and challenges remain in achieving complete removal of antimicrobial residues (Bougnom et al., 2020). This section explores the opportunities and limitations of treated wastewater, focusing on the effectiveness of treatment technologies, the challenges in removing antimicrobial residues, and a comparison of treated versus untreated wastewater in terms of AMR risks.

11.4.1 Conventional Treatment Processes

Conventional wastewater treatment processes, including primary and secondary treatment, are widely employed to remove solids, organic matter, and pathogens from wastewater, but their effectiveness in eliminating ARB and ARGs is limited. Primary treatment, the first stage of wastewater treatment, involves physical processes such as sedimentation and filtration to remove suspended solids and particulate matter. While this step is effective in reducing the bulk of solid waste, it has minimal impact on dissolved antimicrobial residues, ARB, and ARGs, which remain largely unaffected. As a result, a significant portion of these contaminants passes through to subsequent treatment stages (Ranjit et al., 2021).

Secondary treatment, which typically employs biological processes like activated sludge or trickling filters, aims to degrade organic matter and remove pathogens. These processes rely on microbial communities to break down pollutants, and they can achieve a partial reduction in the concentration of ARB and ARGs. For instance, studies have demonstrated that secondary treatment can achieve a 1–2 log reduction (90%–99%) in ARB and ARGs. However, this level of removal is often insufficient to fully eliminate AMR-related contaminants. One reason for this limitation is that secondary treatment is not specifically designed to target antimicrobial residues or the genetic material associated with resistance (Wang et al., 2023). Additionally, the biological processes involved can inadvertently promote the exchange of ARGs among bacteria through HGT, further complicating the issue.

Despite the partial removal of ARB and ARGs during secondary treatment, these contaminants often persist in treated effluent and sludge. Treated effluent discharged into surface water bodies can introduce ARB and ARGs into aquatic ecosystems, while sludge applied as a fertilizer on agricultural land can reintroduce these contaminants into soil environments. This persistence highlights the inadequacy of conventional treatment processes in addressing the growing challenge of AMR.

11.4.2 Advanced Treatment Technologies

Advanced treatment technologies offer enhanced removal of ARB, ARGs, and antimicrobial residues. These technologies include:

11.4.2.1 Membrane Filtration

Techniques such as ultrafiltration (UF) and reverse osmosis (RO) can effectively remove bacteria, viruses, and dissolved contaminants, including antimicrobial residues. Membrane filtration has been shown to achieve up to a 4–6 log reduction in ARB and ARGs. However, the high cost and energy requirements of these technologies limit their widespread adoption.

11.4.2.2 Advanced Oxidation Processes (AOPs)

AOPs, such as ozonation and UV-based treatments, use reactive oxygen species to degrade organic pollutants, including antibiotics. These processes can significantly reduce the concentration of antimicrobial residues and inactivate ARB. However, they may not completely eliminate ARGs, which can persist in the environment even after bacterial cells are destroyed.

11.4.2.3 Constructed Wetlands

Natural treatment systems, such as constructed wetlands, use plants and microorganisms to remove contaminants from wastewater. While cost-effective and environmentally friendly, constructed wetlands are less efficient in removing ARB and ARGs compared to advanced technologies.

The effectiveness of advanced treatment technologies depends on factors such as the type and concentration of contaminants, treatment conditions, and operational parameters.

11.4.3 Disinfection Methods

Disinfection is a critical step in wastewater treatment to ensure the removal of pathogens, including ARB. Common disinfection methods include:

11.4.3.1 Chlorination

Chlorination is widely used due to its low cost and effectiveness in inactivating bacteria and viruses. However, chlorination can lead to the formation of harmful disinfection by-products and may not completely eliminate ARGs.

11.4.3.2 UV Disinfection

UV disinfection is effective in inactivating bacteria and viruses without producing harmful by-products. However, UV disinfection has a limited impact on dissolved antimicrobial residues and ARGs.

11.4.3.3 Ozonation

Ozonation is highly effective in inactivating bacteria and degrading organic pollutants, including antibiotics. However, ozonation is energy-intensive and may not completely remove ARGs.

While disinfection methods can reduce the concentration of ARB, they often fail to address the underlying issue of ARGs, which can persist in the environment and contribute to the spread of AMR (Figure 11.2).

11.4.4 Challenges in Achieving Complete Removal of Antimicrobial Residues

11.4.4.1 Persistence of ARGs

The persistence of ARGs in wastewater treatment systems represents a major challenge in combating AMR. Unlike ARB, which can be inactivated or removed through treatment processes, ARGs

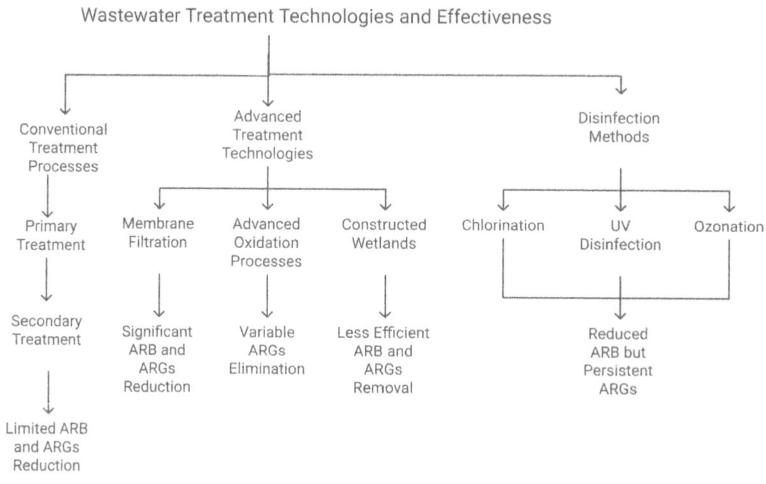

FIGURE 11.2 Wastewater treatment technologies and their effectiveness in ARGs and ARB removal.

are genetic elements that can endure in the environment and continue to pose a threat even after treatment. This persistence occurs because ARGs are often located on mobile genetic elements, such as plasmids, integrons, or transposons, which can be transferred between bacteria through HGT. HGT allows ARGs to spread across different bacterial species, including those that were not originally resistant, thereby perpetuating the problem of AMR (Wang et al., 2025). For example, consider a wastewater treatment plant that employs conventional processes like activated sludge. While these methods can reduce the concentration of ARB by 90%–99% (1–2 log reduction), they do not effectively destroy ARGs. In the activated sludge process, bacteria aggregate into flocs, creating an environment conducive to HGT. Resistant bacteria may be inactivated, but their ARGs can be taken up by other bacteria in the sludge, including environmental or commensal strains. These newly resistant bacteria can then survive and proliferate, even in the absence of antibiotics, due to the selective pressure exerted by other contaminants like heavy metals or biocides. Moreover, treated effluent discharged into rivers or lakes can carry ARGs into aquatic ecosystems. For instance, a study on a river receiving treated wastewater found elevated levels of ARGs downstream of the discharge point, even though the concentration of ARB was significantly reduced. These ARGs can be taken up by aquatic bacteria, which may then enter the food chain through organisms like fish or shellfish, ultimately posing risks to human health. Similarly, when treated sludge is applied to agricultural land as a fertilizer, ARGs can persist in the soil and be transferred to soil bacteria, creating a long-term reservoir of resistance (Grobelak et al., 2024).

11.4.4.2 Complexity of Wastewater Composition

Wastewater is a highly complex mixture containing a diverse array of contaminants, including antibiotics, heavy metals, organic matter, and other chemicals, all of which can contribute to the development and persistence of AMR (Singh et al., 2020a). This complexity poses significant challenges for wastewater treatment processes, as the interactions between these contaminants can reduce the effectiveness of treatment technologies in removing ARB and ARGs. One of the key issues is the phenomenon of co-selection, where the presence of certain contaminants, such as heavy metals, can indirectly promote antibiotic resistance. Heavy metals like copper, zinc, and cadmium, often found in industrial and urban wastewater, can exert selective pressure on microbial communities, favoring the survival of bacteria that possess resistance mechanisms. These resistance mechanisms often involve efflux pumps or detoxification pathways that can also confer resistance to antibiotics, even in the absence of direct antibiotic exposure. For example, bacteria resistant to copper may also exhibit resistance to antibiotics like tetracycline or beta-lactams due to shared genetic determinants. Additionally, organic matter in wastewater, such as proteins, fats, and carbohydrates, can interfere with treatment processes by shielding bacteria from disinfection. For instance, during chlorination or UV disinfection, organic matter can absorb disinfectants or block UV light, reducing their efficacy in inactivating ARB. This protective effect allows resistant bacteria to survive and proliferate, even after treatment. Moreover, organic matter can serve as a nutrient source for microbial growth, further complicating the removal of ARB and ARGs (Brovini et al., 2025). For example, in activated sludge systems, organic matter can promote the growth of biofilms, which provide a protective environment for bacteria and facilitate the exchange of ARGs through HGT. The interplay between these contaminants creates a challenging environment for wastewater treatment. For instance, in a wastewater treatment plant receiving effluent from hospitals, pharmaceuticals, heavy metals, and organic matter may coexist, creating conditions that favor the persistence and spread of AMR. This complexity underscores the need for advanced treatment strategies that can address multiple contaminants simultaneously. Technologies such as AOPs, which generate reactive oxygen species to degrade organic matter and inactivate bacteria, or membrane filtration, which physically removes contaminants, may offer more effective solutions. However, the variability in wastewater composition across different sources and regions requires tailored approaches to ensure the effective removal of ARB, ARGs, and other AMR-promoting contaminants (Graham et al., 2024). Addressing the complexity of wastewater composition is critical to mitigating the environmental and public health risks associated with AMR.

11.4.4.3 Cost, Infrastructure, and Regulatory Gaps

The implementation of advanced wastewater treatment technologies, such as membrane filtration and AOPs, faces significant challenges due to high costs, infrastructure requirements, and the need for technical expertise. These barriers are particularly pronounced in low- and middle-income countries (LMICs), where the use of untreated or inadequately treated wastewater is most prevalent. In these regions, financial constraints and limited access to advanced infrastructure make it difficult to adopt such technologies, despite their effectiveness in removing ARB and ARGs. This disparity highlights the urgent need for cost-effective, scalable, and sustainable solutions that can be implemented in resource-limited settings to address the global challenge of AMR. For example, nature-based solutions like constructed wetlands or solar-based disinfection systems could offer viable alternatives, as they are relatively low cost and require minimal technical expertise (Clemente et al., 2024).

Compounding the issue is the lack of standardized regulations and monitoring protocols for AMR in wastewater. Many countries do not have specific guidelines for the removal of ARB and ARGs during wastewater treatment, making it difficult to assess and mitigate the risks posed by AMR. Without robust regulatory frameworks and consistent monitoring, the effectiveness of treatment processes remains uncertain, and the potential for AMR dissemination through treated wastewater persists. Addressing these gaps requires international collaboration to develop harmonized guidelines, enhance surveillance systems, and promote knowledge-sharing. By integrating cost-effective technologies with stronger regulatory and monitoring frameworks, it is possible to reduce the environmental and public health risks associated with AMR, particularly in regions where the burden of untreated wastewater is highest. This dual approach is essential for achieving global AMR mitigation and ensuring sustainable water management practices.

11.5 PUBLIC HEALTH AND ENVIRONMENTAL RISKS OF AMR IN WASTEWATER-IRRIGATED AGRICULTURE

The use of wastewater in agriculture, particularly untreated or inadequately treated wastewater, has significant implications for both environmental and public health. The introduction of ARB and ARGs into agricultural systems through wastewater irrigation creates multiple pathways for the transmission of AMR to humans and ecosystems. This section explores the transmission pathways of ARB and ARGs from wastewater to humans; the health risks for farmers, consumers, and communities; and the ecological impacts on soil and aquatic ecosystems.

11.5.1 Transmission Pathways of ARB and ARGs from Wastewater to Humans

11.5.1.1 Soil Contamination

Wastewater irrigation introduces ARB and ARGs into the soil, where they can persist and proliferate. Soil serves as a reservoir for these contaminants, creating a long-term source of AMR. The persistence of ARB and ARGs in soil is influenced by factors such as soil type, organic matter content, and microbial activity. Once introduced into the soil, ARB and ARGs can be transferred to crops, water sources, and humans through various pathways (Figure 11.3).

11.5.1.2 Crop Contamination

Crops irrigated with wastewater can become contaminated with ARB and ARGs through several mechanisms, posing significant risks to human health. One primary pathway is root uptake, where ARB and ARGs are absorbed by plant roots and transported to edible parts of the crop. Studies have demonstrated that leafy vegetables like lettuce and spinach, as well as root crops such as carrots and radishes, can internalize ARB and ARGs through their root systems, leading to contamination of the edible portions (Patra and Dubey, 2024). This is particularly concerning because these crops are often consumed raw, increasing the likelihood of human exposure to resistant bacteria and genes.

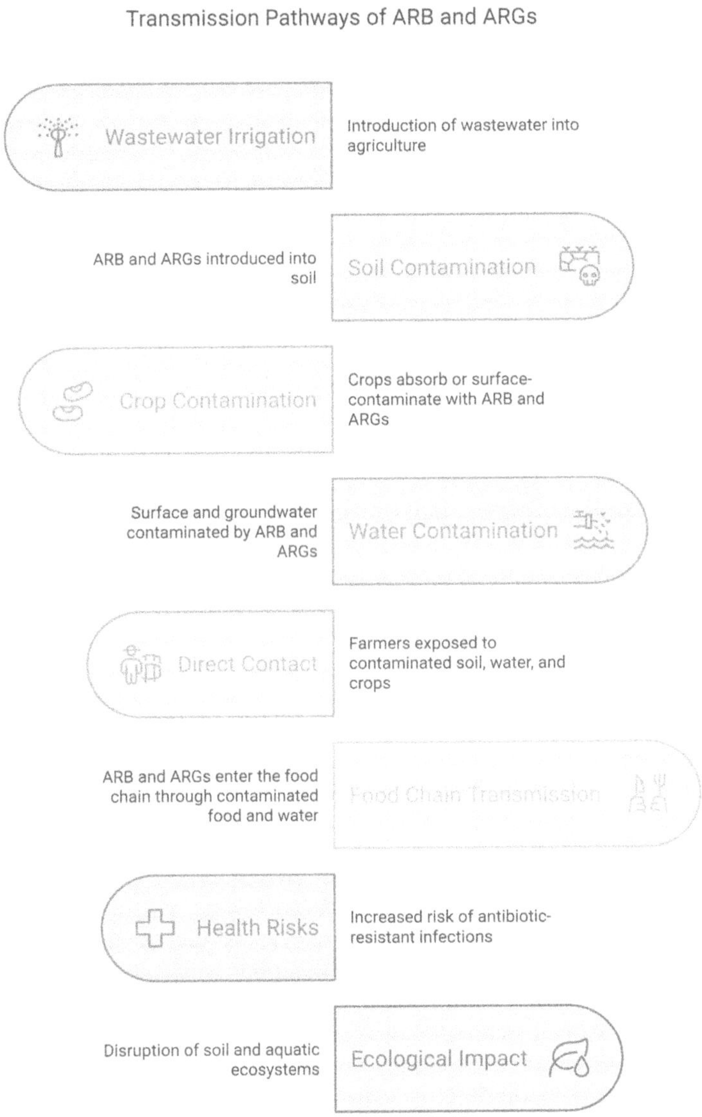

FIGURE 11.3 Transmission pathway of ARB and ARGs.

Another mechanism is surface contamination, where ARB and ARGs adhere to the outer surfaces of crops during irrigation. Leafy greens, such as lettuce and spinach, and fruits like tomatoes and strawberries, are especially vulnerable due to their large surface areas and direct exposure to irrigation water. Research has shown that washing these crops with water alone is often insufficient to remove all contaminants, leaving consumers at risk. Additionally, soil splash—caused by rain or irrigation—can disperse soil particles containing ARB and ARGs onto crops. For instance, a study on tomato plants irrigated with wastewater found that soil splash contributed to the presence of ARB on the surfaces of the fruits, even when the plants were not directly irrigated with contaminated water. These contamination pathways highlight the direct risks to consumers who ingest these crops, as ARB and ARGs can colonize the human gut, potentially leading to infections that are difficult to treat due to antibiotic resistance (Sun et al., 2024). The widespread use of wastewater in agriculture, particularly in water-scarce regions, underscores the urgent need for improved irrigation practices and advanced wastewater treatment technologies to mitigate these risks and ensure food safety.

11.5.1.3 Water Contamination

Wastewater irrigation can lead to the contamination of surface water and groundwater with ARB and ARGs. Surface water bodies, such as rivers and lakes, can become reservoirs for AMR, particularly in regions where wastewater is discharged without treatment. Groundwater contamination occurs when ARB and ARGs leach through the soil and enter aquifers. Contaminated water sources can serve as a pathway for the transmission of AMR to humans through drinking water, recreational activities, and irrigation of other crops.

11.5.1.4 Direct Contact

Farmers and agricultural workers are at risk of direct exposure to ARB and ARGs through contact with contaminated soil, water, and crops. This can occur during activities such as planting, weeding, harvesting, and irrigation. Direct contact with contaminated materials can lead to skin infections, respiratory infections, and gastrointestinal illnesses caused by antibiotic-resistant pathogens.

11.5.1.5 Food Chain Transmission

The contamination of crops and water sources with ARB and ARGs creates a pathway for AMR to enter the food chain. Consumers who ingest contaminated food or water are at risk of acquiring antibiotic-resistant infections. Additionally, livestock that consume contaminated feed or water can become carriers of ARB and ARGs, further amplifying the spread of AMR through the food chain.

11.5.2 Health Risks for Farmers, Agricultural Workers, Consumers, and Communities

Farmers and agricultural workers face significant health risks due to their frequent and direct exposure to wastewater used for irrigation, which often contains ARB and ARGs. Their close contact with contaminated soil, water, and crops increases the likelihood of acquiring antibiotic-resistant infections, making them one of the most vulnerable groups. One common health issue is skin infections, which can occur through direct contact with contaminated soil or water during farming activities. For example, a study in India found that farmers using untreated wastewater for irrigation had a higher incidence of skin infections caused by multidrug-resistant bacteria, such as *Staphylococcus aureus*, compared to those using clean water. Another significant risk is respiratory infections, which can result from inhaling dust or aerosols containing ARB during irrigation or tilling (Verma et al., 2023). In Mexico, researchers observed that agricultural workers exposed to wastewater irrigation had elevated rates of respiratory infections linked to resistant pathogens like *Pseudomonas aeruginosa*. Additionally, gastrointestinal illnesses are a major concern, as farmers may inadvertently ingest contaminated water or food while working in the fields (Adegoke et al., 2018). A study in Pakistan revealed that farmers using untreated wastewater had a higher prevalence of gastrointestinal infections caused by antibiotic-resistant *Escherichia coli* and *Salmonella* species. These findings underscore the occupational health risks associated with wastewater irrigation and highlight the urgent need for protective measures, such as the use of personal protective equipment, improved hygiene practices, and the adoption of safer irrigation methods.

Communities living near wastewater-irrigated fields are at risk of exposure to ARB and ARGs through multiple pathways, including contaminated water sources, airborne dust, and direct contact with contaminated materials. The spread of AMR within communities can lead to increased rates of antibiotic-resistant infections, particularly among vulnerable populations such as children, the elderly, and immunocompromised individuals. The economic burden of AMR on communities is also significant, as the treatment of antibiotic-resistant infections often requires more expensive and prolonged medical care. This can strain healthcare systems and exacerbate health inequalities, particularly in LMICs.

11.5.3 Ecological Impacts on Soil and Aquatic Ecosystems

11.5.3.1 Soil Ecosystems

The introduction of ARB and ARGs into soil ecosystems through wastewater irrigation poses significant ecological risks, disrupting microbial communities and essential ecosystem functions. Soil microbial diversity is particularly vulnerable to these changes, as the presence of ARB and ARGs can alter the composition of microbial populations. Beneficial microorganisms, which play critical roles in maintaining soil health, may decline in abundance, while resistant strains proliferate. This shift can reduce the overall functional diversity of soil microbes, compromising their ability to perform vital ecological processes (Cycoń et al., 2019). Nitrogen fixation, a process essential for converting atmospheric nitrogen into forms usable by plants, can be negatively impacted, as can the decomposition of organic matter, which is crucial for nutrient recycling and soil fertility. These disruptions in nutrient cycling can lead to reduced agricultural productivity and long-term soil degradation, threatening food security and environmental sustainability.

Moreover, ARGs can persist in soil environments for extended periods, even in the absence of selective pressure from antibiotics. This persistence creates a long-term reservoir of AMR, which can be transferred to other bacteria through HGT. HGT facilitates the spread of resistance traits across diverse microbial species, further amplifying the ecological risks. The soil ecosystem, therefore, becomes a potential hotspot for the evolution and dissemination of AMR, with implications for human and animal health. Additionally, the leaching of ARGs and ARB from soil into adjacent aquatic ecosystems can exacerbate the problem, spreading resistance to waterborne microbes and disrupting aquatic biodiversity (Jadeja and Worrich, 2022). The cascading effects of AMR on soil and aquatic ecosystems highlight the urgent need for sustainable wastewater management practices and strategies to mitigate the ecological impacts of AMR, ensuring the preservation of ecosystem functions and long-term environmental health.

11.5.3.2 Aquatic Ecosystems

The contamination of surface water and groundwater with ARB and ARGs through wastewater irrigation poses significant threats to aquatic ecosystems, disrupting microbial communities and ecosystem functions. Aquatic microbial communities play a critical role in maintaining water quality and ecosystem health by driving processes such as nutrient cycling, organic matter decomposition, and pollutant degradation (Anand et al., 2021). The introduction of ARB and ARGs can alter the composition and functionality of these microbial communities, leading to imbalances in ecosystem dynamics. The proliferation of resistant strains can outcompete native microbial species, reducing biodiversity and impairing essential ecological processes. This disruption can have cascading effects on aquatic food webs, as microbial communities form the foundation of these ecosystems, supporting higher trophic levels, including fish, invertebrates, and other aquatic organisms. One of the most concerning ecological impacts is the spread of AMR to aquatic organisms. ARB and ARGs can be transferred to fish, shellfish, and other aquatic species through contaminated water, leading to the bioaccumulation and biomagnification of resistance traits within aquatic food chains. This not only threatens aquatic biodiversity but also poses risks to human health, as the consumption of contaminated seafood can facilitate the transmission of resistant pathogens to humans (Eapen et al., 2024). Additionally, the presence of ARB and ARGs in water bodies can compromise water quality, rendering it unsafe for drinking, recreation, and agricultural use. Contaminated water sources can serve as reservoirs for AMR, enabling the persistence and dissemination of resistance genes across different environments.

11.6 STRATEGIES TO MITIGATE AMR RISKS IN WASTEWATER IRRIGATION

The use of wastewater in agriculture, while addressing water scarcity and supporting food security, poses significant risks related to AMR. To mitigate these risks, a multifaceted approach is required, encompassing improvements in wastewater treatment infrastructure, robust policy and regulatory frameworks, effective monitoring and surveillance systems, and public awareness and education campaigns. This section delves into these strategies in detail, providing a comprehensive roadmap for reducing the risks of AMR in wastewater-irrigated agriculture.

11.6.1 Improving Wastewater Treatment Infrastructure and Technologies

Upgrading conventional wastewater treatment systems is a critical first step in mitigating AMR risks. Advanced treatment technologies, such as membrane filtration (UF and RO) and AOPs (ozonation, photocatalysis, and Fenton reactions), offer higher removal efficiencies for ARB, ARGs, and antimicrobial residues. These technologies are particularly useful for producing high-quality treated wastewater suitable for irrigation. Constructed wetlands, a nature-based solution, use plants and microorganisms to remove contaminants from wastewater. While less efficient than advanced technologies, constructed wetlands are cost-effective and environmentally friendly, making them suitable for low-resource settings. Decentralized treatment systems, including small-scale treatment plants and household-level treatment units, can provide a viable alternative in regions lacking centralized infrastructure (Kalbar and Lokhande, 2023). These systems can incorporate advanced technologies, such as membrane bioreactors (MBRs) and solar-powered disinfection, to enhance treatment efficiency. Continued research and innovation are essential to develop cost-effective and scalable wastewater treatment solutions. Exploring emerging technologies, such as nanotechnology and bioelectrochemical systems (BES) and optimizing existing technologies through process integration can further enhance treatment efficiency and affordability (Prasad and Thirugnanasanbandham, 2019).

11.6.2 Monitoring and Surveillance of AMR in Wastewater and Agricultural Systems

Regular monitoring of wastewater, soil, and crops is essential to assess and mitigate the risks of AMR. Monitoring programs should include sampling and analysis of wastewater, soil, and crops for the presence of ARB, ARGs, and antimicrobial residues, as well as standardized protocols for data collection and reporting to ensure consistency and comparability of results. Surveillance systems can provide early warning of AMR risks and support evidence-based decision-making. These systems should track trends in the prevalence and distribution of ARB and ARGs in wastewater and agricultural systems and identify regions and sectors with high levels of AMR for targeted interventions. Integrating AMR monitoring with existing environmental and public health surveillance systems can enhance efficiency and reduce costs. For example, wastewater surveillance can be linked to national AMR action plans and One Health initiatives. Effective monitoring and surveillance systems are essential for identifying AMR hotspots, assessing the effectiveness of interventions, and informing policy and regulatory decisions.

11.6.3 Public Awareness and Education Campaigns

Raising awareness among farmers about the health risks associated with untreated wastewater and the importance of safe irrigation practices is critical. Farmers should be educated about the risks of direct contact with contaminated soil, water, and crops, and trained on best practices for wastewater reuse, such as drip irrigation and crop selection. Engaging consumers in the safe reuse of wastewater is equally important. Consumers play a critical role in driving demand for safe agricultural products. Awareness campaigns should promote safe food choices, encouraging consumers to choose produce from farms that use treated wastewater or alternative water sources, and highlight the health risks of consuming crops irrigated with untreated wastewater. Effective public awareness campaigns require collaboration among governments, NGOs, academia, and the private sector. Workshops and training sessions for farmers, consumers, and policymakers, as well as media campaigns using mass media, social media, and community outreach, can disseminate information on AMR risks and safe wastewater reuse. Public awareness and education campaigns are essential for fostering a culture of safety and responsibility in the use of wastewater in agriculture.

11.6.4 Policy and Regulatory Frameworks for Safe Wastewater Reuse

Establishing clear guidelines for the safe reuse of wastewater in agriculture is critical to mitigating AMR risks. National and international guidelines should include quality standards for ARB, ARGs, and antimicrobial residues in treated wastewater, specify the types of treatment processes required, and mandate regular monitoring and reporting of wastewater quality. International organizations, such as the World Health Organization (WHO) and the Food and Agriculture Organization, can play a key role in developing and promoting these guidelines. Effective enforcement of regulations is essential to ensure compliance with wastewater reuse standards. This requires capacity building for regulatory authorities and wastewater treatment operators, as well as imposing penalties for noncompliance to deter unsafe practices. Governments can encourage the adoption of safe wastewater reuse practices by providing incentives, such as subsidies and grants for the construction and operation of wastewater treatment facilities, and tax breaks for industries and farmers that adopt safe practices. Robust policy and regulatory frameworks, combined with effective enforcement and incentives, can create an enabling environment for the safe reuse of wastewater in agriculture (Srivastava et al., 2024).

11.6.5 Innovative Approaches to Reducing AMR in Wastewater Irrigation

11.6.5.1 Constructed Wetlands

Constructed wetlands have emerged as a cost-effective and environmentally friendly solution for wastewater treatment, particularly in low-resource settings. These systems use plants and microorganisms to remove contaminants, including ARB and ARGs, from wastewater. A study in China demonstrated that constructed wetlands could achieve significant reductions in ARB and ARGs, making treated water safer for irrigation. The integration of constructed wetlands with conventional treatment processes can further enhance their effectiveness. This approach is particularly suitable for rural and peri-urban areas where advanced treatment infrastructure is lacking.

11.6.5.2 Solar-Powered Disinfection

Solar-powered disinfection (SODIS) is an innovative approach that uses sunlight to inactivate pathogens in wastewater. This method is particularly effective in regions with high solar radiation, such as Sub-Saharan Africa and South Asia. A pilot project in Kenya demonstrated that SODIS could significantly reduce the concentration of ARB in wastewater used for irrigation. The low cost and simplicity of SODIS make it an attractive option for smallholder farmers. However, its effectiveness in removing ARGs and antimicrobial residues requires further research.

11.6.5.3 Biochar Filtration

Biochar, a carbon-rich material produced from organic waste, has been used as a soil amendment and water filtration medium. Studies have shown that biochar can adsorb antibiotics and heavy metals, reducing their concentration in wastewater. Additionally, biochar can improve soil health by enhancing microbial diversity and nutrient cycling. A project in India demonstrated that biochar filtration could reduce the levels of ARB and ARGs in wastewater used for irrigation. The use of biochar in wastewater treatment and soil management offers a dual benefit of reducing AMR risks and improving agricultural productivity.

11.6.5.4 Nanotechnology for AMR Mitigation in Wastewater

Nanotechnology holds significant promise for the removal of ARB and ARGs from wastewater. Nanoparticles, such as silver and titanium dioxide, have antimicrobial properties that can inactivate ARB and degrade ARGs. Silver nanoparticles disrupt bacterial cell membranes and inhibit the replication of ARGs, while titanium dioxide nanoparticles generate reactive oxygen species under UV light, degrading organic pollutants, including antibiotics (Noor et al., 2021). However, the potential environmental and health impacts of nanoparticles require further research to ensure their safe and sustainable use. BES, such as microbial fuel cells and microbial electrolysis cells, use microorganisms to generate electricity or produce hydrogen while treating wastewater. These systems can degrade organic pollutants, including antibiotics, and reduce the concentration of ARB and ARGs. BES also offer the potential for resource recovery, such as energy and nutrients, making them a sustainable option for wastewater treatment. Research is needed to optimize the performance of BES and scale up their application in real-world settings. AOPs, such as photocatalysis, Fenton reactions, and ozonation, use reactive oxygen species to degrade organic pollutants, including antibiotics and ARGs. Emerging AOPs, such as solar-driven photocatalysis and electro-Fenton processes, offer enhanced efficiency and lower energy consumption. Solar-driven photocatalysis uses sunlight to activate catalysts, such as titanium dioxide, for the degradation of pollutants. These technologies are particularly suitable for regions with high solar radiation. Further research is needed to optimize the design and operation of AOPs for large-scale wastewater treatment. MBRs combine biological treatment with membrane filtration to achieve high removal efficiencies for ARB, ARGs, and antimicrobial residues. MBRs are effective in producing high-quality treated wastewater suitable for irrigation (Ahmed, et al., 2022). However, the high cost and energy requirements of MBRs limit their widespread adoption. Research is needed to develop cost-effective and energy-efficient MBR systems, as well as to explore the use of novel membrane materials with enhanced performance. Biochar and activated carbon are effective adsorbents for removing antibiotics and heavy metals from wastewater. These materials can also improve soil health and microbial diversity when used as soil amendments. Research is needed to optimize the production and application of biochar and activated carbon for wastewater treatment and soil management. Additionally, the potential of these materials for reducing ARB and ARGs in wastewater and soil requires further investigation.

11.7 CASE STUDIES: SUCCESS STORIES AND LESSONS LEARNED

The global challenge of AMR in wastewater-irrigated agriculture has prompted several regions to transition from untreated to treated wastewater, adopting innovative approaches to mitigate AMR risks (Bekele et al., 2025; Hounmanou et al., 2025). These case studies highlight successful strategies, lessons learned, and the potential for scaling up these efforts to other regions facing similar challenges. Israel is a global leader in wastewater reuse, with over 90% of its treated wastewater being recycled for agricultural irrigation. The country faced severe water scarcity, prompting the development of advanced wastewater treatment infrastructure. The Shafdan wastewater treatment plant, one of the largest in the Middle East, treats wastewater from the Tel Aviv metropolitan area using a combination of biological treatment, membrane filtration, and disinfection. The treated wastewater, known as "reclaimed water," meets stringent quality standards and is used to irrigate crops in the Negev Desert. Israel's success in wastewater reuse has significantly reduced its reliance on freshwater resources and minimized the risks of AMR in agriculture. Key lessons from Israel include the importance of government commitment, investment in advanced treatment technologies, and strict regulatory frameworks (Ben-Mordechay et al., 2024). Jordan, one of the most water-scarce countries in the world, has embraced treated wastewater as a key resource for agriculture. The As-Samra wastewater treatment plant, which serves the capital city of Amman, treats wastewater using activated sludge and constructed wetlands. The treated water is used to irrigate crops in the Jordan Valley, a major agricultural region. Jordan's approach demonstrates the potential of nature-based solutions, such as constructed wetlands, in low-resource settings. The country has also implemented public awareness campaigns to educate farmers and consumers about the benefits and safety of treated wastewater. Jordan's experience highlights the importance of integrating traditional and advanced treatment methods, as well as engaging local communities in wastewater reuse initiatives (Alshawabka, 2019). California has been at the forefront of wastewater reuse in the United States, driven by recurring droughts and water shortages. The Orange County Water District's Groundwater Replenishment System is a flagship project that treats wastewater using microfiltration, reverse osmosis, and UV disinfection. The treated water is used to recharge groundwater aquifers, which are then used for irrigation. California's stringent regulatory framework ensures that treated wastewater meets high-quality standards, minimizing the risks of AMR. The state has also invested in research and innovation to develop cost-effective treatment technologies. California's success underscores the importance of regulatory oversight, technological innovation, and public–private partnerships in promoting safe wastewater reuse (Duong and Saphores, 2015).

11.8 CONCLUSIONS AND PROSPECTS

Addressing AMR in wastewater-irrigated agriculture requires continuous innovation, interdisciplinary collaboration, and global cooperation. As AMR challenges evolve, mitigation strategies must also advance. An interdisciplinary approach integrating microbiology, engineering, and environmental science is crucial, where microbiologists study AMR mechanisms, engineers develop treatment technologies, and environmental scientists assess ecological impacts. Social and economic aspects are equally important, as social scientists analyze behavioral factors influencing wastewater reuse, and economists evaluate the cost-effectiveness of interventions. Emerging technologies, including AI and machine learning, enhance AMR monitoring and mitigation by predicting the spread of ARB and optimizing wastewater treatment efficiency. While advanced treatment technologies such as membrane filtration and oxidation processes reduce AMR risks, affordability and complete ARG removal remain challenges, particularly in low-resource settings. Scalable solutions like constructed wetlands and solar-powered disinfection show promise. Global collaboration, standardized protocols, and evidence-based policies are essential to ensuring safe wastewater reuse, protecting public health, and strengthening agricultural resilience

ACKNOWLEDGEMENT

Nisha Singh is thankful to University of Lucknow, Lucknow for providing technical support.

CONFLICT OF INTEREST

None.

REFERENCES

Adegoke, A.A., Amoah, I.D., Stenström, T.A., Verbyla, M.E. and Mihelcic, J.R., 2018. Epidemiological evidence and health risks associated with agricultural reuse of partially treated and untreated wastewater: a review. *Frontiers in Public Health*, 6: 337.

Ahmed, Y., Zhong, J., Yuan, Z. and Guo, J., 2022. Roles of reactive oxygen species in antibiotic resistant bacteria inactivation and micropollutant degradation in Fenton and photo-Fenton processes. *Journal of Hazardous Materials*, 430: 128408.

Alam, M.U., Ferdous, S., Ercumen, A., Lin, A., Kamal, A., Luies, S.K., Sharior, F., Khan, R., Rahman, M.Z., Parvez, S.M. and Amin, N., 2021. Effective treatment strategies for the removal of antibiotic-resistant bacteria, antibiotic-resistance genes, and antibiotic residues in the effluent from wastewater treatment plants receiving municipal, hospital, and domestic wastewater: protocol for a systematic review. *JMIR Research Protocols*, 10 (11): e33365.

Alcamo, J., Henrichs, T. and Rosch, T., 2000. World Water in 2025. *World Water Series Report*, 2.

Alshawabka, Z.A., 2019. *Water Availability and Quality in the Jordan River Valley and the Zarqa River Basin: Stakeholders' Perspectives* (Doctoral dissertation, The University of Waikato).

Anand, U., Reddy, B., Singh, V.K., Singh, A.K., Kesari, K.K., Tripathi, P., Kumar, P., Tripathi, V. and Simal-Gandara, J., 2021. Potential environmental and human health risks caused by antibiotic-resistant bacteria (ARB), antibiotic resistance genes (ARGs) and emerging contaminants (ECs) from municipal solid waste (MSW) landfill. *Antibiotics*, 10 (4): 374.

Bangre, J., Sinha, N.K., Mohanty, M., Jayaraman, S., Sahoo, R.N., Dwivedi, A.K., Singh, D., Mishra, A., Kumar, S., Wanjari, R. and Jha, P., 2024. Long-term impact of inorganic fertilizers and farmyard manure on soil quality and productivity in subtropical vertisols under a soybean-wheat cropping system. *Land Degradation & Development*, 35 (14): 4257–4270.

Bekele, E.T., Sintayehu, Y.D., Ravikumar, C.R. and Murthy, H.A., 2025. Green titanium oxide nanoparticles for nanoremediation. In Fernanda, M.P.T., Arpita, R. and Khalid, R.H. (Eds) *Sustainable Nanoremediation* (pp. 515–541). Apple Academic Press, Palm Bay, FL.

Ben-Mordechay, E., Abdeen, Z., Robeen, S., Schwartz, S., Abdeen, A.M., Mordehay, V., Troen, A.M., Chefetz, B. and Tal, A., 2024. Regional water and food security require joint Israeli-Palestinian guidelines for wastewater reuse and food safety. *Food and Nutrition Bulletin*, 45 (4): 113–124.

Bougnom, B.P., Thiele-Bruhn, S., Ricci, V., Zongo, C. and Piddock, L.J.V., 2020. Raw wastewater irrigation for urban agriculture in three African cities increases the abundance of transferable antibiotic resistance genes in soil, including those encoding extended spectrum β-lactamases (ESBLs). *Science of the Total Environment*, 698: 134201.

Brovini, E.M., Dias, M., Teodoro, A., Machado, E., Santos, L., Leal, C., de Aquino, S.F. and de Araújo, J.C., 2025. Effects of UVC doses on the removal of antimicrobial resistance elements from secondary treated sewage. *Environmental Science and Pollution Research*, 32 (7): 1–12.

Chávez-Mejía, A.C., Navarro-González, I., Magaña-López, R., Uscanga-Roldán, D., Zaragoza-Sánchez, P.I. and Jiménez-Cisneros, B.E., 2019. Presence and natural treatment of organic micropollutants and their risks after 100 years of incidental water reuse in agricultural irrigation. *Water*, 11 (10): 2148.

Clemente, E., Domingues, E., Quinta-Ferreira, R.M., Leitão, A. and Martins, R.C., 2024. European and African landfilling practices: an overview on MSW management, leachate characterization and treatment technologies. *Journal of Water Process Engineering*, 66: 105931.

Cycoń, M., Mrozik, A. and Piotrowska-Seget, Z., 2019. Antibiotics in the soil environment—degradation and their impact on microbial activity and diversity. *Frontiers in Microbiology*, 10: 338.

Desai, S., Hasan, M., Sonawane, A., Singh, D., Satpute, S. and Yadav, D., 2023. Soil water and potassium dynamics simulation under drip fertigated Kinnow mandarin in sandy loam soils. *Indian Journal of Soil Conservation*, 51 (3): 219–227.

Duong, K. and Saphores, J.D.M., 2015. Obstacles to wastewater reuse: an overview. *Wiley Interdisciplinary Reviews: Water*, 2 (3): 199–214.

Eapen, J.V., Thomas, S., Antony, S., George, P. and Antony, J., 2024. A review of the effects of pharmaceutical pollutants on humans and aquatic ecosystem. *Exploration of Drug Science*, 2 (5): 484–507.

El-Rawy, M., Fathi, H., Wahba, M. and Negm, A., 2024. An overview of water resource challenges and managed aquifer recharge (MAR) in the MENA region. In El-Rawy, M. and Negm, A. (Eds) *Managed Aquifer Recharge in MENA Countries: Developments, Applications, Challenges, Strategies, and Sustainability* (pp. 1–27). Springer, Cham.

Graham, D.W., Mateo-Sagasta, J., Haile, A.T., Moodley, A., Goshu, G. and Kibret, M., 2024. *Reducing the Emergence and Spread of Waterborne Antimicrobial Resistance (AMR) in Ethiopia from a One Health Perspective*. International Water Management Institute (IWMI), Colombo, Sri Lanka.

Grobelak, A., Całus-Makowska, K., Jasińska, A., Klimasz, M., Wypart-Pawul, A., Augustajtys, D., Baor, E., Sławczyk, D. and Kowalska, A., 2024. Environmental impacts and contaminants management in sewage sludge-to-energy and fertilizer technologies: current trends and future directions. *Energies*, 17 (19): 4983.

Gupta, A.K., Kumar, A., Maurya, U.K., Singh, D., Islam, S., Rathore, A.C., Kumar, P., Singh, R. and Madhu, M., 2023. Comprehensive spatio-temporal benchmarking of surface water quality of Hindon River, a tributary of river Yamuna, India: adopting multivariate statistical approach. *Environmental Science and Pollution Research*, 30 (55): 116804–116830.

Hanna, N., Tamhankar, A.J. and Lundborg, C.S., 2023. Antibiotic concentrations and antibiotic resistance in aquatic environments of the WHO Western Pacific and South-East Asia regions: a systematic review and probabilistic environmental hazard assessment. *The Lancet Planetary Health*, 7 (1): e45–e54.

Hegde, M., Raj, S., Tikadar, D. and Nyamagoud, S.B., 2025. Addressing antimicrobial resistance in India: a 2 stewardship roadmap. *Journal of Health Science and Medical Research*, 43 (2): 20241071.
Hosain, M.Z., Kabir, S.L. and Kamal, M.M., 2021. Antimicrobial uses for livestock production in developing countries. *Veterinary World*, 14 (1): 210.
Hounmanou, Y.M.G., Houefonde, A., I-CRECT Consortium, Nguyen, T.T. and Dalsgaard, A., 2025. Mitigating antimicrobial resistance through effective hospital wastewater management in low-and middle-income countries. *Frontiers in Public Health*, 13: 1525873.
Jadeja, N.B. and Worrich, A., 2022. From gut to mud: dissemination of antimicrobial resistance between animal and agricultural niches. *Environmental Microbiology*, 24 (8): 3290–3306.
Kalbar, P.P. and Lokhande, S., 2023. Need to adopt scaled decentralized systems in the water infrastructure to achieve sustainability and build resilience. *Water Policy*, 25 (4): 359–378.
Khan, M.M., Siddiqi, S.A., Farooque, A.A., Iqbal, Q., Shahid, S.A., Akram, M.T., Rahman, S., Al-Busaidi, W. and Khan, I., 2022. Towards sustainable application of wastewater in agriculture: a review on reusability and risk assessment. *Agronomy*, 12 (6): 1397.
Lofrano, G. and Brown, J., 2017. Wastewater management in. *Wastewater and Biosolids Management*, 12 (6): 1.
Michaelis, C. and Grohmann, E., 2023. Horizontal gene transfer of antibiotic resistance genes in biofilms. *Antibiotics*, 12 (2): 328.
Naghavi, M., Vollset, S.E., Ikuta, K.S., Swetschinski, L.R., Gray, A.P., Wool, E.E., Aguilar, G.R., Mestrovic, T., Smith, G., Han, C. and Hsu, R.L., 2024. Global burden of bacterial antimicrobial resistance 1990–2021: a systematic analysis with forecasts to 2050. *The Lancet*, 404 (10459): 1199–1226.
Noor, Z.Z., Rabiu, Z., Sani, M.H.M., Samad, A.F.A., Kamaroddin, M.F.A., Perez, M.F., Dib, J.R., Fatima, H., Sinha, R., Khare, S.K. and Zakaria, Z.A., 2021. A review of bacterial antibiotic resistance genes and their removal strategies from wastewater. *Current Pollution Reports*, 404: 1–16.
Pant, A., Dwivedi, A.K., Murasingh, S., Singh, D., Mayank, M. and Ojha, C.S.P., 2024. Treatment of pulp and paper mill effluent through combined aerobic and anaerobic suspended fixed-bed bioreactor. *Water Environment Research*, 96 (9): e11133.
Patra, M. and Dubey, S.K., 2024. Understanding the spread of antibiotic resistance in vegetables cultivated with sewage sludge: implications for food safety and human health. *Environmental Systems Research*, 13 (1): 21.
Prasad, R., Thirugnanasanbandham, K. 2019. Advances Research on Nanotechnology for Water Technology. Springer International Publishing. https://www.springer.com/us/book/9783030023805
Ranjit, P., Jhansi, V. and Reddy, K.V., 2021. Conventional wastewater treatment processes. In Maddela, N.R., García Cruzatty, L.C. and Chakraborty, S. (Eds) *Advances in the Domain of Environmental Biotechnology: Microbiological Developments in Industries, Wastewater Treatment and Agriculture* (pp. 455–479). Springer Singapore, Singapore.
Singh, D. and Goyal, V.C., 2018. Planning of agricultural inputs in Ur watershed to maximize net benefit under limited resources. *Indian Journal of Agricultural Sciences*, 88 (2): 326–332.
Singh, D., Mishra, A.K., Patra, S., Dwivedi, A.K., Ojha, C.S.P., Singh, V.P., Mariappan, S., Babu, S., Singh, N., Yadav, D. and Ojasvi, P.R., 2023. Effect of long-term tillage practices on runoff and soil erosion in sloping croplands of Himalaya, India. *Sustainability*, 15 (10): 8285.
Singh, D., Patel, N., Gadedjisso-Tossou, A., Patra, S., Singh, N. and Singh, P.K. 2020a. Incidence of *Escherichia coli* in vegetable crops and soil profile drip irrigated with primarily treated municipal wastewater in a semi-arid peri urban area. *Agriculture*, 10 (7): 291.
Singh, D., Patel, N., Patra, S. and Singh, N., 2020b. Growth and yield of cauliflower under surface and subsurface drip irrigation with primarily treated municipal wastewater in a semi-arid peri-urban area. *Current Science*, 119 (8): 1357–1363.
Singh, D., Patel, N., Patra, S., Singh, N., Roy, T., Caucci, S. and Hettiarachchi, H. 2021a. Efficacy of drip irrigation in controlling heavy-metal accumulation in soil and crop. *Journal of Environmental Engineering and Science*, 16 (3): 109–121.
Singh, D., Patel, N., Singh, N., Rosin, K.G., Yadav, D., Singh, S., Gupta, A.K., Kumawat, A., Sharma, V.K. and Kumar, M., 2021b. Emitter clogging and hydraulic performance of drip system under different water qualities and placement techniques. *Food Science and Engineering*, 2 (2): 112–125.
Srivastava, N., Maurya, V.K., Bee, Z., Singh, N., Khare, S., Singh, S., Agarwal, N. and Rai, P.K., 2024. Green Solutions for a blue planet: harnessing bioremediation for sustainable development and circular economies. In Srivastav, A.L., Zinicovscaia, I. and Cepoi, L. (Eds) *Biotechnologies for Wastewater Treatment and Resource Recovery* (pp. 283–296). Elsevier, Amsterdam, Netherlands.

Sun, Z., Hong, W., Xue, C. and Dong, N., 2024. A comprehensive review of antibiotic resistance gene contamination in agriculture: challenges and AI-driven solutions. *Science of The Total Environment*, 953: 175971.

Tella, T.A., Festus, B., Olaoluwa, T.D. and Oladapo, A.S., 2025. Water and wastewater treatment in developed and developing countries: present experience and future plans. In Olusola, O.A., Azeez, O.I., Sadanand, P. and Peter, A.O. (Eds) *Smart Nanomaterials for Environmental Applications* (pp. 351–385). Elsevier, Amsterdam, Netherlands.

Tucker, K., 2021. *Antibiotic Resistance in Surface Waters and Biofilm-Response to Environmental Contaminants* (Doctoral dissertation, Stellenbosch University, Stellenbosch).

Verma, A., Gupta, A. and Rajamani, P., 2023. Application of wastewater in agriculture: benefits and detriments. In Rai, P.K. (Eds) *River Conservation and Water Resource Management* (pp. 53–75). Springer Nature Singapore, Singapore.

Wang, D., Zhou, X., Fu, Q., Li, Y., Ni, B.J. and Liu, X., 2025. Understanding bacterial ecology to combat antibiotic resistance dissemination. *Trends in Biotechnology* 43 (7): 1566–1582.

Wang, J., Xu, S., Zhao, K., Song, G., Zhao, S. and Liu, R., 2023. Risk control of antibiotics, antibiotic resistance genes (ARGs) and antibiotic resistant bacteria (ARB) during sewage sludge treatment and disposal: a review. *Science of the Total Environment*, 877: 162772.

Wu, X., Nawaz, S., Li, Y. and Zhang, H., 2024. Environmental health hazards of untreated livestock wastewater: potential risks and future perspectives. *Environmental Science and Pollution Research*, 31 (17): 24745–24767.

12 Antimicrobial Resistance in Wastewater Bacteria

Sandhya, Amar P. Garg, and Khushboo

12.1 INTRODUCTION

Bacterial resistance to antibiotics poses a significant public health issue globally. An estimated 700,000 people perish each year as a result of diseases spurred on by bacterial strains that are resistant to antibiotics (Duarte et al. 2022). It is forecasted that this figure will rise to 10 million annual deaths by the year 2050 if thorough and efficient preventive measures are not put in place globally.

The development of resistance to antibiotics in bacterial populations, a consequence of evolutionary adaptation through natural selection, has been officially recognized as a global health crisis by WHO (Duarte et al. 2022). In recent times, bacterial antibiotic resistance has become a major public health issue that affects both infection prevention and the effectiveness of treatment. The selection and spread of antibiotic-resistant bacteria (ARB) and the spread of antibiotic-resistant genes (ARGs) have been greatly aided by the persistence and build-up of environmental antibiotics, particularly in wastewater sources.

This increasing prevalence of antimicrobials and their ineffective removal using traditional disposal techniques are the main causes of this issue. Furthermore, the emergence of antimicrobial resistance has been exacerbated by the poor effectiveness of conventional water treatment methods in completely degrading drugs and eliminating resistant microorganisms, especially ARGs (Duarte et al. 2022). The environmental distribution of antibiotic-resistance bacteria (ARB) and antibiotic resistance genes (ARGs) is influenced by multiple factors, including the indiscriminate usage of antibiotics in humans and veterinary medicine, improper disposal of expired antibiotics, escalating discharge of pharmaceutical industrial effluents, and the insufficient efficacy of wastewater and drinking water treatment systems in removing these contaminants. Notably, merely reducing antibiotic consumption is insufficient to mitigate the persistence and spread of ARGs and ARB in aquatic environments, particularly in drinking water sources. Comprehensive strategies addressing both antibiotic usage and environmental contamination are necessary to control the propagation of antimicrobial resistance (AMR) (Amarasiri et al. 2020). The development and widespread antibiotic resistance among bacterial populations poses a global threat to humans, animals, and environmental health. Extensive research has highlighted the critical role of environmental matrices, such as water and soil, in the persistence and transmission of antibiotic resistance.

Wastewater is a microbial ecosystem that can serve as a reservoir or amplifier of resistant determinants that human pathogens have already acquired and subsequently released into the environment as contaminants, a source of resistance genes, or a dynamic bioreactor that promotes the horizontal gene transfer (HGT) of resistant elements among pathogenic and nonpathogenic bacterial populations. The precise circumstances under which waterborne bacteria contribute to the development of novel antibiotic resistance mechanisms are still unknown, despite a great deal of recent research in this area. It is also unclear how they either directly serve as reservoirs for AMR or indirectly aid in its spread through intricate ecological interactions. Aquatic ecosystems play a critical role in sustaining high levels of biodiversity, supporting livelihoods, and maintaining the overall productivity of the biosphere (Hassan et al. 2020). The occurrence of antimicrobials (e.g., antibiotics), Antibiotic-resistant genes (ARGs), and Antibiotic-resistant bacteria (ARB) in aquatic environments has emerged as a significant concern due to their potential role in the dissemination

DOI: 10.1201/9781003269823-12

and widely spread of antimicrobial drugs, including multidrug-resistance "superbugs." The continuous release of these contaminants into water bodies through anthropogenic activities fosters selective pressure, promoting the persistence and HGT of resistance determinants among microbial communities. This phenomenon poses substantial risks to environmental health and human well-being by compromising the efficacy of antimicrobial therapies and increasing the likelihood of untreatable infections (Zhuang et al. 2021). Every antibiotic used eventually finds its way into the environment, accelerating the development of AMR. Antibiotic resistance has been widely recorded from wastewater systems and freshwater sources worldwide, including but not limited to pharmaceutical companies and wastewater treatment plants (WWTPs) (Praveen Kumar Reddy et al. 2020). The escalation of AMR is occurring at an alarming rate, posing an increasingly critical threat to public health globally and therapeutic effectiveness. Pathogenic microorganisms responsible for common infectious diseases are rapidly acquiring and exhibiting varying levels of resistance to newly developed antibiotics, often within less than 5–10 years of their market introduction. AMR is a widespread phenomenon observed across diverse microbial taxa, including bacteria, fungi, viruses, and parasites, which undergo adaptive evolutionary mechanisms to survive and proliferate in the presence of antimicrobial agents. This growing resistance threatens the effectiveness of existing therapeutics and posing a significant challenge to public health globally (Singh et al. 2022b). Significant evidence is produced by this study that reinforces recommendations for the prudent use of antibiotics and to provide practical preventative measures meant to lessen the AMR threat. This study supports a comprehensive, One Health approach to combating AMR and promoting sustainable antimicrobial stewardship (AMS) by addressing the difficulties in treating infectious diseases across multiple sectors, including humans, animals, and environmental health (Jofre et al. 2023).

12.2 CHARACTERISTICS OF WASTEWATER

Wastewater is a by-product of many company operations and frequently contains contaminants that need to be treated before being disposed of or used again. Whereas the food processing business releases organic matter, oils, and cleaning agents, the textile industry generates effluent that is loaded with chemicals, dyes, and heavy metals. While the oil and gas business creates wastewater tainted with hydrocarbons, heavy metals, and saline water from drilling operations, the chemical manufacturing sector produces toxic wastewater containing solvents and dangerous substances. While the mining business produces acid mine drainage with heavy metals and suspended sediments, the pharmaceutical industry emits wastewater that contains organic contaminants and active medicinal components. The high BOD wastewater from the paper and pulp industries, which contains bleaching chemicals and lignin, is well known. Wastewater containing heavy metals, cyanides, and acids is also released by the metal processing and electroplating industries. While the beverage business, which includes breweries and soft drink producers, produces wastewater containing sugars, alcohol, and organic compounds, the dairy industry discharges effluent that is high in organic matter, fats, and proteins. To reduce environmental contamination and guarantee sustainable water management, these companies must process their wastewater effectively (Qadir et al. 2020).

India is the second most populated country in the world, home to 1.38 billion people. Of the total population, 35% (483 million) reside in urban centers, while 65% (900 million) dwell in rural areas. The amount of wastewater generated is expected to be around 72368 MLD for the 2020–2021 year. Wastewater generation in urban centers is predicted to be 72,368 MLD, whereas 39,604 MLD is generated in rural areas. Over 50% of the global population already lives in metropolitan regions, and by 2050, it is predicted that 70% of people will call large centers home. Although rapid urbanization offers chances for economic growth, it can also increase strain on freshwater resources to satisfy the needs for food and water in regions with limited water supplies and those experiencing unheard-of growth rates. Finding water resources is crucial as the population of these cities grows to support the expansion (Prasad et al. 2024).

The main obstacle to encouraging reuse is making sure that wastewater is sufficiently cleansed before use or release into the environment, which is safe for ecosystem and human health. The necessary paradigm change in water resources management must be accomplished immediately, where wastewater is no longer seen as "waste" but rather as a resource (for water, nutrients, and energy).

Major rivers like the Yamuna and Ganga are severely impacted by this insufficient treatment capacity. For example, the Yamuna River gets more than 500 million liters of untreated industrial effluent per day, which causes excessive levels of contaminants such as hazardous compounds and heavy metals (Naznine et al. 2024). It was a major political issue in the recent Delhi elections held on February 5, 2025. Yamuna is contaminated with a large amount of industrial waste and sewage being discharged without adequate treatment. Similarly, industrial effluents, especially those from tanneries, chemical plants, and the textile industry, seriously pollute the Ganga river (Prasad et al. 2024).

The National River Conservation Plan and the Ganga Action Plan (GAP-I 1985, GAP-II 1993): Cleaning of the Ganga and its principal tributaries, the Yamuna and Gomti, was started under GAP-I and GAP-II. It was enlarged to include other rivers that fall under the National River 1995 Conservation Plan. 1098.31 ML/day (MLD) of sewage treatment capacity was developed under GAP (Dutta et al. 2020). Among the actions adopted to reduce pollution are as follows: (1) intercepting and rerouting sewerage lines for treatment; (2) establishing STPs; (3) providing inexpensive restrooms; (4) developing riverfronts; and (5) raising public knowledge and capability.

Untreated wastewater discharges into aquatic environments release harmful microbial communities, contaminating river ecosystems and posing significant risks to the health of humans, animals, and plants. A notable example of this environmental and public health concern is the presence of toxic foam observed in the Yamuna River. This foam results from the accumulation of industrial pollutants, exacerbating the deleterious effects on both the ecosystem and public health (Prasad et al. 2024) that became a hot topic of debate on TV in the early 2025 elections campaign in Delhi.

High degrees of turbidity, warmth, and odor are features of wastewater that contains ARB. Additionally, wastewater contains traces of pesticides and antibiotics that can encourage the growth of ARB. Wastewater contains various contaminants that aid in the dissemination and widespread of resistant bacteria, making it a major reservoir for AMR. It has been documented that during the past century, human activities fueled by industrialization and agricultural practices have produced chemical and biological contaminants that are released into water bodies. Toxic compounds with antibacterial properties are present in many of these contaminants. In particular, water sources have been shown to contain antibiotic families such as tetracyclines, penicillin, macrolides, quinolones, and aminoglycosides (La Rosa et al. 2025).

Before releasing the treated water into different water bodies like dams, streams, rivers, and seas, WWTPs were initially built to remove physical contaminants like debris, pathogens, and excessive loads of organics. More effective wastewater treatment systems were created to take into consideration the consequences of population development, industrial expansion, and new difficulties. Even though WWTP treatment techniques have significantly improved due to technological advancements, antimicrobial chemicals and other new pollutants cannot be completely removed. The primary goal of this treatment procedure was to keep these water bodies from becoming eutrophic (Sambaza and Naicker 2023). The analysis of wastewater properties identifies several variables affecting the spread of AMR:

12.2.1 Microbial Load

The term "microbial load" describes the entire population of microorganisms in wastewater, including both dangerous pathogens and benign bacteria. Due to the presence of various pollutants, this environment can function as a reservoir for microorganisms exhibiting AMR. The selective pressure imposed by these pollutants favors the survival and proliferation of the most resistant strains, thereby facilitating the development and persistence of resistant microbial populations. It includes bacteria, viruses, fungi, and protozoa found in wastewater. Wastewater serves as a habitat and transmission medium for various microorganisms, including potentially hazardous diseases and ARB. Understanding the microbiological load in wastewater is crucial for determining public health hazards, environmental implications, and the efficacy of wastewater treatment methods. Numerous microbial communities, including both pathogenic and nonpathogenic microorganisms, can be found in wastewater. Pollutant-induced selection acts as a breeding habitat for resistant bacteria (Ryu et al. 2021).

The untreated or insufficiently treated wastewater acts as a breeding ground for resistant microbial strains, AMR in wastewater bacteria represents a serious danger to environmental stability and public health. High quantities of disinfectants, heavy metals, and antibiotics found in wastewater from home, industrial, hospital, and pharmaceutical sources create selective pressure that promotes the evolution of ARB and ARGs. Complex microbial communities that can withstand harsh environments are created due to the persistent introduction of these resistant microorganisms into natural water bodies, which raises the microbial load. These bacteria aid in HGT, which spreads resistance determinants to pathogenic species, in addition to competing with native microbial populations (Qadir et al. 2020).

12.2.1.1 Fecal Matter from both Humans and Animals

Wastewater has significant microbiological loads from human and animal excrement, which contribute to bacterial, viral, and protozoan contamination. Microbes such as *Escherichia coli*, *Salmonella* species, *Shigella* species, and *Enterococcus* species are frequently detected in untreated wastewater (Bonetta et al. 2022).

12.2.1.2 Hospital and Clinical Waste

Hospitals discharge a large number of harmful germs and ARB into wastewater. This source is especially alarming given the high incidence of ARGs and mobile genetic elements (MGEs) (Ramírez-Coronel et al. 2023).

12.2.1.3 Runoff from Industrial and Agricultural Activities

Slaughterhouses, the food processing industry, and livestock farms influence microbial loads in wastewater. Antibiotics used in livestock production encourage the survival of ARB in wastewater (Ramírez-Coronel et al. 2023).

12.2.1.4 Storm Water Runoff

Rainwater attracts microorganisms from the soil, sewage overflows, and animal excrement, which increases the microbial load in the wastewater system.

12.2.2 Wastewater Microorganisms Include Bacteria

Common bacteria discovered in wastewater include *Escherichia coli*, *Salmonella* species, *Klebsiella* species, *Pseudomonas aeruginosa*, and *Enterococcus* species. Many of these bacteria are markers of fecal pollution and pose potential public health problems (Kushkevych 2021).

Viruses include enteric viruses such as Noroviruses, Enteroviruses, Rotaviruses, and Adenoviruses, which are commonly found in wastewater. These viruses can survive in water for long periods, posing a risk to public health.

Fungi, including *Candida albicans*, *Aspergillus* species, and *Cryptococcus* species, can be found in wastewater, especially from hospitals.

Protozoa and helminths include Protozoan pathogens such as *Giardia lamblia* and *Cryptosporidium parvum*, which are resistant to traditional wastewater treatment technologies. Helminth eggs, such as those from *Ascaris lumbricoides,* can live in wastewater sludge for a long time.

12.3 FACTORS INFLUENCING MICROBIAL LOAD IN WASTEWATER INCLUDE POPULATION DENSITY AND COMPOSITION

Increased microbiological loads are caused by increased sewage discharge in densely populated areas (Ryu et al. 2021).

12.3.1 Wastewater Treatment Efficiency

Conventional wastewater treatment lowers microbial burdens; however, certain bacteria survive and persist in the treated effluent.

12.3.2 Temperature and Environmental Conditions

Warm temperatures encourage microbial growth, whereas high pH levels and disinfection procedures can deplete microbial populations.

12.3.3 Implications for Public Health and Environment, Including Disease Transmission

Untreated or insufficiently treated wastewater can transmit diseases such as Cholera, Typhoid, Hepatitis A, and Gastroenteritis.

12.3.4 Antimicrobial Resistance Spreads

Wastewater containing antibiotic residues and resistant microorganisms can help propagate AMR in natural habitats.

12.3.5 Pollution of Water Bodies

Discharging untreated wastewater into rivers, lakes, and oceans contaminates drinking water sources and recreational bodies.

12.4 PRESENCE OF ANTIBIOTICS AND PHARMACEUTICALS

Companies that manufacture pharmaceuticals are major contributors to the antibiotic load in WWTPs worldwide. For instance, large levels of oxytetracycline, lincomycin, or fluoroquinolone antibiotics have been found in treated pharmaceutical effluent from Asian nations like China, India, and South Korea. Furthermore, manufacturing facilities may unintentionally release active pharmaceutical chemicals, which can lead to the evolution of drug-resistant pathogens and have worldwide repercussions. The presence of antimicrobial drugs and pharmaceutical compounds in the ecosystem has raised significant concerns due to their potential adverse effects on ecosystems and human health. The key points are outlined below (Felis et al. 2020). The proper management of nutrients in wastewater requires advanced treatment technologies such as biological nitrification-denitrification to remove nitrogen, enhanced biological phosphorus removal to reduce phosphorus, and chemical precipitation to eliminate heavy metals. Additionally, membrane filtration and adsorption techniques help remove organic matter and trace elements. These technologies work in tandem to reduce the nutrient load in wastewater, mitigate eutrophication, and restore ecological balance. Ultimately, addressing the nutrient contents in wastewater is crucial for ensuring sustainable water resources and maintaining the health of aquatic ecosystems and communities that rely on clean water (Huemer et al. 2020).

12.4.1 Sources of Antibiotics and Pharmaceuticals in Wastewater

12.4.1.1 Human Excretion

A significant number of medications consumed are excreted unmetabolized or as active metabolites through urine and feces, which enter municipal wastewater systems. Various medications have been discovered in WWTP effluents, showing that treatment techniques do not remove them completely (Parra-Saldivar et al. 2021).

12.4.1.2 Improper Disposal

Unused or expired prescriptions are frequently disposed of incorrectly, such as by flushing them down the toilet or throwing them away in the garbage, allowing them to enter wastewater streams. This technique adds to the pollution of water sources with pharmaceutical residues.

12.4.1.3 Agricultural Runoff

The usage of antibiotics in animal farming can cause excretion of these compounds, which can then leach into soil and water systems, contributing to environmental contamination.

12.4.1.4 Pharmaceutical Production Effluents

Effluents from pharmaceutical production plants can contain significant levels of active pharmaceutical compounds, particularly when waste management procedures are poor. These discharges can considerably increase local medication concentrations in aquatic bodies. The primary concern associated with antimicrobial discharge into the ecosystem is the emergence and proliferation of ARGs and ARB, which compromise the efficacy of antimicrobial therapies for humans and animal infections. Resistant microorganisms can develop and persist across diverse environmental reservoirs and ecosystems (Omuferen et al. 2022).

12.4.1.5 Heavy Metal Contaminants

One well-known environmental contaminant is heavy metal ions. Heavy metal ions have been found in wastewater in several recent publications. Through various methods, including co-resistance and cross-resistance, heavy metal ion pollution has been shown to co-select antibiotic and metal resistance. These metal ions, which include Zn^{2+}, Co^{2+}, and Cu^{2+}, among others, combine with antibiotic molecules to create complexes. These complexes might not biodegrade and be persistent. A metal ion's properties and the complex's charge, the complexes are more dangerous because they differ from the parent molecules in certain ways.

Common heavy metals in wastewater are as follows:

1. Lead (Pb): Found in battery manufacturing, paints, and plumbing materials.
2. Mercury (Hg): Released from coal-burning plants, mining, and chemical industries.
3. Cadmium (Cd): Present in electroplating, battery production, and fertilizers.
4. Chromium (Cr): Used in tanneries, textile dyeing, and stainless-steel production.
5. Arsenic (As): Comes from mining, pesticide use, and industrial waste.
6. Nickel (Ni): Discharged from metal processing, battery manufacturing, and electroplating.
7. Zinc (Zn): Found in galvanization, alloy production, and rubber manufacturing.
8. Copper (Cu): Present in electrical industries, pesticides, and plumbing waste (Alalwan et al. 2020)

12.4.2 Impact of Heavy Metal Contamination on Human Health and Ecosystem

1. Aquatic Ecosystems: Heavy metals disrupt aquatic life by accumulating in fish, causing toxicity and affecting biodiversity.
2. Human Health: Consumption of contaminated water leads to neurological disorders, kidney damage, and cancers.

3. Soil Contamination: Metals accumulate in agricultural lands, reducing soil fertility and entering the food chain.
4. Microbial Resistance: Heavy metals encourage the growth of ARB, posing health risks (Bonetta et al. 2022)

12.4.3 Impact of Heavy Metals on Microbial Communities

Heavy metals, including mercury (Hg), lead (Pb), chromium (Cr), cadmium (Cd), nickel (Ni), and arsenic (As), perturb microbial ecosystems by altering community composition, inhibiting key metabolic functions, and exerting selective pressure that facilitates the proliferation of metal- and antibiotic-resistant microorganisms (Ali et al. 2024).

12.4.3.1 Reducing Microbial Diversity

- Many helpful bacteria are poisoned by heavy metals, which reduces the diversity of microorganisms.
- The natural equilibrium of microbial communities is changed as certain bacteria become extinct and others develop resistance mechanisms.

12.4.3.2 Promoting Metal-Resistant Bacteria

- Some bacteria develop resistance genes to survive in metal-contaminated environments.
- These metal-resistant bacteria (MRB) can transfer resistance genes to pathogens, making diseases harder to treat.

12.4.3.3 Enhancing the Spread of Antibiotic Resistance Genes (ARGs)

- Because metal-resistant genes and ARGs are frequently related, research suggests that heavy metals are the primary cause of antibiotic resistance in bacteria. This phenomenon leads to the development of multidrug-resistant (MDR) microorganisms, presenting significant public health challenges.

12.4.3.4 Heavy Metal–Microbial Interactions in Ganga and Yamuna Rivers

12.4.3.4.1 Microbial Changes in the Yamuna River

- According to the previous study**,** harmful metals like lead and chromium decrease the number of helpful microorganisms that break down organic debris.
- Because of their capacity for biodegradation, bacteria like *Pseudomonas* and *Bacillus* become less prevalent in places contaminated by metals.
- The dispersion of heavy metals, with particular emphasis placed on the concentration of dissolved metals. Similar to the total metal concentration, the dissolved metal concentration is observed to decrease downstream, primarily due to dilution by uncontaminated surface water, as well as sorption, sedimentation processes, and bioaccumulation. The overall water toxicity to total metal load is influenced by additional factors, such as water hardness and dissolved oxygen levels (Prabhakar 2021).

12.4.3.4.2 Microbial Impact on the Ganga River

- According to the research, the chromium (Cr) released by the Kanpur tannery waste disturbs the local bacterial communities.
- Increased antibiotic-resistant strains of *Salmonella* and *Vibrio*, which cause gastrointestinal infections, are correlated with heavy metal exposure.
- MRB have been demonstrated to produce biofilms, which exacerbate the levels of contamination and complicate the efficacy of treatment processes (Prabhakar 2021).

12.4.4 Microbial Risks Due to Heavy Metal Contamination

- Environmental conditions high in metals promote the survival of opportunistic diseases, raising the risk of infection in both people and animals.
- *Salmonella* species, which can cause food poisoning and typhoid, are among the examples and *Pseudomonas aeruginosa*, which is connected to respiratory illnesses (Pande et al. 2022).

12.4.4.1 Biofilm Formation

- Bacteria that can withstand heavy metals frequently create protective layers called biofilms.
- These biofilms increase the disease risk by making bacterial colonies resistant to disinfection and cleaning.

12.4.4.2 Spread of Antibiotic-Resistant Genes (ARGs)

- Bacteria can share resistance genes when heavy metals are present because they promote HGT. Antibiotic-resistant infections result from this, making illnesses more difficult to cure (Zhang et al. 2021).

12.4.5 Solutions to Reduce Heavy Metal and Microbial Contamination

12.4.5.1 Better Treatment of Wastewater

- Bioremediation is the process of absorbing and detoxifying heavy metals using bacteria like *Pseudomonas, Bacillus*, and *Rhodococcus.* Using plants, such as water hyacinths, to absorb metal contaminants is known as phytoremediation.
- Membrane filtering: Using sophisticated filtering technologies to eliminate microorganisms and metals (Maqsood et al. 2022).

12.4.5.2 Strict Rules for Industry

- Implement discharge regulations for chemical plants, tanneries, and textile sectors.
- Routine assessment of the levels of metals in rivers and wastewater discharges.

12.4.5.3 Health Monitoring and Public Awareness

- In impacted areas, routine testing of the water quality is necessary. It campaigns to inform the public about the risks associated with antibiotic resistance and heavy metal contamination.

12.4.6 Nutrient Content (Organic Matter and Nitrogen Compounds)

Microplastics released from manufacturing facilities, textiles, medications, personal hygiene items, household sewage, and runoff are found in WWTPs. They persist and interact with many chemical compounds and bacteria, posing health risks, because they are known to break down slowly in the ecosystem. Because of their innate hydrophobic and physical properties, their interactions with hydrophobic substances have been extensively documented. Tetracycline and other antibiotics have recently been found to be carried by these microplastics (Nguyen et al. 2022). Wastewater contains a diverse range of nutrients that can significantly affect water quality and the surrounding ecosystem. Key nutrients in wastewater include nitrogen compounds (such as ammonia, nitrate, and organic nitrogen), which contribute to eutrophication and oxygen depletion in water bodies. Phosphorus, primarily in the form of orthophosphates and polyphosphates, acts as a major nutrient for algae, leading to harmful algal blooms when present in excess. Organic matter, measured by biochemical oxygen demand (BOD), chemical oxygen demand (COD), and total organic carbon (TOC), fuels microbial activity but also depletes oxygen in aquatic environments, harming aquatic life. Trace elements, such as calcium, magnesium, sulfur, sodium, and heavy metals like iron, zinc, and copper, are also present, often derived from industrial effluents and agricultural runoff. These nutrients, if left untreated, can cause serious environmental problems like eutrophication, waterborne diseases, and toxicity in aquatic organisms. Effective wastewater treatment technologies, including biological nitrification-denitrification, chemical precipitation, and filtration, are crucial in managing these nutrient levels and preventing water pollution (Nguyen et al. 2022).

12.4.6.1 Wastewater Containing Organic Matter

Organic Matter Sources: Wastewater organic stuff comes from Domestic sewage, including urine, feces, and food waste industrial effluents from the textile, paper, and food processing industries runoff from agriculture (manure, agricultural wastes).

12.4.6.2 Typical Indicators of Organic Matter

BOD quantifies the amount of oxygen required by microorganisms to degrade organic matter. Elevated BOD values indicate excessive organic pollution, leading to oxygen depletion in water. Untreated wastewater typically exhibits BOD concentrations ranging from 100 to 300 mg/L, as observed in domestic sewage.

COD represents the amount of oxygen required to oxidize organic materials, both biodegradable and nonbiodegradable. For untreated wastewater, COD levels range from 250 to 800 mg/L and are often greater than BOD.

TOC determines how much carbon is present in organic molecules and aids in determining the degree of pollution and the effectiveness of wastewater treatment (Wang et al. 2021).

Organic Matter Effect on the Environment: It causes hypoxia, or an oxygen shortage, by lowering the water oxygen content to promote the development of anaerobic bacteria, which generate hydrogen sulfide, and methane encourages eutrophication, which results in an abundance of algae blooms.

12.4.6.3 Wastewater Nitrogen Compounds

- Ammonia (NH_3) and ammonium (NH_4^+) are two types of nitrogen compounds released by fertilizers, urine, and the breakdown of proteins and are toxic in high doses to aquatic life.
- Common levels in untreated wastewater include: Nitrate (20–50 mg/L) (NO_3^-)) formed by nitrification, which is the microbial oxidation of ammonium extremely soluble and a major cause of eutrophication <10 mg/L is the WHO drinking water limit for nitrite (NO_2^-).
- A by-product of denitrification and nitrification in between poisonous drinking water that interferes with the passage of oxygen in the blood <1 mg/L is the acceptable limit.
- Total Kjeldahl Nitrogen: Ammonia and organic nitrogen are added together and used to evaluate wastewater levels of nitrogen contamination.

12.5 RESISTANCE GENES AND MOBILE GENETIC ELEMENTS (MGEs)

ARGs and MGEs play a crucial role in the spread of AMR in wastewater environments. Understanding their properties and interactions is critical for creating measures to slow the spread of resistance. Managing the spread of AMR requires an understanding of the processes by which genetic material is transferred between ambient and clinical microorganisms. Numerous ARGs are found on MGEs, including integrin-specific gene cassettes, transposons, and plasmids, which serve as carriers for these determinants and facilitate their spread. Microorganisms can successfully adapt to changing environmental conditions and evolutionary forces because this genetic material interchanges. This flexibility is particularly important when dealing with changing conditions and environmental challenges (Mazhar et al. 2021). The three primary mechanisms of horizontal gene transfer (HGT) in bacteria include transformation, involving the direct uptake of exogenous DNA; transduction, mediated by bacteriophages; and conjugation, facilitated by pili-dependent cell-to-cell transfer of genetic material. Conjugation involves the direct transfer of genetic material through physical cell-to-cell contact, transduction is facilitated by bacteriophages that mediate gene transfer between bacterial cells, and transformation entails the uptake and incorporation of free extracellular DNA from the surrounding environment (Singh et al. 2022b). By sharing their genetic components of resistant with another nonresistant bacterial species through HGT, MGEs encouraged the growth and spread of ARGs in Gram-positive and Gram-negative bacteria. There is a significant risk to human health when highly pathogenic superbacteria resistance to the majority of antimicrobial medications when drug-resistance genes in bacteria build up to a particular level (Nadeem et al. 2020).

12.5.1 Horizontal Gene Transfer

The horizontal transfer of ARGs serves as the principal mechanism facilitating the dissemination of antibiotic resistance among bacterial populations. The rapid dissemination of resistance genes is considerably aided by HGT, which is crucial to bacterial evolution. Conjugation, transformation, and transduction are the primary methods of HGT. Furthermore, it has been confirmed that membrane vesicles (MVs) play a part in HGT (Blakely 2024).

The process of conjugation involves direct physical contact between cells to transmit genetic material, including plasmid DNA, from donor bacteria to recipient bacteria. The most significant horizontal transfer process is conjugation, which is extensively found in bacteria. MGEs, such as plasmids and integrative conjugative elements (ICEs), are transferred between proximal bacterial cells through a pilus or membrane-associated pore during the contact-dependent process of conjugation (Nadeem et al. 2020).

Conjugation between members of the same genus or distinct species can spread resistance genes. Both commensal and opportunistic infections have been shown to disseminate mobile genetic components during their colonization of the human gut. The conjugative transfer of plasmid-mediated AMR genes and the subsequent dissemination of resistance to therapeutic agents pose a significant threat to human health. Gram-negative bacteria plasmids with carbapenem resistance genes (such as blaKPC, blaOXA-48, and blaNDM) can quickly conjugate with other susceptible bacteria, posing a major threat to global health. According to reports, additional members of the Enterobacteriaceae family in the gastrointestinal system may conjugately acquire the plasmid from *Enterobacter cloacae* that codes for OXA-48 (carbapenem resistance). Research has shown that Gram-positive bacteria, including *Streptococcus* species, also exhibit ICE- mediated drug resistance transfer pathways (Luo et al. 2022) (Figure 12.1).

12.5.2 Transformation

Transformation is a process in which recipient bacteria uptake extracellular DNA from lysed donor cells and integrate it into their genomes, facilitating the acquisition of novel genetic traits. Plasmid DNA and fragmented DNA released during active secretion or lysis by bacteria, which frequently carry the ARG, make up the majority of extracellular DNA. It is believed that many clinical bacterial species regularly acquire resistance through spontaneous transformation. For instance, *Streptococcus pneumoniae*, *Neisseria gonorrhoeae,* and *Vibrio cholerae* can develop resistance to antibiotics by metamorphosis (Nadeem et al. 2020). It has been reported that sequencing analyses of *Acinetobacter* plasmids have revealed the absence of essential genes required for conjugative transfer, indicating that certain drug-resistant plasmids in *Acinetobacter baumannii* may not be disseminated via conjugation but instead propagate through the natural transformation pathway. This is because *E. coli* can be transformed by plasmid DNA under natural conditions, indicating that *E. coli* can absorb DNA in the gut (Jian et al. 2021).

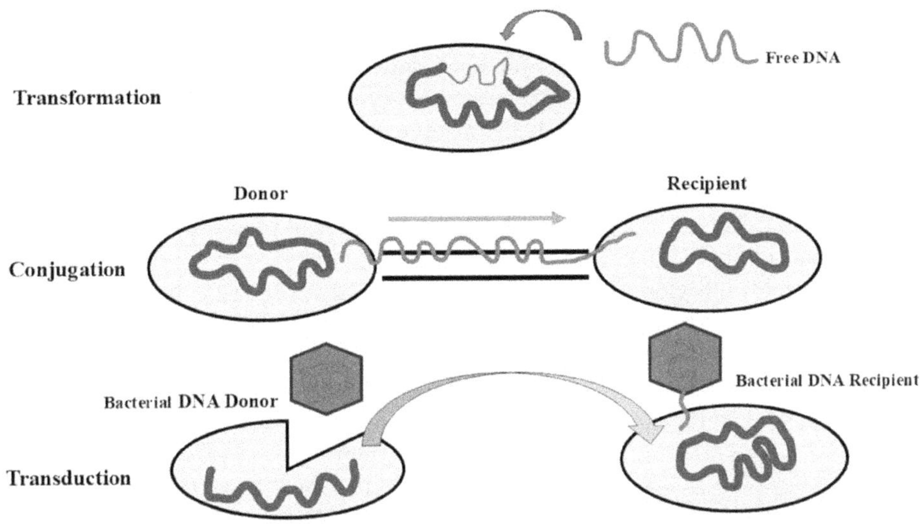

FIGURE 12.1 Schematic representation of horizontal gene transfer by transformation, conjugation, and transduction, enabling the sharing of genetic information between the same or different species.

12.5.3 Transduction

Transduction is a mechanism in which temperate bacteriophages serve as vectors to mediate the transfer of chromosomal and extrachromosomal DNA from donor to recipient bacteria, enabling the acquisition of novel genetic traits. Bacteriophages can coexist with ARGs within the same ecological niche and bacterial host, indirectly suggesting their potential role in the dissemination of resistance determinants. *Staphylococcus aureus* exhibits a high propensity for resistance transduction. Through phage-mediated transduction, methicillin-resistant *Staphylococcus aureus* (MRSA) acquires resistance from other bacterial species harboring the *mecA* gene. Additionally, Phage φ80α has been shown to transfer penicillin and tetracycline resistance genes to the MDR *S. aureus* strain USA300, as well as to *S. aureus* strains that are not inherently phage-sensitive. Transduction is a ubiquitous and continuous process in natural environments, playing a far more complex role in the dissemination of antibiotic resistance than previously understood. Studies using murine models indicate that transduction is a key driver of genetic diversity in gut-colonizing *Escherichia coli* strains and contributes to the evolution of antibiotic resistance in intestinal microbiota (Lee et al. 2022).

12.5.4 Other Mechanisms of the Horizontal Gene Transfer

Protoplast fusion, transposition, and lysogenic conversion are other methods for HGT. Gram-negative bacteria secrete bacterial outer MVs, which have particle sizes between 20 and 400 nm and are involved in various biological processes such as interactions between cells, HGT, and the export of cellular metabolites. Genes that cause antibiotic resistance may be delivered via MVs. It has been established that *Acinetobacter baumannii* can transmit drug-resistant genes via MVs, and more current studies have revealed that MVs can also transfer the β-lactamase gene to *E. coli* (Blakely 2024).

The primary mechanism via which bacterial drug resistance spreads is HGT. In vitro investigations have been the primary focus of HGT of ARGs. Furthermore, it may be far more difficult to transfer ARGs to the recipient bacteria, and the outcomes of in vitro experiments may not match the actual circumstances in vivo. It suggests that in contrast to the in vitro models now in use, in vivo models may be useful for studying the spread of therapeutically relevant ARGs under more realistic circumstances. More and more researchers have recently focused on using in vivo models to examine antibiotics (Duarte et al. 2022).

12.6 MICROBIAL COMMUNITIES IN WASTEWATER

Microbial community diversity represents the taxonomic richness and evenness of species within a defined ecological system. Soil and freshwater ecosystems, including aquifers, groundwater, lakes, rivers, drinking water, and wastewater, constitute natural reservoirs that support the highest abundance and phylogenetic diversity of bacterial lineages. These habitats provide a wide range of ecological niches, fostering extensive microbial diversity and functional adaptability. In biological wastewater treatment systems, microbial diversity plays a crucial role in governing treatment efficiency and process stability (Ali et al. 2024).

Several environmental bacterial taxa, including *Acinetobacter* species, *Aeromonas* species, and *Pseudomonas* species exhibit a high propensity for developing MDR. These opportunistic pathogens, adapted to humid and diverse aquatic environments, are frequently associated with anthropogenic sources and persist within the microbial consortia of municipal WWTPs. Furthermore, wastewater bacterial communities consist of diverse taxa encompassing various biochemical groups, including nitrogen-fixing, nitrifying, denitrifying, and sulfur-oxidizing bacteria, as well as physiological groups such as anaerobic, aerobic, phototrophic, and heterotrophic microorganisms. Many of these microbial populations contribute to water purification processes by facilitating nutrient removal and bioremediation.

Wastewater, particularly in its raw form, represents one of the most enriched aquatic habitats for known ARGs. This environment harbors a characteristic resistome signature, predominantly composed of genes conferring resistance to legacy antibiotics, including tetracyclines (tet), sulfonamides (sul), aminoglycosides (aac), and β-lactams (e.g., class A β-lactamases). A significant proportion of these ARGs are plasmidborne, while others are embedded within variable gene cassettes of integrins, facilitating their potential mobilization and HGT among bacterial populations. Several bacterial taxa are important reservoirs and vectors of tetracycline, macrolide, and MDR genes in wastewater environments (Singh et al. 2022b) (Figure 12.2).

Most common bacteria found in wastewater and surfaces include those belonging to the genera *Curvibacter, Acidovorax, Aeromonas, Acinetobacter, Pseudomonas, Sphingomonas, Rhodococcus,*

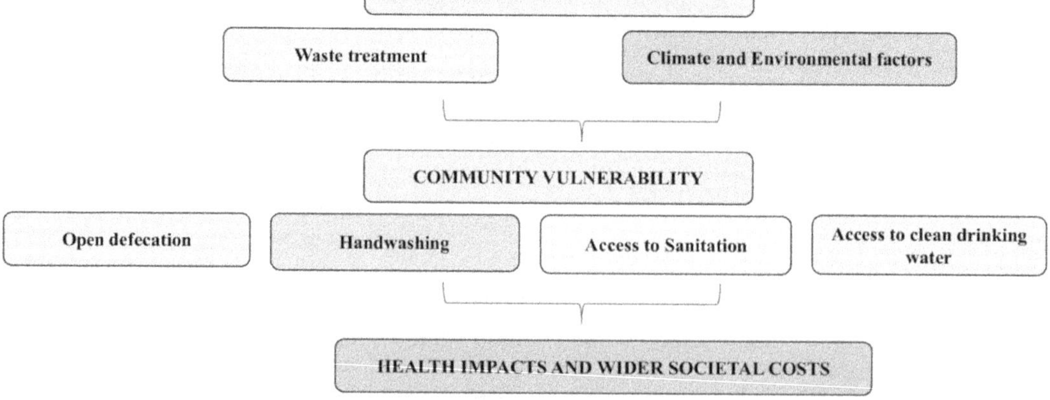

FIGURE 12.2 Origin of discharge and socioeconomic vulnerabilities contributing to waterborne antimicrobial resistance (AMR).

Legionella, Mycobacterium, Gordonia, Bacillus, Clostridium, and *Flavobacterium*. At higher taxonomic levels, such as phylum or class, the dominant bacterial taxa in aquatic ecosystems generally belong to the phyla Proteobacteria (primarily the classes Alphaproteobacteria, Betaproteobacteria, and Gammaproteobacteria), Actinobacteria, Bacteroidetes, and Firmicutes. HAIs are becoming more severe due to antibiotic and MDR spread. This makes treating these bacterial infections more difficult, raising mortality rates and medical expenses (Hayward et al. 2020). The World Health Organization (WHO) has classified multidrug-resistant pathogens, including *E. coli* and ESKAPE (*Enterococcus faecium, Staphylococcus aureus, Klebsiella pneumoniae, Acinetobacter baumannii, Pseudomonas aeruginosa,* and *Enterobacter* species) bacteria, as critical public health threats due to their resistance to most commonly used antimicrobial drugs (Duarte et al. 2022).

In WWTPs, microorganisms are organized into species-rich consortia that facilitate the biodegradation of a broader spectrum of substrates compared to monocultures. The development of microbial aggregates is associated with the production of extracellular polymeric substances (EPS). EPS are biogenic macromolecules synthesized through cellular lysis, excretion, detachment of surface-associated material, and assimilation of environmental compounds. The primary constituents of EPS are proteins and polysaccharides, predominantly in colloidal form, which play a crucial role in defining the structural organization and physicochemical properties of the microbial biomass (Ali et al. 2024)

Activated sludge flocs are structured through the physicochemical adhesion of microbial cells and wastewater-borne pollutants, along with the establishment of microbial colonies. Bacteria account for approximately 5%–20% of the organic fraction within flocs, while the remaining organic content primarily consists of extracellular polymeric substances (EPS).

A biofilm is a highly structured, multilayered microbial consortium in which microorganisms are immobilized on solid surfaces through electrostatic interactions, covalent bonding, and hydrophobic associations, facilitating their adherence and stability within the extracellular matrix. This matrix-embedded consortium provides structural stability and facilitates microbial adhesion, enabling enhanced resistance to environmental stressors and antimicrobial agents.

12.7 HARMFUL EFFECTS OF WASTEWATER ON ECOSYSTEM

Concern over AMR in wastewater bacteria is on the rise. Agricultural runoff, pharmaceutical industry wastewater, and hospital wastewater frequently contain antibiotics and other antimicrobials. These chemicals foster an environment that encourages the development of bacterial resistance. Public health is seriously threatened by these resistant bacteria since they can transfer AMR genes to other bacteria when they get into natural water bodies (Owhonka et al. 2021). There are several ways that wastewater damages ecosystems:

12.7.1 Heavy Metals and Chemical Contamination

A diverse variety of pollutants, such as industrial chemicals, medications, heavy metals, and personal care items, are frequently found in wastewater. These pollutants have the potential to upset ecological equilibrium and harm aquatic life. Water bodies receive hazardous chemicals and heavy metals from industrial effluents. These compounds have the potential to build up in biota and sediments, which could have long-term ecological effects. China Creek in Karachi, for instance, receives about 16 million gallons of untreated wastewater every day, which includes a wide range of hazardous contaminants from both industrial and residential sources. Mangrove ecosystems have deteriorated and biodiversity has decreased as a result of this pollution (Singh et al. 2022a).

Hazardous chemicals, heavy metals, and infections found in wastewater can contaminate water bodies, endangering aquatic life as well as human health. Waste from industrial activities, such as factories, mills, and mining operations, encompasses any material that is rendered unusable during the manufacturing process. The majority of power plants and many factories are situated close to bodies of water in order to access significant volumes of water for equipment cooling or production activities (Karri et al. 2021). The biggest uses of water in the United States are electric power plants. Aluminum smelters, food processing facilities, chemical plants, iron and steel mills, pulp and paper mills, and petroleum refineries are further businesses that need a lot of water. Wastewater that has been partially or completely treated is frequently returned to a nearby body of water. The discharge of metals, pollutants, and sewage into aquatic ecosystems directly affects both human health, particularly for populations dependent on these waterways for food or drinking water, and marine biodiversity. The impact of wastewater contaminants varies based on the nature and concentration of pollutants, potentially leading to the mortality of marine organisms or the bioaccumulation of toxic substances, which pose health risks to humans consuming contaminated seafood (Prabhakar 2021).

12.7.2 Medicinal and Personal Care Products

Research has found various medicines in groundwater and rivers, including painkillers, antidepressants, and antibiotics. For example, a study conducted in England examined 171 river and groundwater catchments and discovered almost 500 distinct pollutants, including the ones that are prohibited. Freshwater biodiversity is seriously threatened by more than half of these substances, which are extremely hazardous to aquatic life (Garg et al. 2022)

12.7.3 Nutrient Loading and Eutrophication

Eutrophication, a condition marked by excessive algal blooms that reduce oxygen in water bodies, can result from an overabundance of nutrients, especially nitrogen and phosphorus from sewage and agricultural runoff. Hypoxia, or the lack of oxygen, creates "dead zones" where aquatic life finds it difficult to live. For example, decades of industrial pollution and nutrient-rich agricultural runoff have severely deoxygenated the Baltic Sea, compromised its ecosystem and caused fish numbers to plummet. Nutrient-rich wastewater (phosphates and nitrates) frequently results in eutrophication, which can wipe out aquatic life.

12.7.4 Introduction of Pathogens

The most common microorganisms in wastewater influents include helminths, bacteria, fungi, viruses, and protozoa. While different microorganisms in water are thought to be important contributors to many waterborne epidemics and have various advantageous functions in wastewater influents, microorganisms have historically been employed in wastewater secondary treatment to eliminate dissolved organic materials. Depending on the treatment plant preferences, the microorganisms are employed in lagoon, suspended, or fixed-film systems (Liguori et al. 2022).

Their presence throughout the various stages of treatment can accelerate solids breakdown and reduce the amount of sludge produced. Water quality can also be determined by microbiological contaminants. The identification, separation, and detection of the various microbiological eliminating contaminants from wastewater are always challenging, costly, and time-consuming. To prevent this, the relative risk of a certain pathogens potential existence in wastewater is always assessed using indicator organisms. For example, faecal *Streptococci, Escherichia coli*, and coliforms are examples of enteric bacteria that are utilized as markers of faecal contamination in water sources (Choudhary et al. 2020).

12.7.5 Physical Habitat Alteration

Aquatic habitats may undergo physical changes as a result of wastewater discharge. For example, sedimentation from runoff from construction sites might disturb breeding grounds and suffocate benthic creatures. A construction company's poor sediment controls in New South Wales,

Australia, caused contaminated runoff to enter Curl Creek, endangering the severely endangered Climbing Galaxias fish. The rural character of many places has also been altered by urbanization (Jofre et al. 2023).

As the population grows, more food must be grown to accommodate the expanding population, and as agriculture uses the majority of the water available, economic growth is putting further strain on the already accessible water supplies.

12.7.6 Microplastic Waste

The researchers elucidate that the fragmentation of plastic waste into micro- and nano-sized particles facilitates their adsorption of environmental contaminants, including antibiotics, heavy metals, and pathogenic bacterial species. Furthermore, recent studies indicate that microplastics serve as a novel adhesive substrate, promoting bacterial colonization on their surfaces. In environmental settings, microplastics function as vectors for microbial attachment and facilitate HGT by providing a stable platform for bacterial communities. This process enables the exchange of genetic material, particularly through plasmid-mediated conjugation, which enhances the dissemination of AMR genes.

This study's findings reveal that the presence of microplastics significantly amplifies the rate of plasmid conjugation, increasing the transfer of antibiotic-resistance genes (ARGs) among bacterial populations by up to 200-fold compared to control conditions devoid of microplastics. "Antimicrobial resistance and environmental pollution represent critical global health challenges with far-reaching ecological and public health implications," the researchers emphasize (https://www.soci.org/news/2025/1/antimicrobial-resistance-how-microplastics-can-increase-the-spread-of-amr).

Plastic pollution in the environment has garnered significant public and scientific attention, with ongoing debates regarding the allocation of financial resources to address this issue in comparison to other global challenges such as climate change. The global impact of plastic pollution and its management strategies are increasingly being examined within the framework of sustainable development.

Environmental plastic pollution arises from the improper disposal of used or discarded plastic waste, leading to ecosystem contamination. A particularly critical concern is the role of municipal wastewater effluents, sewage sludge landfills, and plastic mulch from agricultural activities in contributing to soil pollution. These sources not only introduce microplastics into terrestrial and aquatic ecosystems but also pose long-term environmental and ecological risks, necessitating urgent mitigation strategies (Chaudhary and Garg 2024).

Plastic pollution occurs when synthetic polymer-based materials, regardless of their size, morphology, or composition, are introduced into the ecosystem, posing potential risks to ecosystems, biodiversity, and human health. Once released into natural habitats, plastics exhibit high mobility, transitioning from terrestrial environments to freshwater systems and ultimately reaching marine ecosystems. Throughout this transport process, plastics can exert adverse ecological and toxicological effects on various organisms, including humans (Sarma et al. 2022; Yang et al. 2025).

Furthermore, plastics can act as vectors for persistent organic pollutants, heavy metals, and pathogenic microorganisms, facilitating their dissemination across environmental matrices. Additionally, plastics have been demonstrated to release various chemical constituents, including monomers, oligomers, and additives, into the environment. This phenomenon highlights the potential role of plastics as sources of hazardous substances, contributing to environmental contamination and associated health risks (Chen et al. 2020).

12.8 WASTEWATER TREATMENT PROCESS

Due to the presence of physical, chemical, and biological pollutants, wastewater treatment requires a combination of unit operations and processes designed to remove or neutralize these contaminants, ensuring the safe discharge or reuse of treated water (Jofre et al. 2023).

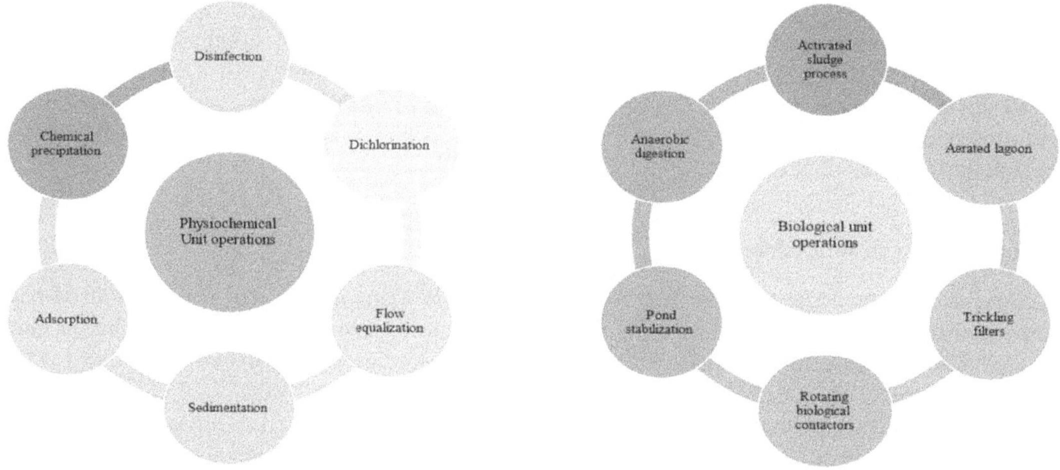

FIGURE 12.3 Processing of wastewater treatment.

Treatment processes can also be categorized accordingly. The following provides a summary of the unit operations and procedures involved in wastewater treatment, as outlined by the Economic and Social Commission for Western Asia (ESCWA) in 2003.

At this point, water can be disinfected using injections of chlorine, ozone, and ultraviolet (UV) radiation to satisfy current worldwide guidelines for urban and agricultural reuse.

Techniques for treating wastewater**:** Both traditional and unconventional wastewater treatment techniques are effective in wastewater treatment. The degree of automation in conventional wastewater treatment techniques is comparatively higher than that of nonconventional techniques requiring power and pumps in most cases. For the system to function and be maintained, skilled labor is needed (Choudhary et al. 2020) (Figure 12.3).

12.8.1 Conventional Method

Conventional wastewater treatment methods include activated sludge, trickling filters, and rotating biological contactors. Trickling filters and rotating biological contactors exhibit lower efficiency in biochemical oxygen demand (BOD) removal, have higher construction costs compared to activated sludge systems, and are sensitive to temperature variations. Additionally, activated sludge systems entail significantly higher operational costs due to the energy requirements for operating pumps and aeration blowers (National Programme on Technology Enhanced Learning [NPTEL], 2010).

12.8.2 Nonconventional Technique

These are inexpensive, less advanced biological treatment operations and upkeep systems for wastewater in cities. If the system is appropriately built and not overloaded, these systems can remove pathogens more effectively and consistently, even if they require more acreage than traditional high-rate biological processes. Stabilization ponds, artificial wetlands, oxidation ditches, and soil aquifer treatment are a few of the unconventional techniques (Jofre et al. 2023).

12.9 IMPACT OF WASTEWATER ON HUMAN HEALTH

Wastewater from many industries is released into nearby waterways and the environment. This effluent is occasionally released partially or untreated. The degradation of the environment has become a big concern for various nations due to recent developments. Wastewater harbors a diverse array of microorganisms, including bacteria, viruses, protozoa, and algae, which pose significant health risks to the public due to their association with various waterborne diseases. The discharge of untreated wastewater adversely affects human health, disrupts trophic levels within aquatic food chains, and deteriorates water quality in receiving bodies of water. Additionally, wastewater may contain emerging pollutants with endocrine-disrupting properties. In some instances, certain chemicals present in the effluent are not effectively removed by the treatment methods employed, further exacerbating environmental and human health impacts (Garg et al. 2022).

A concerning topic is the emergence of bacterial resistance to antibiotics, which is brought about by the excessive and careless usage of antimicrobial drugs. Wastewater treatment facilities have been the focus of much attention lately. The quantity of microbes discharged into the environment is extremely high and hazardous, and these stations serve as a reservoir for resistant bacteria. Reusing treated wastewater for irrigation is a workable way to deal with the water shortage, but there are some environmental and health hazards involved. It might make soil and water more resistant to antibiotics. WWTP effluents have been recognized as significant environmental reservoirs of antibiotic resistance due to the selective pressure exerted by the continuous release of antibiotics into aquatic systems. This selective pressure increases the abundance of resistance genes, thereby enhancing the likelihood of HGT among microbial communities. This review explores the emergence of antibiotic resistance in wastewater, the treatment processes employed, the associated challenges, and their implications for human health.

Wastewater is becoming increasingly scarce worldwide, and in most areas of the world, agriculture relies heavily on the reuse of treated or untreated wastewater due to the lack of freshwater. However, reusing wastewater carries a number of health and environmental concerns. By reducing the amount of freshwater resources used, wastewater irrigation reduces the strain on the aquatic ecosystem. However, this might seriously harm both the environment and human health. A significant amount of nutrients found in recycled wastewater serve as crop fertilizers and support microorganisms' metabolic processes. Increased agricultural output, nutrient recycling, less strain on freshwater, financial support, and the ability to support farmers are the main benefits of wastewater irrigation. However, because of improper irrigation and wastewater management techniques, several detrimental effects of wastewater irrigation are also noticeable (Pratap et al. 2021). These include significant health risks to agricultural workers, contamination of arable land and crops with toxic metals, chemical pollutants, salts, and microbial pathogens. Furthermore, prolonged use of wastewater for irrigation can adversely impact groundwater quality through the seepage of saline and heavy metal-laden effluents, rendering the water unsuitable for human consumption.

Only a portion of wastewater pathogenicity potential and antibiotic resistance may be understood, according to the state of the art. Furthermore, new diseases may arise, and new resistance and virulence genes may evolve and spread more easily in wastewater ecosystems. To provide a more thorough evaluation of the long-term impacts of wastewater releases, more research is also required.

Effects on human health of clinically relevant bacteria in wastewater habitats:

Pathogens (bacteria, protozoa, and viruses) and contaminants of growing concern (pharmaceuticals, insecticides, and personal care items) that affect human and animal health as a result of wastewater irrigation were discussed earlier. Additionally, covered are the vectorborne illnesses associated with water such as Zika, dengue Fever, Yellow fever, and Malaria. There is a discussion of the connection between greenhouse gas emissions (carbon dioxide, methane, and nitrous oxide) and wastewater irrigation and climate. Additionally, case studies of wastewater irrigation from throughout the world are discussed, along with their ramifications. The literature review found that, in addition to increasing the number of infections and microbiological hazards to human and animal health, improperly treated wastewater led to the deposition of heavy metals and other hazardous components in soils and plants. This study aims to establish the long-term impacts of emerging contaminants, as these are relatively novel compounds for which standardized analytical methodologies for precise detection and quantification are currently lacking. Similarly, a comprehensive and reliable long-term assessment of the potential health risks associated with heavy metal exposure requires further investigation (Singh 2021).

The section discusses how wastewater irrigation affects soil and plants by accumulating heavy metals and harmful substances in the soil profile and plant tissues, which has a detrimental effect on health. The section's main advantages and disadvantages are the harmful effects that infections (such as microbes, protozoa, viruses, and helminths) and poisonous chemical compounds have on humans and animal health (Singh et al. 2022b).

This review examines the development of antibiotic resistance in wastewater, wastewater treatment processes, associated challenges, and their implications for human health. Based on the current state of knowledge, we conclude that advancements in wastewater treatment technologies are necessary not only for the removal of solids, organic matter, and nutrients but also for the effective elimination of antibiotic resistance determinants and resistant bacteria.

12.10 ANTIMICROBIAL RESISTANCE IN WASTEWATER BACTERIA

The discovery of antibiotics in 1928 and their subsequent clinical application represented a paradigm shift in medical science. The introduction of antibiotic therapy enabled physicians to implement targeted treatment strategies following precise diagnoses. Even so, the benefits of using these antibiotics are threatened by the consistently increasing levels of AMR. Additionally, there has been a concerning global rise in the number of microbial strains that are resistant to several antimicrobial drugs. It is widely recognized that excessive, improper, and uncontrolled use of antimicrobial drugs contributes to AMR by causing resistant bacteria and their resistance genes to propagate uncontrolled (Amato et al. 2021). When bacteria are exposed to antibiotic molecules, some of their susceptible cells are killed, while the resistant cells live and proliferate (Cheng et al. 2021).

Researchers highlighted that the breakdown of plastic waste into microplastics enables them to absorb pollutants like antibiotics, heavy metals, and bacterial pathogens. Microplastics also serve as adhesive surfaces for bacterial colonization, facilitating HGT. This process enhances the exchange of ARGs via plasmids. Studies show that microplastics significantly increase the frequency of plasmid conjugation, accelerating the transfer of ARGs between bacteria up to 200 times more than in environments without microplastics (Figure 12.4).

12.10.1 WASTEWATER TREATMENT PLANTS (WWTPs)

In the past century, anthropogenic activities associated with industrialization and agricultural practices have been identified as major contributors to the release of biological and chemical

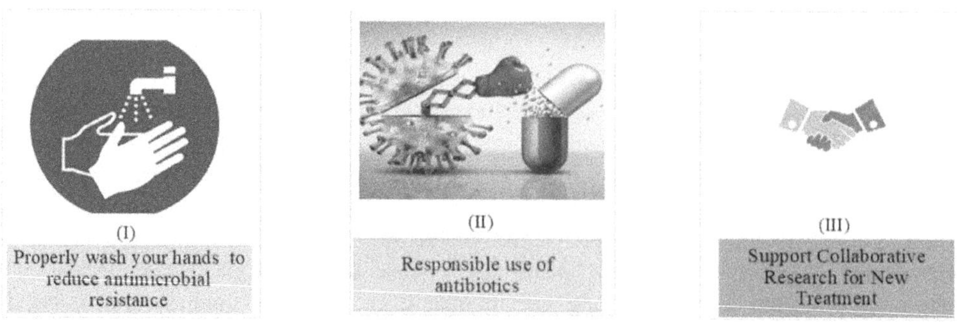

FIGURE 12.4 Major steps required to reduce antimicrobial resistance.

contaminants into aquatic ecosystems. A significant proportion of these pollutants comprises toxic compounds exhibiting antimicrobial properties. Notably, various antibiotic classes, including tetracyclines, penicillin, macrolides, quinolones, and aminoglycosides, have been detected in wastewater sources, raising concerns about their potential ecological and public health implications. WWTPs started to filter out physical contaminants including debris, pathogens, and large amounts of organic matter before releasing the cleaned water into rivers, streams, dams, and the ocean (Hammond et al. 2021).

AMR is caused by common infections, ARGs, and ARB found in wastewater. A few bacteria adhere to organic materials during primary treatment, which promotes genetic exchanges. Increased levels of resistance are caused by several stressors, including pH, temperature, heavy metals, and antibiotic residues. Wastewater is a crucial focus for AMR surveillance since studies have shown an increase in antimicrobial-resistant determinants in WWTPs (Syranidou and Kalogerakis 2022).

12.10.2 ARG and ARB Removal Efficiency in WWTPs

ARGs and ARB are largely produced by WWTPs. Health risk can rise as a result of these resistant bacteria spreading into the environment, especially drinking water. Multidrug-resistant bacteria are promoted through HGT and vertical gene transfer events that occur among wastewater-associated pathogens. The acquisition of extracellular DNA and HGT processes like conjugation, transformation, and transduction are two ways that ARGs might proliferate. This method contributes to the persistence and evolution of ARB by allowing bacteria that are sensitive to antibiotics to develop resistance (Cai et al. 2021).

12.10.3 Antibiotics in WWTPs

Research across the globe has identified antibiotics in the influents and effluents of WWTPs, even at subinhibitory levels, which play a role in AMR. Pathogens can develop resistance due to exposure to antibacterial compounds in wastewater, resulting in the proliferation of ARB within WWTPs. Due to the increased use of antimicrobial drugs to treat secondary infections and their repurposing, the COVID-19 pandemic has exacerbated this issue (Ansari et al. 2021). Furthermore, the presence of antimicrobial agents in cleaning products, like triclosan and benzalkonium chloride, exacerbates the increase in resistance levels. Their ongoing global use, despite prohibitions in certain nations, worsens AMR. Research, such as one conducted in India, has reported a rise in antibiotic-resistant *E. coli* prevalence during the pandemic, underscoring the vital contribution of wastewater to the spread of AMR (Sambaza and Naicker 2023).

12.10.4 Global Impact of Antimicrobial Resistance

The discovery of antimicrobial drugs was a groundbreaking achievement in the 20th century, transforming modern medicine. However, their success is challenged by the development of resistance, which reduces their effectiveness over time. The global public health community acknowledges the urgent need to tackle AMR. In accordance with the World Health Organization (WHO) 2015 Global Action Plan on Antimicrobial Resistance (AMR), nations have been mandated to formulate and implemented national action plans to address the increasing threat of AMR. The WHO has identified a critical gap in the lack of data on the incidence, prevalence, and geographic distribution of AMR, underscoring the urgent need for enhanced surveillance.

The incidence of infections caused by antimicrobial-resistant pathogens, as documented by the European Antimicrobial Resistance Surveillance Network (EARS-Net), served as a basis for estimating the prevalence of resistant infections in regions beyond the EARS-Net surveillance area.

In a study published in *The Lancet Global Health*, Elizabeth Temkin et al. provided global estimates of the incidence and overall burden of infections in the bloodstream and serious patient infections, such as those caused in 2014 by *Klebsiella pneumoniae* and third-generation cephalosporin-resistant and carbapenem-resistant *Escherichia coli.*

12.11 ANTIMICROBIAL STEWARDSHIP PROGRAMS

Stewardship refers to the strategic and responsible management of resources entrusted to an individual or organization. The concept of AMS was first introduced in 1996 by John McGowan and Dale Gerding, who proposed a direct causal relationship between antimicrobial usage and the emergence of resistance. Their work emphasized the necessity of large-scale, controlled investigations to regulate antimicrobial use, incorporating advanced epidemiological methodologies, molecular typing techniques, and precise resistance mechanism analyses.

The National Centre for Disease Control (NCDC), under the Directorate General of Health Services, Ministry of Health and Family Welfare, Government of India, developed and disseminated the National Policy for Containment of Antimicrobial Resistance in 2011. Recognizing the growing threat of AMR, the Government of India prioritized AMR mitigation efforts and launched the *National Action Plan on AMR (2017–2021)* in April 2017, with a subsequent phase due in 2022. Additionally, the Indian Council of Medical Research (ICMR) has developed and published standardized guidelines for the implementation of an *Antimicrobial Stewardship Program (AMSP)* in India. The main objective of this initiative is to advocate for the rational and prudent utilization of antimicrobial agents, thereby mitigating the emergence and dissemination of AMR (https://www.who.int/india/publications/m/item/antimicrobial-resistance-sameeksha-volume-32).

The WHO guides countries in the development and implementation of Antimicrobial Stewardship Programs (AMSPs), recognized as one of the most cost-effective strategies for optimizing antimicrobial use, enhancing patient outcomes, and mitigating both AMR and healthcare-associated infections (HAIs). It includes an evidence-based approach aimed at educating and supporting healthcare professionals in the appropriate prescribing and administration of antimicrobial agents. The continuous education and training of healthcare personnel are critical, as they serve as the first line of defense in preserving the efficacy of antimicrobial therapies. AMR arises when microorganisms, viruses, parasites, and fungi evolve mechanisms to withstand the effectiveness of antimicrobial agents intended to eliminate them.

(https://www.who.int/europe/activities/promoting-antimicrobial-stewardship-to-tackle antimicrobial-resistance).

The WHO has played a pivotal role in advancing global AMS initiatives. There are 24 bacterial pathogens from 15 families of antibiotic-resistant organisms on WHO's Bacterial Priority Pathogens List (BPPL) as of 2024. Notably, drug-resistant *Mycobacterium tuberculosis*, multidrug-resistant Gram-negative bacteria resistant to last-resort antibiotics, and other high-burden resistant pathogens like *Salmonella* species, *Shigella* species, *Neisseria gonorrhoeae*, *Pseudomonas aeruginosa*, and *Staphylococcus aureus* are among the most clinically significant pathogens. The inclusion of these pathogens highlights their substantial global health impact, considering factors such as disease burden, transmissibility, treatment challenges, and available prevention strategies. Additionally, the updated list reflects the current research and development (R&D) landscape for new antimicrobial therapies and emerging trends in AMR, emphasizing the urgent need for targeted interventions and innovation in antimicrobial treatment (WHO BPPL 2024).

Antimicrobial stewardship programs (AMSPs) are structured interventions designed to optimize the utilization of antimicrobial agents, including antibiotics, to mitigate the escalating challenge of AMR. AMS programs are designed to achieve the following objectives (Shrestha et al. 2023).

- **Optimization of Antimicrobial Treatment:** Collaborate with healthcare practitioners to ensure adherence to the five "D"s of antimicrobial treatment: selection of the applicable drug, administration of the correct dose, use of the appropriate drug route, prescription of a proper duration, and accurate de-escalation to treatment directed by the pathogen.
- **Prevention of Antimicrobial Misuse:** Curb the overuse, misuse, and inappropriate prescribing of antimicrobials across inpatient, outpatient, and community settings, including applications within the agricultural sector.
- **Reduction of Antibiotic-Associated Adverse Effects:** Mitigate the incidence of complications such as *Clostridioides difficile* infections, which can arise from unnecessary or inappropriate antimicrobial exposure.
- **Minimization of AMR:** Implement strategies to limit the emergence and spread of resistance organisms through judicious antimicrobial use.
- **Reduction of Healthcare-Associated Costs:** Decrease the financial burden associated with antimicrobial overuse, including costs related to prolonged hospitalizations, treatment failures, and resistance-associated complications.

In 2024, Centers for Disease Control and Prevention reported substantial advancements in AMS across diverse healthcare settings in the United States. In 2023, the overall outpatient antibiotic prescribing rate was 756 prescriptions per 1,000 individuals. Furthermore, in 2022, 70.5% of outpatient antibiotic prescriptions belonged to WHO's "Access" category, exceeding the organization's target threshold of 60%. This trend suggests a shift toward the preferential use of antibiotics associated with a lower propensity for resistance selection, reflecting improved AMS practices. In September 2024, global leaders convened in New York City for the second United Nations High-level Meeting on AMR. During this summit, a political declaration was adopted, outlining international commitments and targets aimed at combating AMR. Key objectives include a 10% reduction in mortality attributable to AMR by 2030 and the enhancement of antibiotic stewardship in human health care to optimize antimicrobial use and mitigate resistance development.

(https://www.cdc.gov/antibiotic-use/hcp/data-research/stewardship-report.html).

In 2019, WHO released a comprehensive toolkit designed to facilitate the implementation of antimicrobial stewardship programs (AMSPs) within healthcare settings, particularly in low- and middle-income countries. This evidence-based resource provides structured guidance on the development and sustainability of effective AMSPs, underscoring the necessity of a multidisciplinary framework that integrates expertise from diverse healthcare sectors to optimizing antimicrobial usage and mitigate AMR.

The emergence of antibiotic resistance is a naturally occurring adaptive mechanism that enables bacterial survival; however, its acceleration has been largely driven by anthropogenic factors. These include the excessive prescriptions and inadequate administration of antibiotics, overuse in livestock as growth promoters, and the limited development of novel antimicrobial agents. Recent literature has emphasized the importance of AMS programs in mitigating the complications associated with antibiotic misuse. Such initiatives are crucial in providing healthcare professionals, including undergraduate medical students, with the essential knowledge and skills required to tackle the antibiotic resistance crisis, which represents a critical global public health threat (Aljeldah 2022).

An important pathogen that contributes to the higher rates of morbidity and mortality from dysentery each year worldwide is *Enterococcus*, a major cause of diarrheal sickness that is brought on by wastewater. To lower the diseases prevalence and fatality rates, *Enterococcus* species must be treated with antibiotics. However, because of the worldwide increase in broad-spectrum antibiotic resistance, treating these illnesses is still difficult. Numerous processes, including a reduction in cellular permeability, drug extrusion by active efflux pumps, overexpression of enzymes that modify and inactivate medicines, or target modification by mutation, can lead to drug resistance in *Enterococcus* species. Consequently, there is a growing need to identify and develop alternative therapy approaches that offer novel approaches.

12.11.1 Implementation of Antimicrobial Stewardship Programs in Primary and Secondary Tier Hospitals in India

India exhibits some of the highest rates of AMR among bacterial pathogens commonly responsible for infections in healthcare settings. Additionally, India was the leading global consumer of antimicrobials in 2010, followed by China and the United States. The most frequently utilized antimicrobial classes included broad-spectrum penicillin, cephalosporins, fluoroquinolones, and macrolides. Notably, the national consumption of cephalosporins demonstrated a fourfold increase between 2000 and 2015, indicating a concerning escalation.

Studies conducted across the country have consistently highlighted these trends. Furthermore, the consumption of last-resort antimicrobials, such as polymyxins and carbapenems—typically reserved for treating multidrug-resistant infections—has risen sharply. These data underscore the alarming and rapidly evolving nature of the AMR crisis in India. However, AMR cannot be attributed solely to increased drug consumption, as it is a multifactorial phenomenon influenced by various biological, environmental, and sociobehavioral determinants (Debnath et al. 2024) (https://www.who.int/india/publications/m/item/antimicrobial-resistance-sameeksha-volume-32).

12.11.2 Strategies to Manage Antimicrobial Resistance

By 2050, AMR is projected to result in approximately 10 million deaths annually worldwide, establishing it as a persistent public health crisis. AMR arises when bacteria, fungi, viruses, and parasites develop resistance to antimicrobial agents in both human and animal hosts, enabling their survival despite treatment. The overuse and misuse of antibiotics, particularly their inappropriate administration, remain the primary drivers of the current AMR crisis, contributing significantly to the global burden of resistance. Consequently, the usage and consumption of antimicrobial drugs are subject to continuous monitoring on a global scale (Tang et al. 2023).

The "One Health Approach" is a method that recognizes the obvious linkages between people, animals, and their common environment while integrating a multidisciplinary effort to obtain the greatest possible health outcome. Global and local action plans still need to emphasize the significance of public awareness and lay audience health literacy. One of the biggest problems facing worldwide public health in the 21st century is still antibiotic resistance. The G7 nations are already contributing significantly politically to this issue, and it is still on the agenda of many political conferences (Figure 12.5).

The inadequate management of AMR is anticipated to lead to a reversion to the pre-antibiotic era, wherein common infections associated with childbirth, surgical procedures, and open fractures could become potentially fatal. Addressing AMR effectively and preventing its further emergence presents a significant challenge, as AMR is a multifaceted issue affected by a complex interplay of factors, including social anthropology, civil unrest, war, migration, ethnic displacement, political frameworks, healthcare systems, economic conditions, societal behaviors at both individual and population levels, health literacy, climate events, international travel, and advancements in pharmaceutical research and investment. AMR is the result of biological changes made by microorganisms, such as bacteria, fungi, parasites, and viruses, that make them resistant to antimicrobial treatments, such as antibiotics, which are meant to cure diseases caused by these pathogens (WHO 2021). The growing prevalence of AMR has made it necessary to design and implement measures to address this issue on a worldwide scale (Godman et al. 2021; Saleem et al. 2022; Kariuki et al. 2022).

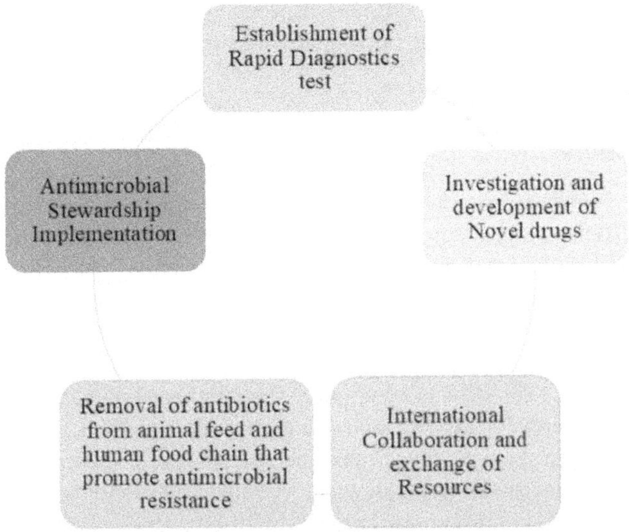

FIGURE 12.5 Diagrammatic representation of important elements that may influence the worldwide battle against AMR.

Countries subsequently pledged to put the NAPs into effect when they were created and assisted in combating AMR. To combat AMR, the GAP focused on five goals:

1. Enhancing awareness and comprehension of AMR through structured education, targeted training programs, and strategic communication initiatives;
2. Strengthening the body of knowledge and evidence through research and surveillance;
3. Reducing the incidence of infections through the implementation of effective infection prevention strategies, improved sanitation, and adherence to strict hygiene protocols;
4. Increasing the efficacy of antimicrobial medication use in both human and animal health; and
5. Developing a comprehensive economic framework for sustainable investment is essential to addressing global AMR. This approach should account for the needs of all countries and promote increased investment in the research and development of novel antimicrobial agents, diagnostic tools, vaccines, and other essential healthcare services (Mudenda et al. 2023).

These objectives underscore the necessities of addressing AMR through a "One Health" approach, which integrates humans, animals, agricultural, and environmental health to develop comprehensive and sustainable AMR mitigation strategies.

AMR is a complicated problem that necessitates a multidimensional and well-coordinated strategy. Fighting AMR necessitates a worldwide effort that includes tactics to maintain the efficacy of currently available antimicrobial drugs, create novel therapeutic drugs, and put in place efficient infection prevention and control mechanisms. Optimizing the use of antimicrobial drugs in humans and animal health and encouraging stewardship initiatives that guarantee these priceless medications are used only when required and appropriately are essential parts of this strategy. Moreover, strong monitoring systems are necessary to monitor the development and dissemination of resistance, offering vital information to guide treatment recommendations and public health initiatives. Putting money into studies to comprehend the mechanisms of resistance and create novel defenses, such as fresh vaccines, antibiotics, and alternative treatment. To lessen the initial need for antibiotics, it is also critical to improve infection prevention and control procedures in communities and hospital settings. This involves advancing immunization, sanitation, and hygiene initiatives. Lastly, promoting a culture of stewardship and encouraging group action to maintain the effectiveness of these life-saving drugs for future generations depends on educating the public, legislators, and healthcare professionals about the dangers of AMR and the significance of responsible usage of antibiotics (Godman et al. 2021).

12.12 CONCLUSION

AMR in wastewater bacterial communities constitutes a substantial environmental and public health risk. WWTPs serve as significant reservoirs and selective environments for ARGs and ARB, promoting their dissemination and persistence. The co-occurrence of heavy metals, antibiotics, and other antimicrobial agents within wastewater matrices exerts selective pressure, enriching for resistant phenotypes. Furthermore, HGT mechanisms facilitate the interspecies and intraspecies transmission of ARGs, contributing to the evolution of multidrug-resistant (MDR) strains.

The escalating prevalence of AMR is compounded by anthropogenic factors, including elevated antibiotic consumption during disease outbreaks (e.g., pandemics), the ubiquitous presence of antimicrobial chemicals in consumer products, and the potential role of microplastics as vectors for ARG transfer. Despite advancements in wastewater treatment methodologies and regulatory restrictions on certain antimicrobial compounds, antimicrobial bacteria and AMR genes are frequently detected in WWTP effluents and receiving aquatic environments.

Mitigation of AMR in wastewater necessitates a comprehensive, multidisciplinary strategy. This includes the emergence and utilization of advanced wastewater treatment technologies optimized for ARG removal, stricter regulatory frameworks governing the production, usage, and disposal of antibiotics and other antimicrobials, and continuous surveillance of AMR prevalence and trends in aquatic ecosystems. International collaboration and data sharing are crucial for effectively containing the global spread of resistance, preserving ecosystem integrity, and safeguarding human health. The role of wastewater in AMR propagation is crucial to develop more effective and innovative treatment methods to reduce the spread of resistant bacteria. This includes optimizing current wastewater treatment technologies, integrating advanced methods such as membrane filtration and UV disinfection, and implementing AMS programs. Furthermore, increased monitoring of wastewater for resistance markers is needed to track trends and inform policy decisions.

The future of managing AMR in wastewater requires a multidisciplinary approach that integrates microbiology, environmental science, and public health initiatives. Key research and policy areas to address include:

12.12.1 Improved Wastewater Treatment Technologies

Research into novel or optimized treatment techniques such as advanced oxidation processes, electrochemical disinfection, and biological treatments will play a significant role in reducing AMR in wastewater. Emphasizing resource recovery, such as using treated wastewater for agriculture, should also be considered to minimize cross-contamination.

12.12.2 AMR Surveillance in Wastewater Systems

Ongoing surveillance and monitoring of ARB and resistance genes in wastewater will help track the spread of AMR and identify high-risk areas. This data will be crucial for decision-making on public health interventions and the development of region-specific strategies.

12.12.3 Policy and Regulatory Frameworks

Governments and international organizations need to create and enforce stricter regulations on the discharge of antibiotics and resistant bacteria into wastewater. Policies encouraging proper pharmaceutical waste disposal and controlling the overuse of antibiotics are essential to curtail the rise of AMR.

12.12.4 Public Awareness and Education

Raising awareness among the general public, healthcare professionals, and industries regarding the importance of reducing antibiotic consumption and ensuring the proper disposal of pharmaceuticals can help reduce the burden of AMR.

12.12.5 Collaborative International Efforts

Since AMR is a global issue, international cooperation will be critical in developing standardized monitoring systems, sharing best practices for wastewater treatment, and coordinating efforts to combat the spread of AMR across borders.

ACKNOWLEDGEMENT

The authors are thankful to the Vice Chancellor, Maj. Gen. Dr. G.K. Thapaliyal for discussions on various occasions while writing this manuscript and we dedicate this paper on his 70th Birthday.

REFERENCES

Alalwan, H. A., Kadhom, M. A., and Alminshid, A. H. (2020). Removal of heavy metals from wastewater using agricultural byproducts. *Journal of Water Supply and Technology- AQUA*, 69(2), 99–112.

Ali, H., Min, Y., Yu, X., Kooch, Y., Marnn, P., and Ahmed, S. (2024). Composition of the microbial community in surface flow-constructed wetlands for wastewater treatment. *Frontiers in Microbiology*, 15, 1421094.

Aljeldah, M. M. (2022). Antimicrobial resistance and its spread is a global threat. *Journal of Antibiotics*, 11(8), 1082.

Amarasiri, M., Sano, D., and Suzuki, S. (2020). Understanding human health risks caused by antibiotic resistant bacteria (ARB) and antibiotic resistance genes (ARG) in water environments: Current knowledge and questions to be answered. *Critical Reviews in Environmental Science and Technology*, 50(19), 2016–2059.

Amato, M., Dasí, D., González, A., Ferrús, M. A., and Castillo, M. Á. (2021). Occurrence of antibiotic resistant bacteria and resistance genes in agricultural irrigation waters from Valencia City (Spain). *Agricultural Water Management*, 256, 107097.

Ansari, S., Hays, J. P., Kemp, A., Okechukwu, R., Murugaiyan, J., Ekwanzala, M. D., and van Dongen, M. B. (2021). The potential impact of the COVID-19 pandemic on global antimicrobial and biocide resistance: An AMR Insights global perspective. *JAC-Antimicrobial Resistance*, 3(2), dlab038.

Blakely, G. W. (2024). Mechanisms of horizontal gene transfer and DNA recombination. *Molecular Medical Microbiology*, 2024, 309–324.

Bonetta, S., Pignata, C., Gasparro, E., Richiardi, L., Bonetta, S., and Carraro, E. (2022). Impact of wastewater treatment plants on microbiological contamination for evaluating the risks of wastewater reuse. *Environmental Sciences Europe*, 34(1), 20.

Cai, Y., Sun, T., Li, G., and An, T. (2021). Traditional and emerging water disinfection technologies challenging the control of antibiotic-resistant bacteria and antibiotic resistance genes. *ACS ES & T Engineering*, 1(7), 1046–1064.

Chaudhary, M., and Garg, A. P. (2024). Global impact of plastic pollution and its management for sustainable development. In Kasturi, S., Yogesh, B. P., Ronald, C. E., and Pedro Antonio, L. H. (Eds) *Quality of Life and Climate Change: Impacts, Sustainable Adaptation, and Social-Ecological Resilience* (pp. 122–152). IGI Global, New York.

Chen, X., Chen, X., Zhao, Y., Zhou, H., Xiong, X., and Wu, C. (2020). Effects of microplastic biofilms on nutrient cycling in simulated freshwater systems. *Science of the Total Environment*, 719, 137276.

Cheng, X., Xu, J., Smith, G., and Zhang, Y. (2021). Metagenomic insights into dissemination of antibiotic resistance across bacterial genera in wastewater treatment. *Chemosphere*, 271, 129563.

Choudhary, M., Peter, C. N., Shukla, S. K., Govender, P. P., Joshi, G. M., and Wang, R. (2020). Environmental issues: A challenge for wastewater treatment. *Green Materials for Wastewater Treatment*, 38, 1–12.

Debnath, F., De, R. G., Chakraborty, D., Majumdar, A., Mukhopadhyay, S., Sarkar, M. D., and Dutta, S. (2024). Antimicrobial stewardship implementation in primary and secondary tier hospitals in India: interim findings from a need assessment study using mixed method design. *Scientific Reports*, 14(1), 28068.

Duarte, A. C., Rodrigues, S., Afonso, A., Nogueira, A., and Coutinho, P. (2022). Antibiotic resistance in the drinking water: Old and new strategies to remove antibiotics, resistant bacteria, and resistance genes. *Pharmaceuticals*, 15(4), 393.

Dutta, V., Dubey, D., and Kumar, S. (2020). Cleaning the River Ganga: Impact of lockdown on water quality and future implications on river rejuvenation strategies. *Science of the Total Environment*, 743, 140756.

Felis, E., Kalka, J., Sochacki, A., Kowalska, K., Bajkacz, S., Harnisz, M., and Korzeniewska, E. (2020). Antimicrobial pharmaceuticals in the aquatic environment-occurrence and environmental implications. *European Journal of Pharmacology*, 866, 172813.

Garg, S., Chowdhury, Z. Z., Faisal, A. N. M., Rumjit, N. P., and Thomas, P. (2022). Impact of industrial wastewater on environment and human health. In Swapnila, R., Alok, G., Shivani, G., and Tien, A. T. (Eds) *Advanced Industrial Wastewater Treatment and Reclamation of Water: Comparative Study of Water Pollution Index During Pre-Industrial, Industrial Period and Prospect of Wastewater Treatment for Water Resource Conservation* (pp. 197–209). Springer, Cham.

Godman, B., Egwuenu, A., Haque, M., Malande, O. O., Schellack, N., Kumar, S., and Seaton, R. A. (2021). Strategies to improve antimicrobial utilization with a special focus on developing countries. *Life*, 11(6), 528.

Hammond, P., Suttie, M., Lewis, V. T., Smith, A. P., and Singer, A. C. (2021). Detection of untreated sewage discharges to watercourses using machine learning. *NPJ Clean Water*, 4(1), 18.

Hassan, B., Qadri, H., Ali, M. N., Khan, N. A., and Yatoo, A. M. (2020). Impact of climate change on freshwater ecosystem and its sustainable management. *Fresh Water Pollution Dynamics and Remediation*, 8, 105–121.

Hayward, C., Ross, K. E., Brown, M. H., and Whiley, H. (2020). Water as a source of antimicrobial resistance and healthcare-associated infections. *Pathogens*, 9(8), 667.

Huemer, M., Mairpady Shambat, S., Brugger, S. D., and Zinkernagel, A. S. (2020). Antibiotic resistance and persistence—Implications for human health and treatment perspectives. *EMBO Reports*, 21(12), e51034.

Jian, Z., Zeng, L., Xu, T., Sun, S., Yan, S., Yang, L., and Dou, T. (2021). Antibiotic resistance genes in bacteria: Occurrence, spread, and control. *Journal of Basic Microbiology*, 61(12), 1049–1070.

Jofre Bartholin, M., Barrera Vega, B., and Berrocal Silva, L. (2023). Antibiotic-resistant bacteria in environmental water sources from Southern Chile: A potential threat to human health. *Microbiology Research*, 14(4), 1764–1773.

Kariuki, S., Kering, K., Wairimu, C., Onsare, R., and Mbae, C. (2022). Antimicrobial resistance rates and surveillance in sub-Saharan Africa: Where are we now?. *Infection and Drug Resistance*, 15, 3589–3609.

Karri, R. R., Ravindran, G., and Dehghani, M. H. (2021). Wastewater—sources, toxicity, and their consequences to human health. In Karri, R. R., Ravindran, G., and Dehghani, M. H. (Eds) *Soft Computing Techniques in Solid Waste and Wastewater Management* (pp. 3–33). Elsevier, Amsterdam, Netherlands.

Kushkevych, I. (2021). The application of microorganisms in wastewater treatment. *Processes*, 9(11), 1914.

La Rosa, M. C., Maugeri, A., Favara, G., La Mastra, C., Magnano San Lio, R., Barchitta, M., and Agodi, A. (2025). The impact of wastewater on antimicrobial resistance: A scoping review of transmission pathways and contributing factors. *Antibiotics*, 14(2), 131. https://doi.org/10.3390/antibiotics14020131.

Lee, A. R., Park, S. B., Kim, S. W., Jung, J. W., Chun, J. H., Kim, J., and Jung, T. S. (2022). Membrane vesicles from antibiotic-resistant Staphylococcus aureus transfer antibiotic-resistance to antibiotic-susceptible *Escherichia coli*. *Journal of Applied Microbiology*, 132(4), 2746–2759.

Liguori, K., Keenum, I., Davis, B. C., Calarco, J., Milligan, E., Harwood, V. J., and Pruden, A. (2022). Antimicrobial resistance monitoring of water environments: A framework for standardized methods and quality control. *Environmental Science & Technology*, 56(13), 9149–9160.

Luo, A., Wang, F., Sun, D., Liu, X., and Xin, B. (2022). Formation, development, and cross-species interactions in biofilms. *Frontiers in Microbiology*, 12, 757327.

Maqsood, Q., Hussain, N., Mumtaz, M., Bilal, M., and Iqbal, H. M. (2022). Novel strategies and advancement in reducing heavy metals from the contaminated environment. *Archives of Microbiology*, 204(8), 478.

Mazhar, S. H., Li, X., Rashid, A., Su, J., Xu, J., Brejnrod, A. D., and Rensing, C. (2021). Co-selection of antibiotic resistance genes, and mobile genetic elements in the presence of heavy metals in poultry farm environments. *Science of The Total Environment*, 755, 142702.

Mudenda, S., Chabalenge, B., Daka, V., Mfune, R. L., Salachi, K. I., Mohamed, S., and Matafwali, S. K. (2023). Global strategies to combat antimicrobial resistance: A one health perspective. *Pharmacology & Pharmacy*, 14(8), 271–328.

Nadeem, S. F., Gohar, U. F., Tahir, S. F., Mukhtar, H., Pornpukdeewattana, S., Nukthamna, P., and Massa, S. (2020). Antimicrobial resistance: More than 70 years of war between humans and bacteria. *Critical Reviews in Microbiology*, 46(5), 578–599.

Naznine, F., Ansari, M. I., Aafreen, U., Singh, K., Verma, R., Dey, M., and Khubaib, M. (2024). Prevalence of antibiotic resistance genes in bacteria from Gomti and Ganga rivers: Implications for water quality and public health. *Environmental Monitoring and Assessment*, 196(10), 992.

Nguyen, L. N., Aditya, L., Vu, H. P., Johir, A. H., Bennar, L., Ralph, P., and Nghiem, L. D. (2022). Nutrient removal by algae-based wastewater treatment. *Current Pollution Reports*, 8(4), 369–383.

Omuferen, L. O., Maseko, B., and Olowoyo, J. O. (2022). Occurrence of antibiotics in wastewater from hospital and convectional wastewater treatment plants and their impact on the effluent receiving rivers: current knowledge between 2010 and 2019. *Environmental Monitoring and Assessment*, 194(4), 306.

Owhonka, A., Fubara, E. F., and Justice, O. B. (2021). Wastewater quality-it's impact on the environment and human physiology: A review. *International Journal of Advance Research and Innovation*, 9, 43–58.

Pande, V., Pandey, S. C., Sati, D., Bhatt, P., and Samant, M. (2022). Microbial interventions in bioremediation of heavy metal contaminants in agroecosystem. *Frontiers in Microbiology*, 13, 824084.

Parra-Saldivar, R., Castillo-Zacarías, C., Bilal, M., Iqbal, H. M., and Barceló, D. (2021). Sources of pharmaceuticals in water. In Sandra, P. S., Nicola, M., Serge, C., and Damià, B. (Eds) *Interaction and Fate of Pharmaceuticals in Soil-Crop Systems: The Impact of Reclaimed Wastewater* (pp. 33–47). Springer, Cham.

Prabhakar, P. (2021). The physico-chemical study of heavy metals in water of Yamuna River. *International Journal of Multidisciplinary Research Education Analysis and Development-IJMREAD*, 7, 40–44.

Prasad, D. K., Shukla, R., and Ahammad, S. Z. (2024). Antibiotic resistance in the Ganga River: Investigation of antibiotic resistant bacteria and antibiotic resistance genes, and public health risk assessment. *Journal of Environmental Chemical Engineering*, 12(6), 114931.

Pratap, B., Kumar, S., Purchase, D., Bharagava, R. N., and Dutta, V. (2021). Practice of wastewater irrigation and its impacts on human health and environment: A state of the art. *International Journal of Environmental Science and Technology*, 20, 1–16.

Praveen Kumar Reddy, Y., Akiba, M., Guruge, K. S., Balakrishna, K., Vandana, K. E., and Kumar, V. (2020). Occurrence of antimicrobial-resistant *Escherichia coli* in sewage treatment plants of South India. *Journal of Water, Sanitation and Hygiene for Development*, 10(1), 48–55.

Qadir, M., Drechsel, P., Jiménez Cisneros, B., Kim, Y., Pramanik, A., Mehta, P., and Olaniyan, O. (2020). Global and regional potential of wastewater as a water, nutrient and energy source. *Natural Resources Forum*, 44(1), 40–51.

Ramírez-Coronel, A. A., Mohammadi, M. J., Majdi, H. S., Zabibah, R. S., Taherian, M., Prasetio, D. B., and Sarkohaki, S. (2023). Hospital wastewater treatment methods and its impact on human health and environments. *Reviews on Environmental Health*, 39(3), 423–434.

Ryu, H., Addor, Y., Brinkman, N. E., Ware, M. W., Boczek, L., Hoelle, J., and Villegas, E. N. (2021). Understanding microbial loads in wastewater treatment works as source water for water reuse. *Water*, 13(11), 1452.

Saleem, Z., Godman, B., Cook, A., Khan, M. A., Campbell, S. M., Seaton, R. A., and Moore, C. E. (2022). Ongoing efforts to improve antimicrobial utilization in hospitals among African countries and implications for the future. *Antibiotics*, 11(12), 1824.

Sambaza, S. S., and Naicker, N. (2023). Contribution of wastewater to antimicrobial resistance: A review article. *Journal of Global Antimicrobial Resistance*, 34, 23–29.

Sarma, H., Hazarika, R.P., Kumar, V., Roy, A., Pandit, S., and Prasad, R. (2022). Microplastics in marine and aquatic habitats: Sources, impact, and sustainable remediation approaches. *Environmental Sustainability* 5, 39–49. https://doi.org/10.1007/s42398-022-00219-8

Shrestha, J., Zahra, F., and Cannady, Jr P. (2023). *Antimicrobial Stewardship*. StatPearls Publishing, Treasure Island, FL. https://www.ncbi.nlm.nih.gov/books/NBK572068/

Singh, A. (2021). A review of wastewater irrigation: Environmental implications. *Resources, Conservation and Recycling*, 168, 105454.

Singh, A. K., Kaur, R., Verma, S., and Singh, S. (2022a). Antimicrobials and antibiotic resistance genes in water bodies: Pollution, risk, and control. *Frontiers in Environmental Science*, 10, 830861.

Singh, A. K., Kumar, A., and Chandra, R. (2022b). Environmental pollutants of paper industry wastewater and their toxic effects on human health and ecosystem. *Bioresource Technology Reports*, 20, 101250.

Syranidou, E., and Kalogerakis, N. (2022). Interactions of microplastics, antibiotics and antibiotic-resistant genes within WWTPs. *Science of the Total Environment*, 804, 150141.

Tang, K. W. K., Millar, B. C., and Moore, J. E. (2023). Antimicrobial resistance (AMR). *British Journal of Biomedical Science*, 80, 11387.

Wang, M., Zhu, J., and Mao, X. (2021). Removal of pathogens in onsite wastewater treatment systems: A review of design considerations and influencing factors. *Water*, 13(9), 1190.

World Health Organization. (2024). *WHO Bacterial Priority Pathogens List, 2024*. World Health Organization, Geneva, Switzerland.

Yang, Q. E., Lin, Z., Gan, D., Li, M., Liu, X., Zhou, S., and Walsh, T. R. (2025). Microplastics mediates the spread of antimicrobial resistance plasmids via modulating conjugal gene expression. *Environment International*, 195, 109261.

Zhang, T., Lv, K., Lu, Q., Wang, L., and Liu, X. (2021). Removal of antibiotic-resistant genes during drinking water treatment: A review. *Journal of Environmental Sciences*, 104, 415–429.

Zhuang, M., Achmon, Y., Cao, Y., Liang, X., Chen, L., Wang, H., and Leung, K. Y. (2021). Distribution of antibiotic resistance genes in the environment. *Environmental Pollution*, 285, 117402.

Subject Index

For Product Safety Concerns and Information please contact our EU
representative GPSR@taylorandfrancis.com
Taylor & Francis Verlag GmbH, Kaufingerstraße 24, 80331 München, Germany

www.ingramcontent.com/pod-product-compliance
Ingram Content Group UK Ltd.
Pitfield, Milton Keynes, MK11 3LW, UK
UKHW052220190926
475702UK00006B/1128
9781032217420